Soviet and Russian Military Aircraft in Africa

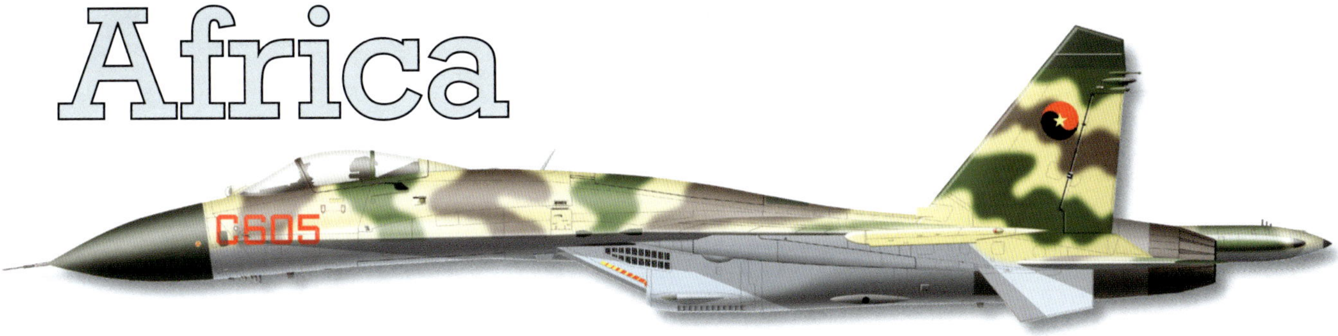

Soviet and Russian Military Aircraft in Africa

Yefim Gordon & Dmitriy Komissarov

HIKOKI PUBLICATIONS

First published in 2013 by
Hikoki Publications Ltd
1a Ringway Trading Est
Shadowmoss Rd
Manchester
M22 5LH
England

Email: enquiries@crecy.co.uk
www.crecy.co.uk

Layout by Polygon Press Ltd., Moscow

Colour profiles: Mikhail Bykov, Aleksandr Gavrilov, Sergey Ignat'yev, Viktor Mil'yachenko, Aleksander Rusinov, Sergey Yershov, Andrey Yurgenson.

ISBN 9 781902 109275

Printed in China

Contents

Acknowledgements 6
Introduction 7
Algeria 9
Angola 44
Benin 80
Burkina Faso 81
Burundi 83
Cameroun 83
Cape Verde 84
Chad 84
Congo (Congo-Brazzaville) 89
Democratic Republic of Congo (Congo-Kinshasa) 95
Côte d'Ivoire 101
Djibouti 105
Equatorial Guinea 107
Eritrea 110
Ethiopia 117
Gambia 137
Ghana 137
Guinea 138
Guinea-Bissau 141
Kenya 143
Lesotho 143
Libya 144
Madagascar (Malagasy Republic) 189
Mali 191
Mauritania 193
Morocco 194
Mozambique 195
Namibia 200
Niger 202
Nigeria 202
Rwanda 212
Senegal 214
Sierra Leone 217
Somalia 219
South African Republic 227
Sudan 230
South Sudan 249
Tanzania 249
Uganda 253
Zambia 265
Zimbabwe 267
Map and Index 270

Acknowledgements

Special thanks go to Chris Lofting, Patrick Roegies, Peter Davison, Sergey Burdin and the Russian Aviation Research Trust (especially Nigel Eastaway), who have contributed invaluably to the selection of illustrations for this book. The authors would also like to apologise, and express their gratitude at the same time, to the persons whose names are unknown to the authors but whose photos have been included due to their obvious value for the reader.

This book is illustrated with photos by Sergey Burdin, Yegor Burkov, Viktor Drushlyakov, Sergey Krivchikov, Alfred Matusevich, Sergey Popsuyevich, M. Slupitskiy, Artyom Starkov, Malcolm Bezzina, Philippe Colin, Peter Davison, Jan Kalina, Chris Lofting, Martin Novak, Patrick Roegies, Royi Sher and T. Laurent.

Additionally, photos from the following sources have been used: www.planes.cz, www.irna.ir, www.elhangardetj.blogspot.com, www.siregar.deviantart.com, www.forcesdz.com, www.xairforces.com, www.benwilhelmi.com, www.lietadla.com, www.defence-arab.com, www.irandefence.net, www.photozagros.com, www.russianplanes.net, www.spotters.net.ua.

Six Libyan *Floggers*, including MiG-23MLAE-2s '132', '5397', '6117', '6130' and '26474', being refurbished at Alrcraft Repair Plant No.121 at Kubinka AB near Moscow.

Introduction

From the outset, the export of revolution and Communist ideology had been one of the cornerstones of Soviet Russia's (and later the Soviet Union's) foreign policy; as the lines from the poem '12' by Aleksandr Blok go, *'we will set the world on fire to the woe of every bourgeois'*. In line with this policy the Soviet Union started exporting arms in the 1930s, even before the Second World War. The first recipients of Soviet weaponry and other military aid were Spain, where a civil war had erupted in 1936, and China, which was hard pressed to fight back the Japanese aggression. Deliveries to these two countries included combat aircraft – the Polikarpov I-15, I-15*bis* and I-16 fighters, the Tupolev SB (ANT-40) fast bomber and TB-3 heavy bomber, the Polikarpov R-5 and R-Z reconnaissance biplanes.

After the Second World War, a new world order was instituted when socialism was introduced in the Soviet-influenced countries of Eastern Europe. The western world reacted by establishing the NATO, to which, in turn, the Eastern Bloc responded by creating the Warsaw Pact. The Soviet Union began exporting weapons, including military aircraft, *en masse* to its socialist allies (which were not limited to WarPac member states). The outbreak of the Korean War in 1950 led to increased military sales to the People's Republic of China and the commencement of arms deliveries to North Korea. It has to be said that the capabilities of Soviet weaponry (first and foremost aircraft) earned recognition not only in the Eastern Bloc but in the West as well. Press articles and analytic reports by western aviation experts gave high acclaim to the Mikoyan/Gurevich MiG-15 jet fighter, whose capabilities the western pilots were to discover – the hard way – in the skies of Korea. Later, these capabilities were confirmed when the Americans evaluated a captured MiG-15*bis*, comparing it with the best contemporary US fighter – the North American F-86 Sabre.

In the mid-1950s the scope of Soviet arms exports began widening, expanding to Africa, which was swept by wars against colonialism. Egypt and Algeria were the first African countries to receive Soviet combat aircraft; Egypt had a chance to put them to the test in the second Arab-Israeli War (aka the Suez Crisis) in 1956. In the 1960s and 1970s, when a large number of African states had gained independence, the Soviet Union began massive supplies of weaponry to some of them. Quite often the weaponry was supplied at a negligible price (for the proverbial '10,000 goatskins') or on credit as military assistance to 'friendly nations' – that is, those who embraced socialism – in the hope of gaining a strategic foothold on the African continent. Actually the Soviet Union was not the only supplier; Czechoslovakia, which built Soviet aircraft under licence, was an active exporter of them, and some WarPac nations (notably East Germany) resold obsolescent aircraft to third-world countries after re-equipping their air forces with up-to-date Soviet hardware. Occasionally East Germany upgraded the aircraft being sold to suit the new operator's requirements.

This situation persisted until the early 1990s – that is, until the break-up of the Soviet Union and the dissolution of the Warsaw Pact, when most of the former Soviet allies shed socialism and switched allegiance to the West. Still, this did not spell the end for Soviet (and then Russian) arms exports; on the contrary, the ongoing wars and insurgencies in many African countries led to an increase in Russian arms deliveries. Today, Soviet/Russian military aircraft can be seen in Algeria, Libya, Angola, Congo (actually *both* Congos), Ethiopia, Eritrea, Uganda and so on. Moreover, China, which had been copying Soviet aircraft (legally or illegally) and creating indigenous versions of them since the mid-1950s, also gained a foothold on the African arms market in the 1990s, supplying several countries with aircraft based on Soviet designs.

In this book – the first volume in a mini-series – the authors have endeavoured to analyse, verify and present to the reader all the available information on Soviet/Russian combat and military transport/utility aircraft delivered to the African continent. These include quasi-civil aircraft (that is, aircraft wearing civil registrations for the sake of convenience only) and occasionally commercial transports which are operated jointly with the military on an on-demand basis. There is, however, a notable exception. Egypt, though certainly an African country, is not included here; instead, it will be included in the forthcoming second volume *Soviet/Russian Aircraft in the Middle East and Iran* – for the simple reason that all the wars involving Egypt (and hence Soviet aircraft in Egyptian service) have been fought in the Middle East. The planned third volume will deal with Soviet/Russian military aircraft in Asia and the Pacific Region.

The authors would like to comment on how the materials are arranged in this book. The aircraft are grouped by category – fixed-wing aircraft first (starting with fighters and their trainer versions, followed by fighter-bombers and/or attack aircraft, then bombers, transports and pur-

This view of an Algerian Air Force Mi-24 SuperHind Mk III shows how the quadruple anti-tank guided missile launch tubes can be tilted down to facilitate engagement.

pose-built trainers), followed by helicopters. In each category, aircraft types are listed in alphabetical order of types (by manufacturer). Next, information is given on the combat use of these aircraft in chronological order. Photos and colour profiles are arranged with the relevant text insofar as possible. Importantly, insignia are provided for each air force, and in cases where the insignia changed over time it is not always possible to show them in chronological order; sometimes the authors have been forced to keep them close to photos of aircraft wearing the respective markings.

Lastly, a fleet list table is given (unless the Soviet/Russian aircraft operated in a given country are too few to warrant a reasonable table). In the tables the aircraft types/versions are listed in alphabetical order, and in serial/registration order for each major version.

In preparing this book the authors have made use of all available unclassified sources and information published in the media, previous books and the Internet. The book includes fleet list tables of all known Soviet/Russian military aircraft (and their Chinese derivatives, as appropriate) operated by each respective African country. It is illustrated with numerous colour profiles which may be of interest to the reader, especially scale modellers.

The authors will appreciate any information submitted by the readers which will be useful for the preparation of subsequent volumes in the series or correct the errors which may occur in this book for objective reasons.

Sudanese Government An-74 ST-PRB awaits delivery at Moscow-Ostaf'yevo in 2005.

Algeria

The Soviet Union began providing military assistance to the Republic of Algeria in 1962, right after the latter had gained independence from France. This included large-scale deliveries of military aircraft to the Algerian Air Force (*Force Aérienne Algérienne*, or *al Quwwat al Jawwiya al Jaza'eriya*).

The first Soviet-built jets to reach the AAF were five Mikoyan/Gurevich UTI-MiG-15 *Midget* trainers; deliveries took place via Egypt in November 1962. 20 MiG-15*bis Fagot-B* fighters were delivered from Egypt in 1964-65; these aircraft equipped the 15ᵉ *Escadron de Chasse* (EdeCh – 15th Fighter Squadron) at Dar el-Beida – Algeria's first jet squadron. More MiG-15s arrived later; some were former East German Air Force aircraft. 20 *Fagot-Bs* and *Midgets* were still operational in Algeria as trainers in 1984.

In 1965, when more fighter pilots had returned to Algeria after completing their training course abroad, the Algerian Air Force took delivery of 30 Mikoyan/Gurevich MiG-17F *Fresco-C* fighters, which equipped a second squadron; this was most probably the 19ᵉ EdeCh, known as the 'Snake Squadron' because of the distinctive unit insignia (a reared-up cobra). Another 30 MiG-17Fs followed in 1966, making it possible to equip three squadrons. According to Egyptian sources, up to 50 more *Frescos* were delivered by June 1967 but these were not reassembled because there was no immediate requirement. By 1979 the MiG-17s were relegated to the ground attack role and replaced by newer MiGs. According to *Interavia*, the AAF had a handful of MiG-17Fs in use as advanced trainers alongside MiG-21U *Mongols* in early 1987; the total number of both types was stated as 20 but the proportion is unknown. However, *Flight International* claims that more than 50 MiG-17Fs were still operated as trainers as late as 1991 (!).

The Algerian Air Force joined the 'supersonic club' in 1965 when the first six Moscow-built Mikoyan/Gurevich MiG-21F-13 (*izdeliye* 74) *Fishbed-C* fighters were supplied; these were operated by a squadron based on the military apron of Algiers International airport. 12 more followed in 1966; some sources say a total of 37 were eventually delivered. 20 MiG-21F-13s were 'wet-leased' to Egypt in 1967 to help in the Six-Day War against Israel; very few survived the conflict to return to Algeria.

12 (some sources say around 30!) MiG-21PF (*izdeliye* 76) *Fishbed-D* fighters were delivered in 1966-67, serving with the 12ᵉ EdeCh. Some sources call them MiG-21FL but that was a version built under licence by Hindustan Aeronautics Ltd. (HAL) solely for the Indian Air Force. The more refined MiG-21PFM (*izdeliye* 94A) *Fishbed-F* followed shortly afterwards. By the beginning of 1967 there were two squadrons of MiG-21 interceptors. The exact number is a matter of conjecture, as official Algerian Air Force figures concerning the number of MiG-21s delivered cite 110 for the 'MiG-21F-13', apparently including the MiG-21PF/MiG-21PFM variants in this total. According to this hypothesis about 40 MiG-21PFMs were received.

The MiG-21PFMs were reported to have served with seven units: the 11ᵉ EdeCh and 19ᵉ EdeCh at Colomb AB,

Five Algerian Air Force Su-30MKAs on the flight line. The ones serialled KF-24 and KF-22 wear a two-tone camouflage, while KF-16, KF-39 and KF-12 are grey overall.

Algerian Air Force MiG-15*bis* fighters, including 6054, wearing old-style winged roundels.

A camouflaged Algerian UTI-MiG-15 (CS-102) being readied for a sortie. The unusually large serial is noteworthy.

Béchar-Oukda, the 140ᵉ EdeCh at Ouargla, the 153ᵉ EdeCh and 630ᵉ EdeCh at Bou Sfer, the 14ᵉ EdeCh at Tindouf, and the 120ᵉ EdeCh at Aïn Oussera (probably a mixed reconnaissance squadron). The 14ᵉ and 19ᵉ EdeCh were mixed squadrons – as more modern variants arrived the MiG-21PFMs were mixed with MiG-21MFs

Six MiG-21R (*izdeliye* 94RA) *Fishbed-H* tactical reconnaissance aircraft were purchased in the 1970s but no details have been released. They were probably based at Aïn Oussera.

Algerian Air Force MiG-15*bis* '6054' with old-style roundels

Algerian Air Force UTI-MiG-15 (Aero CS-102) '1755'

It appears that both the MiG-21M (*izdeliye* 96) *Fishbed-J* and the MiG-21MF (*izdeliye* 96F) *Fishbed-J* were delivered from 1978 onwards, although both versions were known locally as the MiG-21MF. This deduction derives from information published on the official Algerian Air Force website; the latter quotes the 'MiG-21MF' as having an RP-22 radar (which was indeed fitted to the MiG-21MF) but states it has the Tumanskiy R11F2S-300 engine, which powered the MiG-21M but not the MiG-21MF. It is believed about 60 MiG-21Ms and MiG-21MFs were procured.

The ultimate variant supplied to Algeria was the MiG-21*bis* (*izdeliye* 75A) *Fishbed-L/N*, of which about 60 were delivered in the second half of the 1970s. The 110^{e}, 140^{e} and 190^{e} EdeCh were entirely equipped with the MiG-21*bis*, and the 630^{e} EdeCh was a 'mixed bag' squadron which also had MiG-21MFs. The procurement of 120 MiG-21MF and MiG-21*bis* fighters allowed a similar number of older variants to be replaced.

At first, Algerian Air Force aircraft wore two- or three-digit numeric serials; interestingly, these were in European characters, not Arabic. In the 1980s Algeria introduced a new alphanumeric serial system with two letters followed by two or three digits. Each aircraft type was assigned its own letter series (or several if large numbers were delivered); the digits did not run in sequence from 01 but were allocated in batches with gaps to create the impression there were more aircraft than Algeria actually had. At the same time the last four digits of the aircraft's construction number were usually painted on the tail.

Up to 1993, the fighter squadrons reported directly to the Headquarters of the Algerian Air Force but thereafter 3e *Escadre de Défense Aérienne* (3rd Air Defence Wing) was formed as a controlling unit. Subsequently the surviving MiG-21s were organised within four squadrons: the 113e EdeCh at Tindouf, 143e EdeCh at Ouargla, 153e EdeCh at Béchar and 193e EdeCh at Bou Sfer. In 1997 a number of surplus MiG-21MF and MiG-21*bis* fighters were traded-in to the Ukraine in exchange for 36 MiG-29s; similar deals may have been struck with Belarus from whom another 36 MiG-29s were acquired.

Very little information is available on the MiG-21U/MiG-21UM *Mongol-A/B* trainer variants delivered to Algeria, except that the total number is reported as being around 30. From the 1980s their serials were prefixed FE but it is unlikely that the same numbers were retained.

The last MiG-21 fighters and trainers were finally retired in 2003, having been steadily replaced by more modern types since 1997.

An obvious publicity shot of an Algerian MiG-17F, '24 Black', wearing the 'attacking cobra' emblem of the 19^{e} EdeCh. Note that the pilot is getting out on the wrong side of the aircraft.

MiG-21F-13s from the first shipment delivered to Algeria, with Algerian Air Force An-12Bs in the background. The fighters are equipped with UB-16-57 rocket pods. At that time the national insignia was carried on the fin.

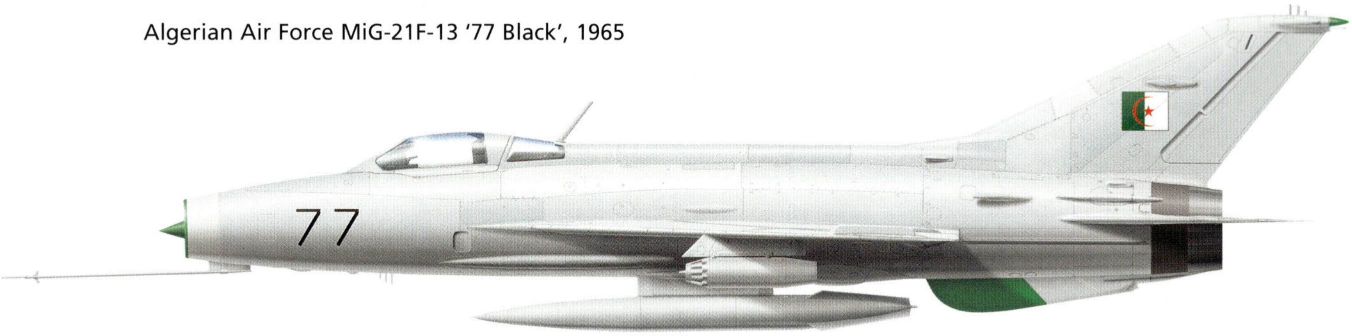

Algerian Air Force MiG-21F-13 '77 Black', 1965

In 1979 the Algerian Air Force received about 40 Mikoyan MiG-23MS *Flogger-E* fighters which were later augmented by some 30 MiG-23MF *Flogger-B*. The latter version served with the 28ᵉ *Escadron d'Interception* (28th Interceptor Squadron) of the 21ᵉ *Escadre de Chasse-Bombardement* (21st Fighter-Bomber Wing) at Laghouat AB in the centre of northern Algeria. Starting in 2001, the type was progressively withdrawn and replaced by the more modern MiG-29. No records of the service life of the MiG-23MS have been released but none were reported to be in service in 2004. It has been reported that only one squadron of MiG-23MFs based at Laghouat was still operational in 2004.

About ten MiG-23UB *Flogger-C* trainers were also delivered. Originally purchased to train the MiG-23MS pilots, these aircraft were used also to train personnel for the MiG-23MF, MiG-23BN (see below) and even MiG-29 – that is, until the Algerian Air Force took delivery of MiG-29UBs.

In 1978 the Algerian Air Force purchased a number of Mikoyan MiG-25 heavy supersonic jets. These included MiG-25PD *Foxbat-E* interceptors, MiG-25RBK and MiG-25RBV *Foxbat-B* reconnaissance/strike aircraft, and MiG-25PU *Foxbat-C* trainers; the total number delivered is reportedly 48 but the proportion of the subtypes is unknown. The *Foxbats* saw service with the 10ᵉ *Escadre de Chasse* (10th Fighter Wing) at Aïn Oussera AB, the interceptors and MiG-25PUs being operated by the 110ᵉ (and, according to some sources, 120ᵉ) *Escadron de Chasse* while the MiG-25RBK and MiG-25RBV versions served with the 510ᵉ *Escadron de Reconnaissance* (EdeR – Reconnaissance

Non-airworthy ex-Algerian Air Force MiG-21MF FD-19 sits at the Odessa aircraft repair plant, awaiting refurbishment and resale (a sticker below the cockpit reads 'Export'). The Algerian roundels on the rear fuselage have been painted out.

Algerian Air Force MiG-21*bis* fighters sit in storage at Blidah beside an Algerian Mi-17 helicopter. FA-65 is a *Fishbed-N*, while FA-89 and FA-96 are *Fishbed-Ls*.

Algerian Air Force MiG-23MF FP-27 shows off its desert camouflage and the last four digits of the c/n on the tail.

Squadron). In the 1990s the MiG-25s were refurbished by the Zaporozhye Aircraft Repair Plant (ZARZ) in the Ukraine.

One of the latest Soviet/Russian types in Algerian Air Force service – and the most controversial one, too – is the Mikoyan MiG-29 light fighter. These aircraft were obtained from three sources, and reports on MiG-29 operations in Algeria are highly contradictory; there is no consensus even on when the first *Fulcrums* were delivered, some sources saying 1994 and others 1997. Algeria purchased its first 36 MiG-29S (*izdeliye* 9.13S) *Fulcrum-C* fighters from the Ukraine – as already mentioned, via a trade-in arrangement; this number probably included several MiG-29UB *Fulcrum-B* combat trainers. Another 36 (or 31?) MiG-29Ss/MiG-29UBs were acquired from Belarus under a similar arrangement; the first 16 of these arrived in 1999, followed by eleven in 2000, two in 2001 and the final two in 2002. The *Fulcrums* were reportedly exchanged for 120 MiG-21s which Beltechexport, the Belorussian arms export agency, hoped to resell to other African countries. Other sources say 36 MiG-29s were received from the Ukraine in 1999-2000 – or 28 *Fulcrum-Cs* from Belarus and eight MiG-29UBs from Russia. (Here it should be noted that in the 1990s Algeria bought second-hand military hardware quite a lot – possibly because the civil war against Muslim fundamentalists was eating up resources and insufficient money was available to buy new hardware.)

However, the second-hand MiG-29s lacked in-flight refuelling (IFR) capability and had only limited air-to-ground capability. Therefore, in March 2006 Algeria signed a contract with Russia for the delivery of a weapons package worth no less than US$ 8 billion, which included 28 MiG-29SMT multi-role fighters and six MiG-29UBT combat trainers with a total worth of US$ 1.3 billion. The MiG-29SMT (in its latest *izdeliye* 9.19 version) featured increased internal fuel capacity, plus a semi-retractable IFR probe on the port side of the nose to extend range further, and the ability to carry precision-guided munitions; the two-seat MiG-29UBT had a similarly modified fuel system. Unlike the second-hand MiG-29s already in service, which had an overall light grey colour scheme, the new-build MiG-29SMTs and MiG-29UBTs wore a grey/light blue camouflage.

Deliveries began that same year, the RSK MiG (MiG Russian Aircraft Corporation) plant in Lukhovitsy near Moscow reinstating production of the type (the mothballed airframes of late-production *Fulcrum-Cs* from a deferred

The badge worn by Algerian Air Force personnel.

MiG-23BN FM-17 is loaded with American Mk 82 low-drag bombs (four under the fuselage and four under the wing gloves on MBD3-U2T-1 tandem racks).

Algerian Air Force MiG-23MF FP-27 in two-tone desert camouflage with the 800-litre 'banana' centrelilne drop tank

Algerian Air Force MiG-23MLD FP-84 in three-tone camouflage armed with R-23R and R-60M missiles

This MiG-23BN wearing three-tone camouflage has the old-style serial '383 Black' and the Algerian Air Force fin flash

'389 Black', another Algerian MiG-23BN in old colours/markings; note the absence of the roundel on the fuselage

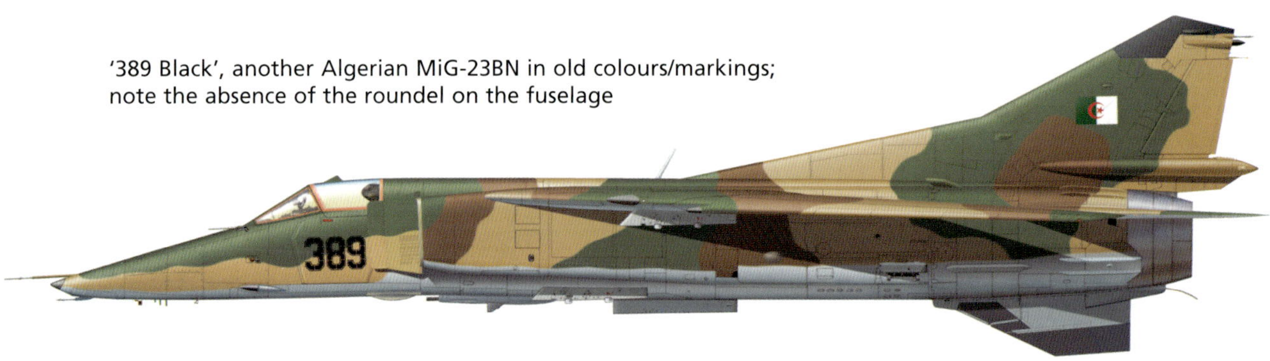

MiG-23BN FM-14 fitted with MBD3-U2T-1 bomb racks wears the new desert camouflage and AAF roundels but no fin flash

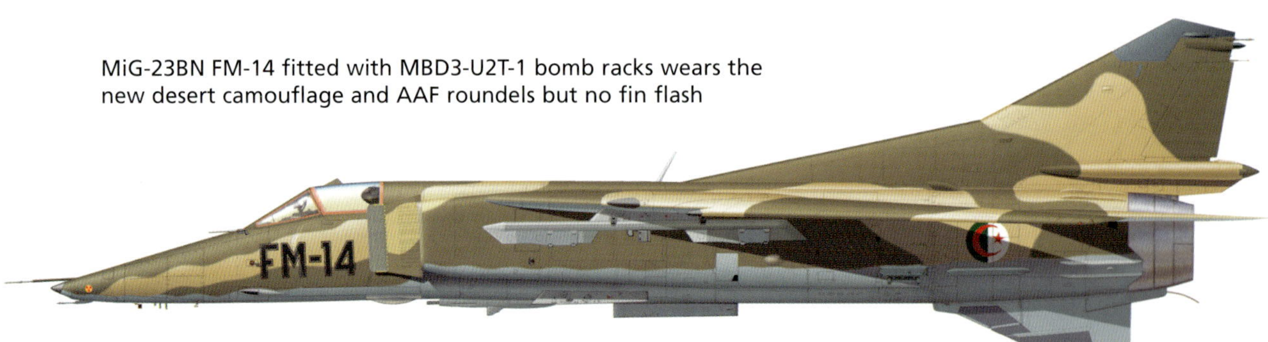

A trio of Algerian MiG-25PDs, including FU85 and FU56. Note the 'Danger, air intake' stencils.

An Algerian Air Force MiG-25PD is prepared for the formal handover to the customer in the Ukraine.

Refurbished by the Zaporozhye Aircraft Repair Plant, MiG-25PD FU-31 looks immaculate in Algerian Air Force markings

Algerian Air Force MiG-25PD FU-75; note the emblem of the Zaporozhye AIrcraft Repair Plant (ZARZ) on the fins

Russian MoD order that had been sitting at Lukhovitsy-Tret'yakovo airfield were used for this).

However, in 2007, after taking delivery of 15 aircraft, Algeria raised hell, refusing to accept any more MiG-29SMTs/UBTs and stating that it would return the ones it had already bought – an unprecedented gesture. The Algerians cited substandard quality of the fighters as the reason, claiming that the MiGs contained used or counterfeit components. The scandal gained momentum as Algeria suspended payments on the other Russian defence contracts until the fighters had been returned. Not wishing the matter to be blown out of proportion (which might indeed have resulted in termination of other contracts and even complete loss of the Algerian market), in February 2008 the Rosoboronexport arms trading agency agreed to take the fighters back by the end of April. On the other hand, the Russian Federal Agency for Defence Technology Co-operation stated that the Algerian advance payment of US$ 250 million for the fighters would not be refunded. The re-export MiG-29SMTs/UBTs plus the ones already completed but still undelivered (24 aircraft in all) were delivered to the Russian Air Force instead, thus becoming Russia's first IFR-capable *Fulcrums*.

Some Russian sources say that the MiGs had indeed been delivered with substandard parts, and there had been prior cases in Russia when unscrupulous businessmen had sold counterfeit spares to Russian aviation enterprises. On the other hand, Russian defence observers maintain that the reason behind the Algerian demarche is clan strife within the Algerian political and military establishment. At that time the Algerian President Abdulaziz Bouteflika was running for his third term of office. Whereas arms procurement for the Army was controlled by his loyal Chief of General Staff Ahmed Salah Gaïd, aircraft procurement for the Air Force was the domain of an opposing clan which included some influential persons. It's just possible that someone saw his chance to throw a spanner in the works, 'exposing' the opposition as incompetent or corrupt.

Of course, such a scandal was bound to cause repercussions in Russia. The Federal Security Service (FSB) investigated the matter; as a result, criminal charges were brought against the management of the Russian company Aviaremsnab (= Aviation Repair Supplies)

Algeria also bought a few reconnaissance *Foxbats*, including MiG-25RBK FG-78. Again, the emblem of the Zaporozhye AIrcraft Repair Plant is visible on the tail.

Algerian Air Force MiG-25RBK FG-78; again it carries the ZARZ emblem on the fins

MiG-25RBV FG-34, showing the starboard camera window

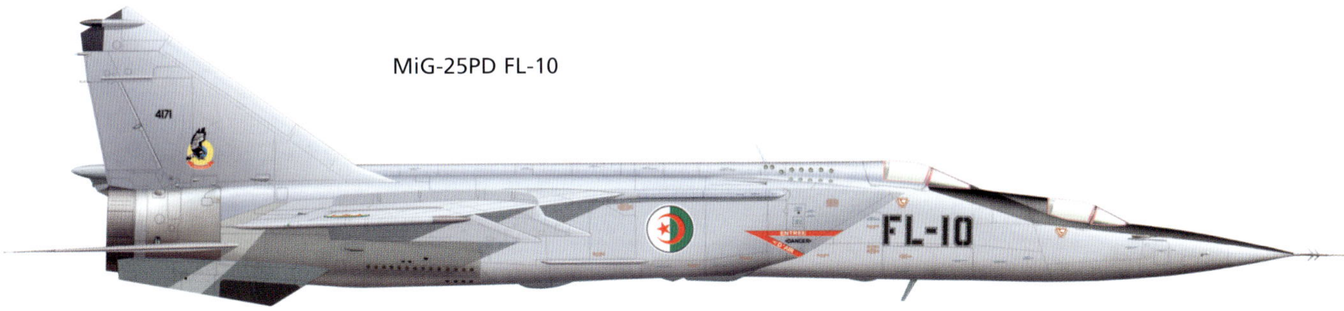

MiG-25PD FL-10

which had furnished the defective parts (landing gear struts, radio altimeters etc.). The parts had been manufactured between 1982 and 1996 and their shelf life had expired; yet the parts were supplied to RSK MiG with forged papers telling they had been manufactured in 2005-06. Eventually the director and vice-director of the company were found guilty as charged and jailed for five and a half years.

In spite of the scandal, the Algerian Air Force did not withdraw the MiG-29 completely. Fact is, Algeria needs fighters of this class in order to oppose the permanently hostile Morocco, which is procuring Lockheed Martin F-16C/D Fighting Falcons (paid for by Saudi money), and the MiG-29SMT's western counterparts are pricey. 25 MiG-29s remained in service as of 2010. The fighters are operated by the 3rd Air Defence Wing comprising the 113ᵉ EdeCh at Tindouf AB, the 143ᵉ EdeCh at Ouargla AB, the 153ᵉ EdeCh at Béchar-Oukda AB and the 193ᵉ EdeCh at Bou Sfer AB.

Two Algerian *Fulcrum-Cs* were lost in accidents. On 10th October 1999 a 193rd Sqn MiG-29SE serialled FC-84 and piloted by Capt. Bouafara Abdel-Malek while flying a daytime air combat training sortie. 16 minutes after take-off from Bou Sfer AB (at 1643 hrs 39 sec CET), when the fighter was climbing through 1,476 m (4,842 ft) and flying at 550 km/h (341 mph), strong vibration of the starboard

FC-61, one of the MiG-29Ss from the early deliveries to Algeria, with a full load of missiles; drop tanks and R-73 AAMs are arrayed in front.

Algerian Air Force MiG-29S (*izdeliye* 9.13S) FC-93, 3rd Air Defence Wing

Algerian Air Force *Fulcrum-Cs*, such as FC-53, have an overall grey finish.

engine and the accessory gearbox set in. (The MiG-29 has a common accessory gearbox serving both engines – *Auth.*) One second later the affected engine ran down abruptly; another two seconds later a fire warning went off. The pilot shut down both engines and then, at 1643 hrs 53 sec, attempted a relight. The port engine came alive but the starboard one would not; the fire warning extinguished at 1643 hrs 55 sec but went off again at

MiG-29SMT (*izdeliye* 9.19) FC-12, 3rd Air Defence Wing

FC-57, one of the MiG-29SMTs built for Algeria, shows off its blue/grey camouflage and IFR probe.

MiG-29SMT FC-32 departs on a training sortie, carrying wingless dummy AAMs.

Algerian Air Force MiG-29UBT FB-05 at Zhukovskiy prior to delivery. Note the IFR probe and the formation lights.

A fine study of Algerian Air Force MiG-29SMT FC-14 cruising above the clouds.

Algerian Air Force Su-7BMK '333' with old-style insignia (fin flash and numeric serial), Blidah

Su-7BMK 'KR-57/1957' preserved with new-style insignia (fuselage roundels and alphanumeric serial)

Algerian Air Force Su-7BMK '333' and Su-7UMK '1303' in storage at Blidah in 2008 in company with two MiG-21MFs.

1644 hrs 32 sec and stayed on until the end. Eight seconds later the wingman confirmed that his leader's aircraft was on fire and told Abdel-Malek to use the fire suppression system. At 1644 hrs 58 sec the pilot ejected after being told to do so by his wingman; the aircraft crashed and burned on a rocky outcrop without causing any collateral damage. The investigation established that the starboard engine had suffered an uncontained failure of both high-pressure and low-pressure turbines, the blade fragments knocking out the accessory gearbox as well.

On 19th April 2000 (not 19th January as sometimes reported) the same pilot had an accident in another 193rd Sqn MiG-29SE

This Su-7BMK is preserved at a road junction in Algeria, wearing the serial KR-57. This time the number 1957 on the fin bears no relation to the c/n, being much too low for a Su-7BMK.

Su-24MK KX-99 illustrates the early 'dark top' camouflage worn by Algerian *Fencers*. Note the slight difference in the rendering of the letters and the digits; the latter are in heavier type, apparently having been applied after delivery.

(FC-84) under similar circumstances. 8 minutes 43 seconds into the mission (at 1718 hrs CET) the port engine failed and a fire developed; the fighter was then at 560 m (1,840 ft) and 512 km/h (318 mph), pulling 2.5 Gs. Cancelling the afterburners, Capt. Bouafara Abdel-Malek advised the tower at Bou Sfer AB of the engine failure; however, he was unaware of the fire because the appropriate audio warning signal had been drowned out by another voice message; also, he failed to perceive the chip detector and engine fire warning lights. The tower instructed the pilot to abort the mission, throttle back the affected engine to idle and make a right turn in order to execute an emergency approach to

Su-24MK KX-11 shows off the new desert camouflage applied after refurbishment and upgrade. Note the different rendition of the serial.

KX-12, another upgraded *Fencer-D* in the new colours.

Algerian Air Force Su-24MK KX-11 as originally delivered (with medium brown/dark brown camouflage on the upper surfaces)

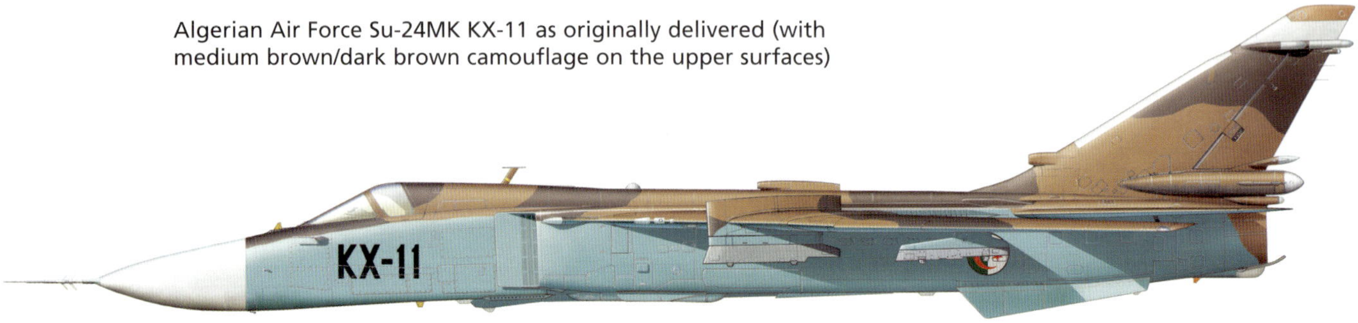

Su-24MK KX-99 in old colours; note that there was no standard pattern

This Su-24MK, KX-29, wears an unusual camouflage scheme with two shades of green on the upper surfaces

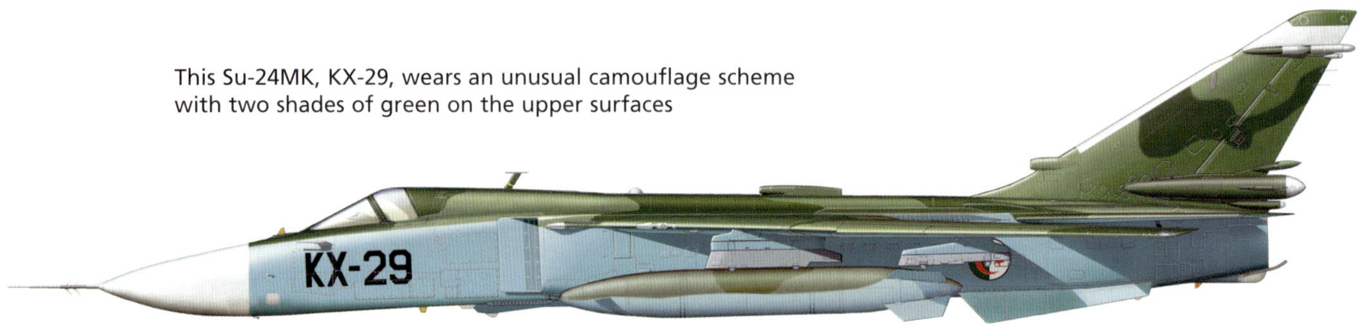

Su-24MK KX-12 in post-refurbishment/upgrade desert camouflage with Kh-25ML laser-guided missiles and UB-32A rocket pods

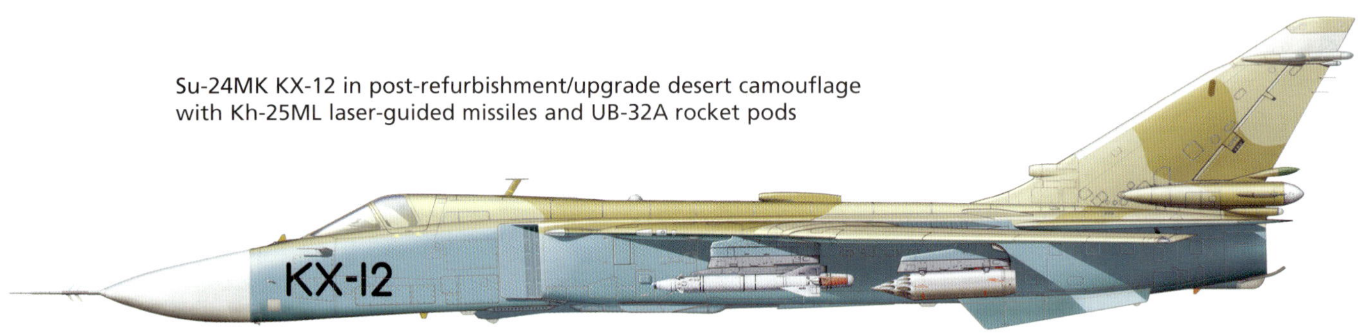

Algerian Air Force Su-24MRK KG-65 in old colours, with a mission equipment pod on the centreline station

Here, Su-24MK KX-11 is shown in its new colours and with MBD3-U6-68 multiple ejector racks on all hardpoints

runway 06. Seeing that the engine had stopped, the pilot retarded the port throttle to 'shut down' but failed to lock it there and the throttle moved forward again. Realising he was too close to runway 06 and would miss the approach, the pilot gave full military power to the good engine and initiated a circuit of the field in order to try an approach to runway 24. The traffic controller advised the pilot that the aircraft was on fire, instructing him to close the port engine's fuel shut-off cock and use the fire suppression system. The pilot complied, but apparently the shut-off cock malfunctioned and stayed open, fuel flowing freely to the burning engine. After passing the threshold of runway 24 on the downwind leg Abdel-Malek initiated a turn onto final approach, the tower informing him that the fighter was still on fire. As it turned onto finals, the MiG-29 spontaneously rolled, the nose 'falling through'; at 1720 hrs 32 sec the pilot ejected safely after sending an appropriate 'Will eject, eject, eject' call. Unfortunately the stricken aircraft fell squarely on a residential building in Ayoun-el-Turk township 3-4 km (1.86-2.48 miles) from the runway, killing one occupant and inuring another.

Apparently the Algerian Air Force had begun sizing up the more modern (Generation 4+) and heavier Sukhoi Su-30MK multi-role fighter in the late 1990s. However, reports that ten Su-27UBK *Flanker-C* combat trainers had been delivered to Algeria in 2002-03 were pure fiction. It was not until May 2006 that Algeria placed a firm order for 28 Su-30MKs as part of the aforementioned eight-billion arms deal. The aircraft were to be manufactured by the Irkut Corporation in Irkutsk and be similar to the Su-30MKI *Flanker-E* version for India (with short tails and canard foreplanes). In keeping with the new practice of including customer designators the Algerian version was designated Su-30MKA; some sources, though, call them Su-30MKI(A).

Deliveries commenced on 27th December 2007, the first two Su-30MKAs arriving in a

A row of Algerian Su-24MKs in new and old colours. Note the Kh-58U anti-radar missile on the dolly in the foreground.

Su-24MKA KF-24 displays its upper surface in a banked flypast. Note the asymmetric wing markings (the roundel to port and the serial to starboard).

Volga-Dnepr Cargo Airlines Antonov An-124-100 Ruslan (*Condor*) freighter, another two following by the end of the year. Four more fighters were supplied in March and June 2008; by 2010 all 28 aircraft of the initial order had been delivered. At the Farnborough International 2010 airshow it was announced that Algeria had placed an additional order for 16 Su-30MKAs, thus bringing the total to 44. The *Flankers* serve with three unidentified squadrons of a likewise unidentified fighter wing based at Aïn Beida.

Yet again, the fighters became the subject of an inquiry – again associated with 'improper' components. This time the components were *politically* improper; it turned out that the Su-30MKAs were equipped with an Elbit Systems SU967 head-up display and an Elta Electronics EL/M8222 ECM system, both companies being divisions of Israel Aircraft

Algerian Air Force Su-30MKA KF-24 in two-tone camouflage

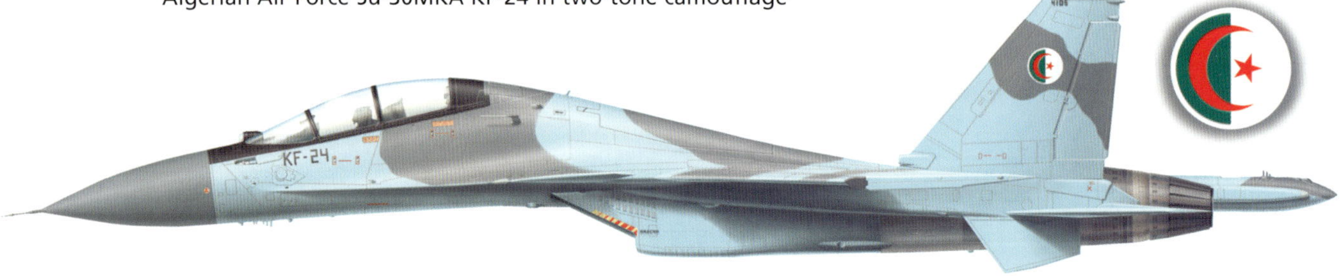

Algerian Air Force Su-30MKA KF-12 in uniform grey camouflage

Grey-painted Su-30MKA KF-94 formates with a foreign transport aircraft.

Here, Su-30MKA KF-48 is the lead ship during formation aerobatics. Note the white ceramic tiles on the insides of the engine nozzles.

Industries. Of course, Algeria was not happy about having critical avionics components of their fighters manufactured by an unfriendly nation, as Israel supports both Morocco and the Berber separatists in Algeria. There have already been cases when the Algerian Air Force had rejected aircraft on the grounds that they have Israeli components.

Interestingly, the 'discovery' of the Israeli components in the Su-30MKAs was not made until all the fighters had been delivered – that is, someone was waiting for the deal to come through in order to use the potential scandal to the full. So far, however (touch wood), there have been no claims and statements about returning the Su-30s.

There have been reports that a reconnaissance version designated Su-30MKR is to be fielded by the Algerian Air Force as a successor to the MiG-25RBSh. The aircraft will carry a podded reconnaissance system supplied by the French Thales Group.

Soviet strike aircraft were initially represented in the Algerian Air Force by the MiG-17F (outfitted as a fighter-bomber) and the specialised Sukhoi Su-7BMK *Fitter-B* fighter-bomber. The Algerian Air Force took delivery of 20 Su-7BMKs and eight Su-7UMK *Moujik* trainers in 1971-72; this was enough to equip two squadrons. Later, some 30 Mikoyan MiG-23BN *Flogger-H* fighter-bombers were delivered to Algeria. Initially the

Partly dismantled, Algerian Air Force Yak-130 NY-11 is manoeuvred into position by a ZiL-131 6x6 lorry at Irkutsk-2 airfield while Volga-Dnepr Airlines An-124-100 RA-82074 prepares to uplift it.

Here, the final preparations are in hand to load an Algerian Yak-130 into the An-124 for the delivery flight.

MiG-23BNs equipped two squadrons: the 28e *Escadron d'Attaque* (28th Attack Squadron) of the 21e *Escadre de Chasse-Bombardement* (21st Fighter-Bomber Wing) at Laghouat and an unidentified squadron of an unidentified wing based at Ouargla. Detachments were also stationed at Béchar-Oukda and Tindouf. Algerian MiG-23BNs used both Soviet and western munitions. The type still equipped at least one squadron in 2004.

An initial batch of twelve Il'yushin IL-28 *Beagle* tactical bombers was delivered to the Algerian Air Force via Egypt, with which Algeria was closely allied, in 1964-65; direct deliveries

Algerian Air Force Yak-130 NY-61

might have caused unfavourable political consequences for the Soviet Union. A further 12 *Beagles* were delivered directly from the USSR after the 1965 military coup when the new government (which was even more pro-Soviet) requested additional military aid. The bombers saw service with the 27e *Escadron*. Nearly all the IL-28s were unserviceable by 1979, and the few which remained operational were relegated to secondary duties.

A successor to the IL-28 was obtained in the 1990s when Algeria took delivery of ten or twelve Sukhoi Su-24MK *Fencer-D* tactical bombers (sources differ). The bombers were converted from ex-Russian Air Force Su-24Ms and served with the 274e Escadron d'Appui (Strike Squadron) of the 4e Escadre d'Appui (penetration wing); though originally at Aïn Oussera, they are now based at Laghouat AB in the centre of northern Algeria. Additionally, five Su-24MRK *Fencer-F* electronic intelligence aircraft (the export version of the Su-24MR) were delivered to the 510e Escadron de Reconnaissance of the 10e EdeCh at Aïn Oussera from Russian Air Force stocks; Algeria is the sole non-CIS operator of the ELINT version. Algeria also received at least one Su-24MPK *Fencer-E* electronic countermeasures aircraft (the export version of the rare Su-24MP) from the Russian Air Force; it is likewise operated by the 510e EdeR.

In 2000 a deal worth US$120 million was signed with the Novosibirsk Aviation Production Association (NAPO), which is the *Fencer*'s original equipment manufacturer (OEM), to upgrade all existing Algerian Su-24MKs/ Su-24MRKs and deliver a further 22 ex-Russian Air Force Su-24Ms to be refurbished and upgraded. The upgrade had been developed by the Russian private enterprise Gefest & T (*Ge**fest*** is the Russian rendering of Hephaestus, the Greek god of technology and fire), and was broadly similar to the Russian Air Force Su-24Ms upgraded by the same company that had proved their much-improved combat potential as compared to the standard bomber. In addition, three Su-24Ms were purchased from the Ukraine, presumably also to be upgraded. By the end of 2001 twenty-five aircraft had been (re)delivered.

In 2004 the Algerian Air Force had 37 Su-24MKs and four Su-24MRKs on strength. These have been occasionally referred to as Su-24MK2s and Su-24MRK2s respectively, although the Su-24MK2 (the export equivalent of the Su-24M2) is a different upgrade developed by the Sukhoi Design Bureau itself (and reportedly inferior to the Gefest & T version). The bombers are in service with three squadrons – the 274e, 284e and 294e *Escadron d'Appui et Pénétration* (Strike and Pénétration Squadron), all based at Laghouat; together they form the 4e *Escadre d'Appui et Pénétration*.

Algeria became the first foreign customer for the Yakovlev Yak-130 *Mitten* advanced trainer. The order for 16 such aircraft was placed in March 2006 and again was part of that big weapons deal. As was the case with the Su-30MKAs, all 16 aircraft were built by the Irkut Corporation. The first three Yak-130s were delivered in November 2011 by a Volga-Dnepr Airlines An-124-100.

The first Soviet transports delivered to the Algerian Air Force were six Il'yushin IL-14T *Crate* piston-engined twins delivered in 1962; by 1970 the number had risen to 14. In early 1966 they were followed by eight Voronezh-built Antonov An-12B *Cub* medium transports in military configuration with a tail gunner's station. The aircraft were operated by the 35e Escadron de Transport (35th Transport Squadron) at Boufarik AB. Interestingly, the *Cubs* wore civil registrations along with Algerian Air Force insignia and military serials. In the early 1970s President Col. Houari Boumedienne switched allegiance to the West and the Algerian Air Force started re-equipping with Western hardware. The An-12s were placed in storage and supplanted by Lockheed C-130H and 'stretched' C-130H-30 Hercules transports. At least four of the aircraft were ultimately sold (to Moldova and possibly to Angola).

In 1990, however, the Soviet Union/Russia managed a comeback in this field; starting that year, the Algerian Air Force purchased 12 Il'yushin IL-76 heavy transports. The order was placed after an IL-76MD captained by Il'yushin Aircraft Complex test pilot Stanislav G. Bliznyuk made several demo flights in Algeria on 15th-25th March 1988. This was something of a breakthrough for the Soviet Union/Russia, as Algeria had not purchased Soviet equipment since the early 1970s. The first three aircraft delivered were very military IL-76MD *Candid-Bs*; the other nine represent the IL-76TD *Candid-A* commercial version. Only five were factory-fresh examples; the others are second-hand but quite new aircraft acquired from various Russian airlines. Again, the *Candids* are based at Boufarik AB, serving with the 347e *Escadron de Transport Tactique* (347th Tactical Transport Squadron). They

7T-WAA, the first An-12B delivered to Algeria. Note the small Arabic 'Algerian Air Force' titles, the registration painted on the tail and the roundel superimposed on the heavy cheatline.

In contrast, sister ship 7T-WAE/550 has a large Algerian flag but no roundel or AAF titles, a differently placed registration and a thinner cheatline.

replaced the last surviving An-12BPs, some of which were reportedly traded in for used IL-76TDs. Like the *Cubs,* the *Candids* wear civil registrations along with Algerian Air Force insignia; unlike the An-12s, however, they have no military serials and carry Algerian Air Force titles (in Arabic to port and in English to starboard).

The Algerian Air Force also purchased six Il'yushin IL-78 *Midas-A* convertible tanker/transports to support its fighter fleet (primarily the Su-30s). All six were bought from the Ukrainian Air Force, which had no use for them as tankers after scrapping its heavy bomber fleet in keeping with its non-nuclear status. Until then the *Midasae* had been used as transports but they had their mission equipment reinstated before the sale to Algeria. They are likewise based at Boufarik and are operated by the 357[e] *Escadron de Ravitaillement en Vol* (Aerial Refuelling Squadron) which, together with the 347th Squadron, forms the 7[e] *Escadre de Transport Tactique et Ravitaillement en Vol* (7th Tactical Transport/Aerial Refuelling Wing).

In 1963 the Algerian Air Force's government flight purchased a single Il'yushin IL-18V *Coot* medium-haul airliner from Aeroflot Soviet Airlines, taking delivery of the aircraft on 6th January 1964. Reregistered 7T-VRA and

Algerian Air Force An-12B 7T-WAB/566

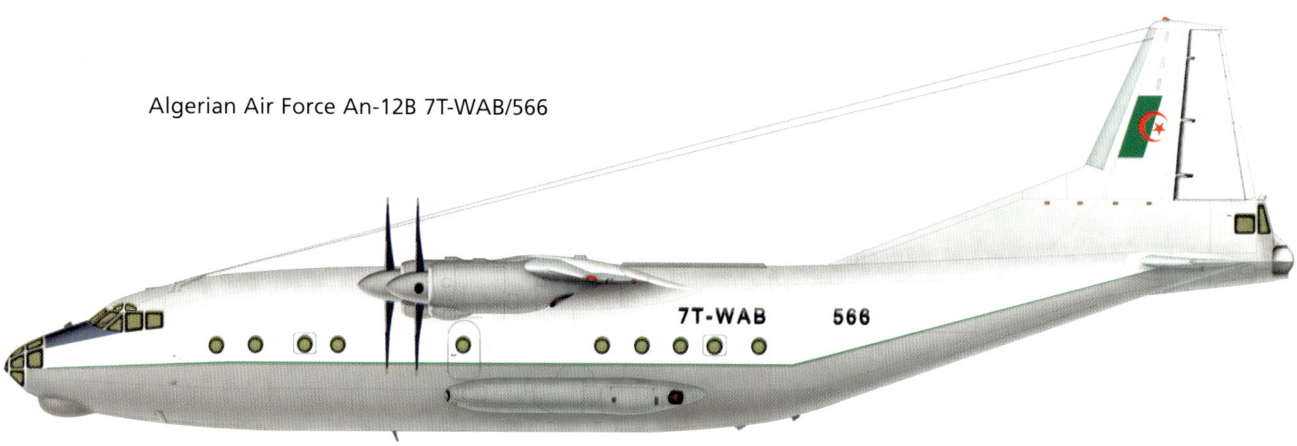

Immaculate-looking An-12B 7T-WAC/514. Note that only the last four digits of the c/n (402605) are painted on, and in much smaller type than hitherto.

refitted with a VIP interior, this was the sole IL-18 delivered to Algeria, despite persistent reports of at least four being operated. The *Coot* carried *République Algérienne* titles and an airline-style colour scheme; the red cheatline curved down to the radome, giving the aircraft a 'smiling' appearance in a head-on view. Some sources reported that 7T-VRA crashed near Tamanrasset in southern Algeria on 12th June 1971 but this was not so.

Soviet helicopters in Algeria are represented by both Mil' and Kamov types. Ten Mil' Mi-4 *Hound* utility helicopters were the first to be delivered in 1962 (a total of 40 was eventually delivered). Three Mil' Mi-1 *Hare* light helicopters were also used in the liaison role. The Mil' Mi-2 *Hoplite* is operated by a squadron of the 9e *Regiment d'Hélicoptères d'Entrainement* (RHE – 9th Training Helicopter Regiment) at Setif/Aïn Arnat since 1988; at least 41 have been delivered. Interestingly, their construction numbers commencing 51 identify them as baseline utility Mi-2Ts rather than dual-control Mi-2U trainers whose c/ns should begin with 54. Equally oddly, almost all the Mi-2s have odd-numbered serials.

The Algerian Air Force received its first 12 Mil' Mi-8 medium helicopters in the late

Algerian Air Force IL-28 '024 Black' with old-style insignia

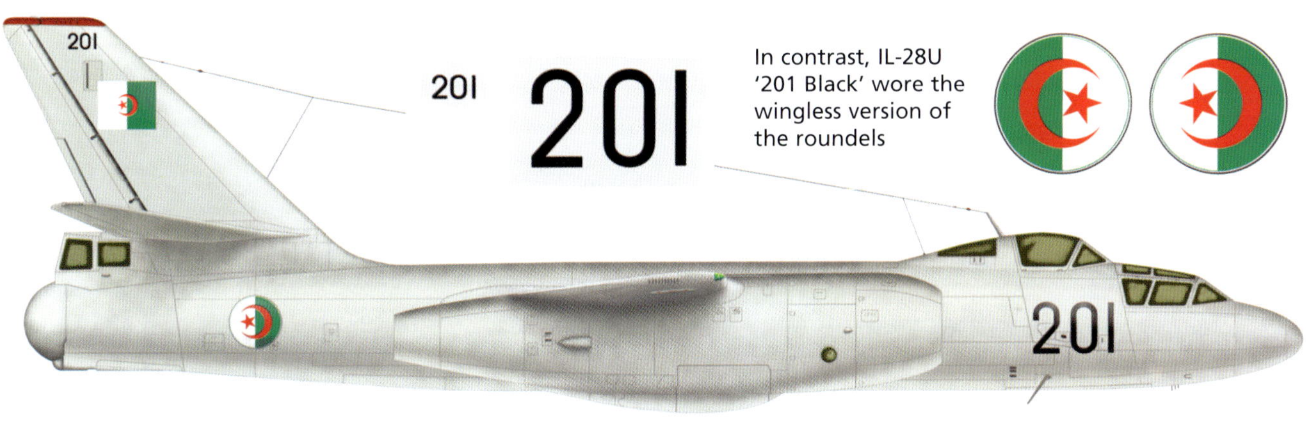

In contrast, IL-28U '201 Black' wore the wingless version of the roundels

Algerian Air Force IL-78 7T-WIF in pure transport configuration (minus UPAZ-1A hose drum units); note the absence of titles

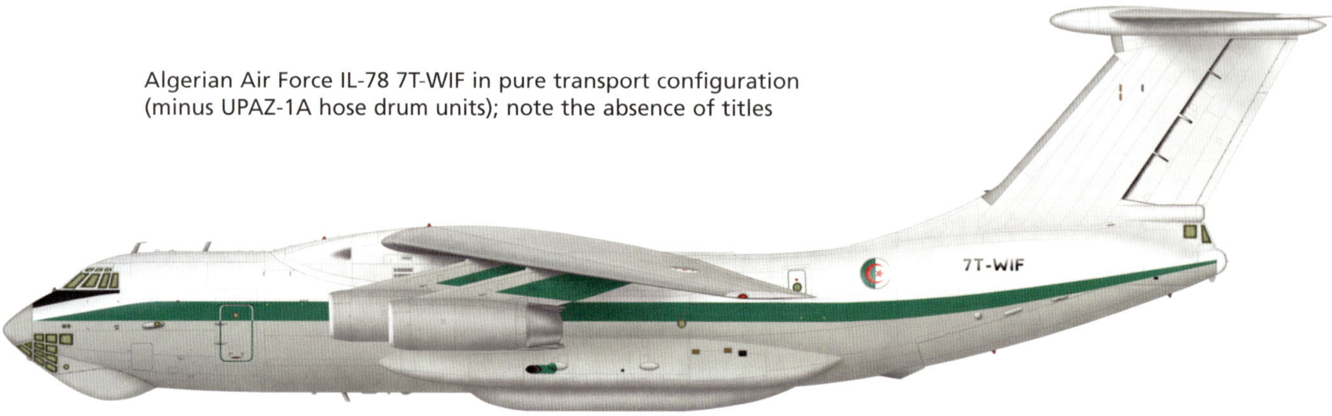

1980s; these included both Mi-8T *Hip-C* utility helicopters and Mi-8PS VIP helicopters. To protect the engines against wear and tear caused by sand ingestion they were retrofitted with boxy western-made intake filters serving both engines (similar to those of the Egyptian *Hip-Cs*). Later deliveries consisted of 'second-generation' Mi-8MT/Mi-17 *Hip-Hs*. In 2002 the Algerian Air Force placed an order for 42 Ulan-Ude-built Mi-171E utility helicopters and Mi-171Sh transport/assault helicopters worth US$ 200 million. Again, most of them had the standard individual intake filters replaced with a common boxy air filter for the two engines produced by the Pall Corporation. Some of the Mi-171Sh's have been upgraded to feature a

Algerian Air Force IL-76TD 7T-WID tucks up its undercarriage on take-off.

Gleaming with fresh paint after an overhaul at the Russian Air Force's Aircraft Repair Plant No.360, an Algerian IL-78 (minus HDUs) shares the hardstand at Dyagilevo AB near Ryazan' with a Russian AF sister ship, '80 Blue'/RF-94283.

ALGERIAN AIR FORCE

Algerian Air Force IL-76TD 7T-WID; the titles on the port side are in Arabic

port starboard

centreline with a Denel Argus 550 optoelectronic targeting system 'turret' and carry four-packs of Kentron ZT35 Ingwe tube-launched anti-tank guided missiles (both of South African design). At least two *Hip-Hs* are specialised medical helicopters based on the airframe of the Mi-19R airborne command post.

The Mi-171s and Mi-8MTs serve with the 436e and 456e *Escadron d'Hélicoptères* of the 6e *Regiment d'Hélicoptères de Manoeuvre* (6th Mobility Helicopter Regiment) at Ech Cheliff and an unidentified *Escadron d'Hélicoptères de Specialisation* (Special Mission Helicopter Squadron) of the 9e RHE at Setif/Aïn Arnat; the Mi-17s are operated by an unknown squadron of the 1e *Regiment d'Hélicoptères de Combat* (RHC) stationed separately from the others at Ouargla, while the older Mi-8s have been relegated to the training role with the same 9th Regiment.

Four Mil' Mi-6 *Hook* heavy-lift helicopters were delivered to Algeria. Originally they wore registrations in the quasi-civil 7T-W** series, though at least one survived long enough to receive a post-1980 alphanumeric serial.

Algeria was among the first foreign customers for the Mil' Mi-24 attack helicopter, starting out with the Mi-24A *Hind-A* (the 'squarehead' initial production model) in 1979. As of early 1995 the Algerian Air Force operated 38 Mi-24As and Mi-24D *Hind-Ds* (or rather Mi-25s, the export version); 35 of these remained by November 2000. In the 1990s the Algerian Air Force obtained a number of second-hand Mi-24Ds and Mi-24V *Hind-Es* from Russian and Ukrainian surplus stocks.

In 2000 the Algerian Air Force contracted the South African company Advanced Technologies and Engineering (ATE) to upgrade its Mi-24Ds and Mi-24Vs. The work was performed on site, and the end result was similar to ATE's SuperHind III demonstrator (see South African section). The upgraded choppers could receive tactical information via data link from Raytheon Beech 1900D regional airliners converted into signals intelligence (SIGINT) aircraft. Predictably, this caused problems with the OEM – the Mil' Design Bureau, which had not authorised the conversion.

Later, Algeria received a number of new-build Mil' Mi-35 *Hind-Es* (the export version of

An Algerian Air Force Mi-2 in training colours with Dayglo orange stripe; note the addition of Mi-2 Plus style air intake filters.

Three AAF Mi-2s, including ST-67, in ordinary camouflage.

Algerian Air Force Mi-8TV SV-87 with western-made air intake filter

SN-12, a Mi-17-1V with the armament removed, flies with the clamshell cargo doors detached for a paradrop sortie. The standard globular intake filters have been replaced with a common filter of western origin.

Mi-171Sh SM-90 (note the wide starboard side door, high-set cabin heater and appropriate nose titles) creates a minor dust storm. This machine represents the early standard with 'glass nose' and clamshell cargo doors. Again, a common western-made intake filter is fitted.

Mi-17s SH-35 and SM-77 sport very different camouflage schemes. SH-35 is actually a medevac version based on the airframe of the Mi-19R airborne command post (note the small clamshell rear doors/airstair) and sports appropriate Red Crescent markings.

the Mi-24V) from Russia. They serve with an unidentified *Escadron d'Hélicoptères d'Assaut* (Assault Helicopter Squadron) at Ech Cheliff, while the older Mi-24s are operated by all three squadrons of the 1e RHC (including the 421e and 441e *Escadron d'Hélicoptères de Combat* at Biskra).

Naval tasks are performed by a small number of Kamov Ka-27 *Helix-A* anti-submarine warfare helicopters and Ka-32 *Helix-D* utility helicopters. These belong to a detachment of the *Escadre de Liaison et de Patrouille Maritime* stationed at Bou Sfer.

Algerian Air Force at war

Algeria used its MiG-15s on a small scale during the brief border conflict with Morocco in 1963. The seeds of conflict had been planted when both countries were still French colonies; France changed the border between the two at will, which led to territorial disputes after Algeria and Morocco gained independence. The main apple of discord was the area near Tindouf in western Algeria which is rich in iron ore. In October 1963 the Moroccan forces launched an offensive, intruding 100 km (62 miles) into Algerian territory near Colomb-Béchir. An urgently convened session of the Organisation of African Unity resulted in a ceasefire being brokered, the belligerents withdrawing to their original positions, and the pre-war status quo was restored. Later, in 1972, Algeria and Morocco agreed to extract the ore deposits in the disputed area jointly.

20 Algerian MiG-17Fs, 20 MiG-21F-13s and 12 IL-28s were dispatched to Egypt on 6th June 1967 to assist in the Six-Day War. However, the first six MiG-21F-13s were lost before they had a chance to fire a single missile. The fighters landed at El-Arish airbase on the Sinai Peninsula which, unbeknownst to the pilots, had recently been seized by the Israelis. The pilot of one MiG managed to destroy his aircraft but the other five jets were captured intact. Of these, four were sent to America for evaluation.

Algeria also lent a hand in the Holy Day War (Yom Kippur War) of 6th-24th October 1973, since Egypt and Syria needed all the help they could get. 23 MiG-17Fs, 13 MiG-21s and 12 Su-7BMKs were deployed to Egypt for operations against Israel, the *Fitters* operating from Bilbeis in the estuary of the Nile. The base was well equipped and, to give credit to the Arab strategists (or their advisors), it was conveniently located close to the Suez Canal, less than 100 km (62 miles) from the action; this

SH-58, another Algerian Air Force medevac helicopter based on the Mi-19R

Mi-171Sh SM-58 has been upgraded by the South African company ATE, featuring a Denel Argus 550 opto-electronic system and Kentron Ingwe anti-tank guided missiles.

allowed the Algerian *Fitters* to dispense with drop tanks, lugging a full ordnance load.

The Su-7s took part in the very first air strikes on 6th October when, according to Israeli estimates based on communications intelligence, up to 240 Arab aircraft pounded the Israeli positions on the Sinai Peninsula. On 7th October they, along with other Arab jets, attacked Israeli troops at El Taza and the Israeli Defence Force/Air Force (IDF/AF) base at Refidim located opposite Bilbeis across the canal. 21 aircraft were lost in so doing, including several Algerian ones. When the Egyptian offensive petered out and Israel launched a counteroffensive, the Algerian *Fitters* were used in the efforts to halt the advance; on 14th October they took part in a crucial battle against Israeli armoured convoys alongside Libyan Dassault Mirage 5Ds, again suffering heavy losses. However, Miss Fortune had turned her back on the Arabs; crossing the Suez Canal, the Israelis seized a bridgehead on the Egyptian side from which they started systematically destroying Egyptian air defences.

Interestingly, unlike the other Egyptian airbases, Bilbeis was not bombarded at all by the

SB-51, an Algerian Mi-24V upgraded by ATE to SuperHind Mk III standard. Note the 'proboscis' carrying a cannon and an optoelectronic system and the same four-packs of Ingwe ATGMs.

Most Algerian Mi-24s, including SuperHind Mk III SB-56, wear this sand/black camouflage scheme.

IDF/AF. Perhaps the reason was that the Israelis had underestimated the importance of this base, knowing that only a training unit had been stationed there before the war; meanwhile, the majority of the Arab powers' fighter-bomber fleets stationed elsewhere had already been destroyed by then. The Algerian aircraft fared no better than the Egyptian and Syrian ones – virtually the entire fleet was wiped out in the course of the conflict.

Albeit details of the Algerian Air Force's MiG-23MS, MiG-23MF and MiG-23BN operations are scarce, they would undoubtedly have been used to support the POLISARIO Front (Popular Front for the Liberation of Segiet-al-Hamra and Rio del Oro) in neighbouring Morocco. Whereas the POLISARIO campaign began as a movement to free Spanish Morocco from colonial rule, as soon as this was accomplished in 1978, Morocco unilaterally took over the Western Sahara and the objective changed to wrestling that area from Moroccan rule. Aided and abetted by Algeria and Libya, the POLISARIO Front had grown to a major army of about 15,000 troops. The Algerian Air Force based fighter-bombers at Tindouf near the Moroccan border, where the newly refurbished and extended runway facilitated air support to the POLISARIO guerrillas.

In 1991, after the military had taken command of the Algerian government, Muslim fundamentalists started a civil war in Algeria. It was precipitated by the government's cancel-

This view of SB-56 shows well the features of the SuperHind Mk III.

lation of the election results due to fears that the Islamic Salvation Front (FIS) party, which had won the first round, would not uphold the concepts of democracy. There were many different Islamic groups involved in the war, frequently fighting among themselves. The Armed Islamic Group (GIA – *Groupe Islamique Armée*) was especially notorious for murdering non-supportive civilians, threatening to kill not only anyone voting for its opponents but also all foreign residents in Algeria. No details of MiG-21 activities in this conflict have been released. However, it would be very surprising if MiG-21 fighter-bombers were not used to attack rebel groups in the countryside as the war officially continued until 2002, although small-scale fighting still continues. What is certain is that MiG-23BN fighter-bombers and Mi-24 attack helicopters were used in the campaign to crush the GIA rebels.

Known Algerian Air Force aircraft of Soviet/Russian origin

Type	Serial/ registration	C/n	F/n	Notes
An-12B (+)*	7T-WAA/560	402607		Serial not always carried visibly!
An-12B (+)	7T-WAB/566	402606		Sold to Mikma (?) as ER-ACC
An-12B (+)	7T-WAC/514	402605		C/n applied as 2605. Sold to Mikma as ER-ACD by 5-95
An-12B (+)	7T-WAD/?	?		No roundels and no c/n on. C/n sometimes reported in error as 402608, which is Aeroflot CCCP-11378!
An-12B (+)	7T-WAE/550	402810		Reportedly sold to an unknown Russian operator as RA-11119 by 12-95? (Or sold to ALADA, Angola, as D2-FAR.
An-12B (+)	7T-WAF/?	402811		
An-12B (+)	7T-WAG/591	402812		Sold to Mikma as ER-ACE by 6-95
An-12B (+)	7T-WAH/516	402809?		
IL-18V 'Salon'	7T-VRA	181003402		Ex-Aeroflot CCCP-75764, D/D 6-1-1964. Retired 1989
IL-28	024 Black	?		
IL-28	211 Black	?		
IL-28U	201 Black	?		
IL-76MD (+)	7T-WIA	0083489674	6709	
IL-76MD (+)	7T-WIB	0093493803	7101	
IL-76MD (–)	7T-WIC	1003405154	7909	No tail gunner's station
IL-76TD	7T-WID	1023414470	8708	Ex-Arkhangel'sk Airlines RA-76419, bought 1995
IL-76TD	7T-WIE	1023414463	8706	Ex-TransSuper Airlines RA-76406, bought by 7-96
IL-76TD	7T-WIG	1023413435	8609	Ex-Polis-Air RA-76407, bought 2003
IL-76TD	7T-WIM	1023411363	8501	Ex-East Line RA-76833, bought 2002
IL-76TD	7T-WIP	1043419636	9109	
IL-76TD	7T-WIR	1013408244	8201	Ex-IL-Avia RA-76835, bought 2003
IL-76TD	7T-WIT	1013408264	8206	Ex-EXPARC RA-76818, bought 2003
IL-76TD	7T-WIU	1023413423	8606	Ex-Arkhangel'sk Airlines RA-76440, bought 1993
IL-76TD	7T-WIV	1043419649	9203	
IL-78	7T-WIF	0043453597	4010	Ex-Ukrainian Air Force UR-76609
IL-78	7T-WIH	0053461837	4610	Ex-Ukrainian Air Force UR-76646
IL-78	7T-WIL	0043454640	4110	Ex-Ukrainian Air Force UR-76610
IL-78	7T-WIN	0063469080	5210	Ex-Ukrainian Air Force UR-76690
IL-78	7T-WIQ	0053462879	4710	Ex-Ukrainian Air Force UR-76653
IL-78	7T-WIS	0053464919	4810	Ex-Ukrainian Air Force UR-76662
MiG-15*bis*	522	?		
MiG-15*bis*	6053 Black	?		Natural metal finish, old 'winged' roundels
MiG-15*bis*	6054 Black	?		Natural metal finish, old 'winged' roundels
MiG-15*bis*	6059 Black	?		Natural metal finish, old 'winged' roundels
UTI-MiG-15	1755 Black	?		Czechoslovak-built (Aero CS-102)? Sand/green camouflage
UTI-MiG-15	?	622974		Czechoslovak-built (Aero CS-102); ex-East German Air Force 15 Black

UTI-MiG-15	?	922330		Czechoslovak-built (Aero CS-102); ex-East German Air Force 138 Black
UTI-MiG-15	?	022529		Czechoslovak-built (Aero CS-102); ex-East German Air Force 112 Black
UTI-MiG-15	?	022534		Czechoslovak-built (Aero CS-102); ex-East German Air Force 137 Black
MiG-17F	02 Black	?		Natural metal finish, new 'wingless' roundels
MiG-17F	24 Black	?		19 EdeCh; overall grey finish, new roundels
MiG-17F	38 Black	?		Seen Algiers 2-1980
MiG-17F	231 Black	?		Seen Algiers 2-1980
MiG-17F	275 Black	2523?		Preserved Army Museum, Algiers
MiG-17F	1768 Black	?		Seen Algiers 2-1980
MiG-17F	?	7019		
MiG-21F-13	09 Black	74****		Silver overall, 'lightning bolt' on nose
MiG-21F-13	66	74****		Silver overall
MiG-21F-13	69 Black	74****		Silver overall
MiG-21F-13	75 Black	74****		Silver overall
MiG-21F-13	77 Black	74****		Silver overall
MiG-21F-13	87 Black	74****		Silver overall
MiG-21F-13	231 Black	74****		
MiG-21F-13	232 Black	74****		
MiG-21F-13	233 Black	74****		
MiG-21F-13	234 Black	74****		
MiG-21F-13	235 Black	74****		
MiG-21F-13	236 Black	74****		
MiG-21F-13	237 Black	74****		
MiG-21F-13	238 Black	74****		
MiG-21F-13	239 Black	74****		
MiG-21F-13	240 Black	74****		
MiG-21F-13	241 Black	74****		
MiG-21F-13	242 Black	74****		
MiG-21F-13	243 Black	74****		
MiG-21F-13	244 Black	74****		
MiG-21F-13	245 Black	74****		
MiG-21F-13	246 Black	74****		
MiG-21F-13	247 Black	74****		
MiG-21F-13	248 Black	74****		
MiG-21F-13	249 Black	74****		
MiG-21F-13	250 Black	74****		
MiG-21F-13	251 Black	74****		
MiG-21F-13	252 Black	74****		
MiG-21F-13	253 Black	74****		
MiG-21F-13	254 Black	74****		
MiG-21F-13	255 Black	74****		
MiG-21F-13	256 Black	74****		
MiG-21F-13	257 Black	74****		
MiG-21F-13	259 Black	74****		
MiG-21F-13	259 Black	74****		
MiG-21F-13	260 Black	74****		
MiG-21F-13	261 Black	74****		
MiG-21F-13	272 Black	74****		190e EdeCh? Ground instructional airframe at Biskra
MiG-21F-13	280 Black	742316		Ex-Hungarian Air Force 2316 Red. 190e EdeCh? Derelict at Tangiers Airport
MiG-21F-13	1395	74****		190e EdeCh? Wfu by 1984
MiG-21FL	31 Black	?		
MiG-21MF	FD-16	964712		Withdrawn from use
MiG-21MF	FD-19	96...		Stored Odessa, Ukraine, 2005
MiG-21MF	FD-26	963811		Stored Odessa, Ukraine, 2005
MiG-21MF	FD-43	96...		
MiG-21MF	FD-59	96...		Stored Odessa, Ukraine, 2005
MiG-21MF	?	964702		Stored Odessa, Ukraine, 2005

MiG-21MF	'FD'	96...		Gate guard at Bou Sfer AB
MiG-21*bis*	132	N75037821	75-24-**?	Noted 1984 withdrawn from use
MiG-21*bis*	142	N750*****	?	Noted 1984 withdrawn from use
MiG-21*bis*	162	N750*****	?	Noted 1984 withdrawn from use
MiG-21*bis*	167	N75037807?	75-24-**?	Stored Blida
MiG-21*bis*	171	N750*****	?	Noted 1984 withdrawn from use
MiG-21*bis*	191	N750*****	?	Noted 1984 withdrawn from use
MiG-21*bis*	FA-41	N750*****	?	Tan/brown camouflage. Stored Odessa, Ukraine, 2005
MiG-21*bis*	FA-51	N750*****	?	Noted stored 2005 Odessa, Ukraine
MiG-21*bis*	FA-52	N75036401	75-23-**	Noted 1984 withdrawn from use
MiG-21*bis*	FA-62	N75036316	75-23-**	Noted withdrawn from use
MiG-21*bis*	FA-63	N750*****	?	Noted stored 2005 Odessa, Ukraine
MiG-21*bis*	FA-65	N750*****	?	Tan/brown camouflage. Stored Odessa, Ukraine, 2005
MiG-21*bis*	FA-73	N750*****	?	Tan/brown camouflage. Stored Odessa, Ukraine, 2005
MiG-21*bis*	FA-74	N750*****	?	Tan/brown camouflage. Stored Odessa, Ukraine, 2005
MiG-21*bis*	FA-89	N750*****	?	Tan/brown camouflage. Stored Odessa, Ukraine, 2005
MiG-21*bis*	FA-92	N75038399	75-25-**?	Noted 1984 withdrawn from use
MiG-21*bis*	FA-94	N75038210?	75-25-**?	See next line!
MiG-21*bis*	FA-96	N750*****	?	Tan/brown camouflage. Noted stored Odessa, Ukraine, 2005; c/n reported as N75038210! Either this aircraft or FA-94 to Yemen Air Force as 2232
MiG-21*bis*	?	N75037857?	75-24-**?	Noted withdrawn from use
MiG-21*bis*	?	N75036406	75-23-**	To Yemen Air Force as 2236
MiG-21*bis*	?	N75038147	75-25-**?	To Yemen Air Force as 2230
MiG-21*bis*	?	N75038206	75-25-**?	To Yemen Air Force as 2231
MiG-21*bis*	?	N75038314	75-25-**?	To Yemen Air Force as 2233
MiG-21*bis*	?	N75038327	75-25-**?	To Yemen Air Force as 2234
MiG-21*bis*	?	N75038399	75-25-**?	To Yemen Air Force as 2235
MiG-21UM	209	?	?	Noted wfu 1984
MiG-21UM	231	?	?	
MiG-21UM	271	662304		
MiG-21U	272	662305		
MiG-21U	273	662311		
MiG-21U	274	662307		
MiG-21U	275	662312		
MiG-21U	276	662314		
MiG-21U	277	662309		
MiG-21U	278	662306		
MiG-21U	279	662319		Crashed 9-1982
MiG-21U	280	662310		
MiG-21U	281	662317		
MiG-21U	282	662315		
MiG-21U	283	662320		Became, see next line
	FC-80			
MiG-21UM	FE-05	?	?	Tan/brown camouflage
MiG-21UM	FE-40	?	?	Tan/brown camouflage
MiG-21UM	FE-209	?	?	Noted wfu 1984, see 209 above
MiG-21UM	FN-70	?	?	Stored Blida
MiG-23MF	FP-11	03902*****	?	
MiG-23MF	FP-27	0390224605	14408	Sand/tan camouflage
MiG-23MF	FP-47	03902*****	?	
MiG-23MF	?	0390220107	13***	
MiG-23MLD	FP-84	03903*****	?	Green/sand/tan camouflage
MiG-23BN	347	03932*****	?	28 Sqn. Became, see next line
	FM-47			
MiG-23BN	357	03932*****	?	28 Sqn. Became, see next line
	FM-57			
MiG-23BN	361?	03932*****	?	28 Sqn. Became, see next line
	FM-61			

MiG-23BN	370 FM-70	03932*****	?	28 Sqn. Became, see next line
MiG-23BN	372 FM-72	03932*****	?	28 Sqn. Became, see next line
MiG-23BN	376? FM-76	03932*****	?	28 Sqn. Became, see next line
MiG-23BN	383 FM-83	03932*****	?	28 Sqn. Sand/green/brown camouflage. Became, see next line
MiG-23BN	387? FM-87	03932*****	?	28 Sqn. Became, see next line
MiG-23BN	388 FM-88	03932*****	?	28 Sqn. Became, see next line
MiG-23BN	389 FM-89	03932*****	?	28 Sqn. Sand/green/brown camouflage. Became, see next line
MiG-23BN	390 FM-90	03932*****	?	28 Sqn. Became, see next line
MiG-23BN	392 FM-92	03932*****	?	28 Sqn. Became, see next line
MiG-23BN	394 FM-94	03932*****	?	28 Sqn. Became, see next line
MiG-23BN	FM-13	03932*****	?	29 Sqn
MiG-23BN	FM-14	03932*****	?	29 Sqn
MiG-23BN	FM-15	03932*****	?	29 Sqn
MiG-23BN	FM-16	03932*****	?	29 Sqn
MiG-23BN	FM-17	0393216837		
MiG-23UB	302? FT-02	B103****?	?	Became, see next line
MiG-25PD	FU-31	N840*3701	?	3701 on tails
MiG-25PD	FU-36	N840*****	?	Serial previously painted with no hyphen!
MiG-25PD	FU-45	N840*****	?	
MiG-25PD	FU-48	N840*****	?	
MiG-25PD	FU56	N840*****	?	No hyphen
MiG-25PD	FU-71	N840*****	?	
MiG-25PD	FU-75	N840*2207	?	2207 on tails
MiG-25PD	FU85	N840*****	?	
MiG-25PD	FU-94	N840*****	?	
MiG-25RBV	503	N020*****	?	
MiG-25RBV	689	N020*****	?	
MiG-25RBV	FG-34	N020*3217?	?	3217 on tails?
MiG-25RBK	FG-62	N020*****	?	
MiG-25RBK	FG-78	N020*3310	?	3310 on tails
MiG-25PU	FL-10	N220*4142?	?	4142 on tails?
MiG-29SMT	FC-12	2960733029	5816?	To Russian Air Force ?? Red
MiG-29SMT	FC-14	2960733017	5726?	To Russian Air Force ?? Red
MiG-29S	FC-15	29607*****	?	
MiG-29...	FC-16	29607*****	?	
MiG-29S	FC-17	29607*****	?	
MiG-29S	FC-18	2960733030	5817?	To Russian Air Force ?? Red
MiG-29S	FC-19	29607*****	?	
MiG-29S	FC-25	29607*****	?	
MiG-29S	FC-31	2960707727	?	Ex-04 White (787th Fighter Regiment). 3.EDA/193 Sqn. Crashed 10-10-1999
MiG-29SMT	FC-32	2960733333	5818?	D/D 3-2007, returned 4-2008. To Russian Air Force ?? Red
MiG-29S	FC-34	29607*****	?	
MiG-29...	FC-35	29607*****	?	Dark camouflage scheme
MiG-29S	FC-45	29607*****	?	
MiG-29S	FC-48	29607*****	?	
MiG-29...	FC-53	29607*****	?	

MiG-29SMT	FC-57	2960734104	?	To Russian Air Force 08 Red
MiG-29...	FC-60	29607*****	?	
MiG-29S	FC-61	29607*****	?	
MiG-29...	FC-64	29607*****	?	
MiG-29S	FC-71	29607*****	?	
MiG-29S	FC-84	2960714935	?	Ex-76 White (787th Fighter Regiment). 3.EDA/193 Sqn. Crashed 19-4-2000
MiG-29S	FC-93	29607*****	?	
MiG-29S	FC-98	29607*****	?	
MiG-29S	FC-99?	29607*****	?	
MiG-29SMT	FC-??	2960731684	5623?	To Russian Air Force ?? Red
MiG-29SMT	FC-??	2960731688	5624?	To Russian Air Force 31 Red
MiG-29SMT	FC-??	2960733021	5727?	Possibly undelivered. To Russian Air Force 23 Red
MiG-29SMT	?	2960733024	5728?	Possibly undelivered. To Russian Air Force 22 Red
MiG-29SMT	FC-??	2960733027	5729?	To Russian Air Force ?? Red
MiG-29SMT	FC-??	2960733028	5730?	To Russian Air Force ?? Red
MiG-29SMT	?	2960733334	5819?	D/D 3-2007, returned 5-2008. To Russian Air Force 26 Red
MiG-29SMT	?	2960733335	5820?	Possibly undelivered. To Russian Air Force 04 Red
MiG-29SMT	?	2960733337	5821?	Possibly undelivered. To Russian Air Force 09 Red/RF-92929
MiG-29SMT	?	2960733340	5822?	Possibly undelivered. To Russian Air Force 05 Red
MiG-29SMT	FC-??	2960733344	5823??	D/D 5-2007, returned 4-2008. To Russian Air Force 05 Red/RF-92925
MiG-29SMT	FC-??	2960733345	5824??	D/D 5-2007, returned 4-2008. To Russian Air Force ?? Red
MiG-29SMT	?	2960734100	?	Possibly undelivered. To Russian Air Force 20 Red/RF-92233
MiG-29SMT	?	2960734101	?	Possibly undelivered. To Russian Air Force 21 Red/RF-92232
MiG-29SMT	?	2960734105	?	Possibly undelivered. To Russian Air Force 10 Red
MiG-29SMT	?	2960734106	?	Possibly undelivered. To Russian Air Force 02 Red
MiG-29SMT	?	2960734110	?	Possibly undelivered. To Russian Air Force 03 Red
MiG-29SMT	?	2960734120	?	Possibly undelivered. To Russian Air Force 07 Red/RF-92927
MiG-29SMT	?	2960734124	?	Possibly undelivered. To Russian Air Force 01 Red
MiG-29SMT	?	2960734130	?	Possibly undelivered. To Russian Air Force 22 Red
MiG-29SMT	?	2960734135	?	Possibly undelivered. To Russian Air Force 23 Red/RF-95811
MiG-29UB	FB-20	N509030*****	?	
MiG-29UB	FB-30	N509030*****	?	
MiG-29UB	FB-60	N509030*****	?	
MiG-29UB	FB-80	N509030*****	?	Dark camouflage scheme
MiG-29UBT	FB-02	N50903029437	?	3.EDA. To Russian Air Force ?? Red
MiG-29UBT	FB-04	N509030*****	?	3.EDA. D/D 5-2007, returned 5-2008, to Russian Air Force ?? Red; c/n N50903029760?
MiG-29UBT	FB-05	N509030*****	?	3.EDA. D/D 5-2007, returned 5-2008, to Russian Air Force ?? Red; c/n N50903029872?
MiG-29UBT	FB-06	N509030*****	?	3.EDA. D/D 5-2007, returned 5-2008, to Russian Air Force ?? Red; c/n N50903029983?
Mi-2T	ST-11	5110747118		
Mi-2T	ST-15	5110802118		
Mi-2T	ST-17	5110803118		
Mi-2T	ST-19	5110804128		
Mi-2T	ST-21	5110805128		
Mi-2T	ST-22	5110748118		
Mi-2T	ST-23	5110806128		
Mi-2T	ST-25	5110807128		
Mi-2T	ST-27	5110808128		Gate guard at Aïn Arnat
Mi-2T	ST-29	5110809128		
Mi-2T	ST-31	5110810128		
Mi-2T	ST-35	5110811128		
Mi-2T	ST-37	5110812128		
Mi-2T	ST-39	5110930069		
Mi-2T	ST-41	5110931069		
Mi-2T	ST-43	5110933069		
Mi-2T	ST-45	5110932069		

Mi-2T	ST-48	5110750118		
Mi-2T	ST-49	5110934069		
Mi-2T	ST-51	5110935069		
Mi-2T	ST-53	5110936069		
Mi-2T	ST-57	5110937069		
Mi-2T	ST-59	5110938069		
Mi-2T	ST-61	5110939069		
Mi-2T	ST-63	5111017109		
Mi-2T	ST-65	5111018109		
Mi-2T	ST-67	5111019109		
Mi-2T	ST-69	5111011089		
Mi-2T	ST-71	5111012089		
Mi-2T	ST-73	5111013089		
Mi-2T	ST-75	5111014099		
Mi-2T	ST-77	5111015099		
Mi-2T	ST-79	5110749118		
Mi-2T	ST-83	5111016099		
Mi-2T	ST-87	5111020109		
Mi-2T	ST-89	5111021119		
Mi-2T	ST-91	5111022119		
Mi-2T	ST-93	5111023119		
Mi-2T	ST-95	5111024119		
Mi-2T	ST-97	5111025119		
Mi-2T	ST-99	5110801118		
Mi-6	7T-WMA	?		
Mi-6	7T-WMB	?		
Mi-6	7T-WMC	?		
Mi-6	7T-WMD	?		
Mi-6	SZ-45	?		Ex-7T-WMA/B/C/D. Seen 1986
Mi-8...	7T-WUX/693	?	?	
Mi-8MT	SV-12	?	?	
Mi-8PS	SV-18	?	?	
Mi-8TV	SV-19/1105	1156		Additional serial as reported (does not match the c/n)
Mi-8S	SV-21	?	?	White with red/white/green trim, Pall intake filter; probably 9ᵉ RHE/Special Mission Squadron
Mi-8T	SV-32	?	?	
Mi-8S	SV-38	?	?	Probably 9ᵉ RHE/Special Mission Squadron
Mi-8TV	SV-43	?	?	
Mi-8TV	SV-53	?	?	
Mi-8TV	SV-64	?	?	
Mi-8T	SV-65	?	?	
Mi-8T	SV-86	?	?	9ᵉ RHE
Mi-8TV	SV-87	?	?	Sand/tan camouflage, Pall air filter
Mi-8T	SV-93	?	?	
Mi-8TV	SV-97	?	?	
Mi-8MT?	SV-36	?	?	Type as reported but may be Mi-8T
Mi-17 (Mi-19 mod?)	SH-35	?	?	Sand/tan camouflage, Red Crescent markings (medevac)
Mi-17 (Mi-19 mod?)	SH-58	?	?	Green/brown camouflage, Red Crescent markings (medevac)
Mi-171	SM-12	?	?	
Mi-171	SM-18	?	?	
Mi-171	SM-23	?	?	
Mi-171	SM-24	?	?	
Mi-171	SM-26	?	?	
Mi-171	SM-27	?	?	Sand/tan camouflage; armed
Mi-171	SM-29	?	?	Sand/tan camouflage; armed
Mi-171	SM-32	?	?	
Mi-171	SM-34	?	?	

Mi-171	SM-35	?	?	
Mi-171	SM-36	?	?	
Mi-171	SM-38	?	?	Green/brown camouflage; armed
Mi-171	SM-42	?	?	
Mi-171	SM-43	?	?	Green/brown camouflage; armed
Mi-171	SM-47	?	?	
Mi-171	SM-48	?	?	
Mi-171	SM-49	?	?	
Mi-171	SM-51	?	?	
Mi-171	SM-53	?	?	
Mi-171	SM-55	?	?	Green/brown camouflage
Mi-171Sh	SM-58	?	?	Modified by ATE, with Atlas 550 optoelectronic turret and Ingwe ATGMs
Mi-171	SM-63	?	?	
Mi-171	SM-68	?	?	Sand/tan camouflage; armed
Mi-171	SM-69	?	?	Sand/tan camouflage; armed
Mi-171	SM-74	?	?	
Mi-171	SM-76	?	?	
Mi-171Sh	SM-77	?	?	
Mi-171	SM-84	?	?	Sand/tan camouflage; armed
Mi-171Sh	SM-85	?	?	
Mi-171	SM-87	?	?	
Mi-171Sh	SM-88	?	?	
Mi-171Sh	SM-90	?	?	Glass nose, two-tone green camouflage
Mi-171	SM-93	?	?	
Mi-171	SM-95	?	?	
Mi-171	SM-96	?	?	Glass nose, sand/tan camouflage
Mi-171	SM-99	?	?	
Mi-171Sh	SN-12	?	?	Glass nose, sand/tan camouflage
Mi-171	SN-16	?	?	Sand/tan camouflage
Mi-171	SN-36	?	?	
Mi-171	SN-42	?	?	9e RHE/Special Mission Squadron
Mi-171	SN-99	?	?	
Mi-24A	634	3532454202034?	?	C/n reported as 4202034 but there cannot be 34 aircraft to a batch under the Mi-24A's original c/n system. Became, see next line
	SU-97			To South Africa for evaluation by ATE; preserved SAAF Museum, Swartkop, since 2001 as 'Soviet AF 34 Yellow'
Mi-24...	SB-12	?	?	ATE SuperHind Mk III upgrade
Mi-24D?	SB-16	070116	?	ATE SuperHind Mk III upgrade, sand/tan camouflage
Mi-24D?	SB-17	070117	?	
Mi-24D?	SB-19	?	?	
Mi-24D	SB-21	070129	?	ATE SuperHind Mk III upgrade. To ATE, South Africa, in 2002 as ZU-GAL
Mi-24...	SB-23	?	?	ATE SuperHind Mk III upgrade, sand/tan camouflage
Mi-24V	SB-25	?	?	
Mi-24V	SB-28	?	?	ATE SuperHind Mk III upgrade, sand/black camouflage
Mi-24V	SB-43	?	?	
Mi-24V	SB-46	?	?	
Mi-24V?	SB-51	?	?	ATE SuperHind Mk III upgrade, sand/black camouflage
Mi-24V	SB-53	070154	?	
Mi-24V?	SB-56	**7537	?	ATE SuperHind Mk III upgrade, sand/black camouflage, 7537 on tail rotor pylon
Mi-24V?	SB-58	?	?	ATE SuperHind Mk III upgrade, sand/black camouflage
Mi-24V?	SB-63	?	?	ATE SuperHind Mk III upgrade, sand/black camouflage
Mi-24V?	SB-65	?	?	
Mi-24V?	SB-72	?	?	ATE SuperHind Mk III upgrade, sand/black camouflage
Mi-24V?	SB-79	?	?	ATE SuperHind Mk III upgrade
Mi-24V?	SB-83	?	?	ATE SuperHind Mk III upgrade
Mi-24V?	SB-92	?	?	ATE SuperHind Mk III upgrade, sand/black camouflage
Mi-24V?	SB-96	?	?	ATE SuperHind Mk III upgrade, sand/black camouflage
Mi-24V	SU-**	3532421809222	?	To South Africa, 1998; became ATE SuperHind Mk II prototype ZU-BOI

Su-7BMK	333	?		Stored Blidah 12-2008
Su-7BMK	335	?		
Su-7BMK	337	?		
Su-7BMK	339	?		
Su-7BMK	341	?		
Su-7BMK	KR-57	?		Not HR-57 as sometimes reported; '1957' on tail. Preserved somewhere in Algeria, serial probably fake
Su-7UMK	1313	?		Stored Blidah 12-2008
Su-24MK	KX-11	?		Sand/tan top, pale blue bottom; upgraded
Su-24MK	KX-12	?		Sand/tan top, pale blue bottom; upgraded
Su-24MK	KX-15	?		Sand/tan top, pale blue bottom; upgraded
Su-24MK	KX-16	?		
Su-24MK	KX-22	?		Sand/tan top, pale blue bottom; upgraded
Su-24MK	KX-25	?		Sand/tan top, pale blue bottom
Su-24MK	KX-29	?		Dark green/dark earth top, pale blue bottom
Su-24MK	KX-32	?		Dark green/dark earth top, pale blue bottom
Su-24MK	KX-33	?		
Su-24MK	KX-37	?		
Su-24MK	KX-40	?		Dark green/dark earth top, pale blue bottom
Su-24MK	KX-41	?		
Su-24MK	KX-42	?		
Su-24MK	KX-43	?		Dark green/dark earth top, pale blue bottom; upgraded
Su-24MK	KX-45	?		Dark green/dark earth top, pale blue bottom
Su-24MK	KX-46	?		Dark green/dark earth top, pale blue bottom
Su-24MK	KX-48	?		
Su-24MK	KX-49	?		
Su-24MK	KX-51	?		Sand/tan top, pale blue bottom
Su-24MK	KX-53	?		Sand/tan top, pale blue bottom; upgraded
Su-24MK	KX-55	?		Sand/tan top, pale blue bottom; upgraded
Su-24MK	KX-57	?		
Su-24MK	KX-62	?		Sand/tan top, pale blue bottom; upgraded
Su-24MK	KX-65	?		Dark green/dark earth top, pale blue bottom
Su-24MK	KX-67	?		Dark green/dark earth top, pale blue bottom
Su-24MK	KX-68	?		Dark green/dark earth top, pale blue bottom
Su-24MK	KX-71	?		Dark green/dark earth top, pale blue bottom; later sand/tan top, pale blue bottom
Su-24MK	KX-72	?		Dark green/dark earth top, pale blue bottom
Su-24MK	KX-73	?		Dark green/dark earth top, pale blue bottom
Su-24MK	KX-75	?		Upgraded
Su-24MK	KX-76	?		
Su-24MK	KX-83	?		Dark green/dark earth top, pale blue bottom
Su-24MK	KX-85	?		
Su-24MK	KX-86	?		
Su-24MK	KX-88	?		Sand/tan top, pale blue bottom; upgraded
Su-24MK	KX-89	?		
Su-24MK	KX-93	?		Dark green/dark earth top, pale blue bottom
Su-24MK	KX-99	?		Dark green/dark earth top, pale blue bottom
Su-24MK	KX-?	4160451002312	02-12	Converted from Soviet Air Force Su-24M c/n 0215312; one of the above 38 aircraft
Su-24MK	KX-?	4160454002318	02-18	Converted from Soviet Air Force Su-24M c/n 0215318; one of the above 38 aircraft
Su-24MK	KX-?	4160454002319	02-19	Converted from Soviet Air Force Su-24M c/n 0215319; one of the above 38 aircraft
Su-24MK	KX-?	4160451810656	10-56	Converted from Soviet Air Force/302nd Bomber Regiment Su-24M c/n 1041656; one of the above 38 aircraft
Su-24MK	KX-?	416045**12*09	12-09	Converted from Soviet Air Force Su-24M c/n 1241609; one of the above 38 aircraft
Su-24MK	KX-?	416045**12*22	12-22	Converted from Soviet Air Force Su-24M c/n 1241622; one of the above 38 aircraft
Su-24MK	KX-?	416045**12*25	12-25	Converted from Soviet Air Force Su-24M c/n 1241625; one of the above 38 aircraft
Su-24MRK	KG-24	?		510.EdeR. Dark green/dark earth top, pale blue bottom
Su-24MRK	KG-46	?		510.EdeR. Dark green/dark earth top, pale blue bottom
Su-24MRK	KG-65	?		510.EdeR. Dark green/dark earth top, pale blue bottom
Su-24MRK	KG-87	?		510.EdeR

Su-24MRK	KG-93	?		510.EdeR
Su-24MRK	KG-87	?		510.EdeR
Su-24MRK	KG-?	0941647		Converted from Soviet Air Force Su-24MR; one of the above five aircraft
Su-24MRK	KG-?	1041604		Converted from Soviet Air Force Su-24MR; one of the above five aircraft
Su-24MRK	KG-?	1041606		Converted from Soviet Air Force Su-24MR; one of the above five aircraft
Su-24MPK	KY-51	416046*******	?	510.EdeR
Su-30MKA	KF-11	10MK4***		
Su-30MKA	KF-12	10MK4101		Overall grey c/s
Su-30MKA	KF-16	10MK4***		Overall grey c/s
Su-30MKA	KF-21	10MK4303		Ex-39 Black (test serial)
Su-30MKA	KF-22	10MK4104?		Two-tone camouflage; had additional Russian Air Force insignia and code '224' prior to delivery
Su-30MKA	KF-24	10MK4105		Two-tone camouflage
Su-30MKA	KF-33	10MK4***		
Su-30MKA	KF-34	10MK4***		Overall grey c/s
Su-30MKA	KF-39	10MK4***		Overall grey c/s
Su-30MKA	KF-45	10MK4***		Overall grey c/s
Su-30MKA	KF-44	10MK4***		
Su-30MKA	KF-45	10MK4***		Overall grey c/s
Su-30MKA	KF-48	10MK4203		Overall grey c/s
Su-30MKA	KF-53	10MK4***		
Su-30MKA	KF-56	10MK4206		Overall grey c/s
Su-30MKA	KF-59	10MK4***		
Su-30MKA	KF-94	10MK4***		Overall grey c/s
Su-30MKA	KF-...	10MK4106		Two-tone camouflage
Yak-130	NY-11	01**		D/D 11-2011
Yak-130	NY-12	01**		D/D 11-2011
Yak-130	NY-13	01**		D/D 11-2011
Yak-130	NY-16	01**		
Yak-130	NY-61	0117		Completed 12-2011

* Note: Hereinafter 'An-12... (+)' in fleet list tables means that the aircraft was built in the military version with a tail gunner's station, while 'An-12... (–)' means the aircraft was built in the commercial version with no tail gunner's station. The same applies to the IL-76MD.

Angola

On 11th November 1975 the former Portuguese colony of Angola proclaimed independence, with a left-wing government headed by José Eduardo dos Santos. The Soviet Union was one of the first to recognise the Republic of Angola and establish political and economic ties with it – in the hope of securing another bridgehead in Africa, of course. When the Angolan People's Air Force/Air Defence Force (FAPA-DAA – *Força Aérea Popular de Angola – Defesa Anti-Aviões*) was established in January 1976, at first it used whatever aircraft had been left behind by the Portuguese. Soon, however, a Soviet-Angolan friendship treaty was signed in October 1976, opening the door for Soviet weapons deliveries to the republic.

Angolan Air Force MiG-17F C24 in two-tone camouflage.

Angolan Air Force MiG-17F C23 as delivered in natural metal finish

Angolan Air Force MiG-17F C24 in sand/green camouflage; note the different insignia and the extra bomb pylons

In 1992, after the First Angolan Civil War, the FAPA was reorganised and became the FANA (*Força Aérea Nacional Angolana*). This continued to operate Soviet aircraft types.

The first Soviet fighter to enter service with the FAPA was the Mikoyan/Gurevich MiG-17F. Some reports say the Soviet Union supplied ten *Fresco-Cs* to Angola by 1976; they reportedly had maintenance stencilling in French, suggesting they were diverted from an order intended for a French-speaking African country. Other sources state that it was Cuba who donated the MiG-17Fs. The latter and the MiG-21F-13s formed the backbone of the FAPA until the early 1980s when they were succeeded by more modern types; however, at least eight were reportedly still operational in late 1991. The MiG-17Fs were based near the Angolan capital of Luanda. Initially they wore their natural metal factory finish (some sources suggest a rather improbable light blue overall finish!) but gained a sand/olive drab camouflage around 1980. They were the only FAPA fighters to wear the old version of the roundels (black with a heavy red surround and a yellow pentastar in the middle).

Two (some sources say three) Mikoyan/Gurevich UTI-MiG-15 trainers were delivered to the FAPA by Cuba in December 1975 or early 1976. Two of the *Midgets* were still operational in the mid-1980s.

The first Mikoyan/Gurevich MiG-21 supersonic fighters to arrive in Angola were 12 MiG-21MF (*izdeliye* 96F) *Fishbed-J* fighters transferred from the Cuban Air Force; they are sometimes reported as MiG-21PFM (*izdeliye* 94) *Fishbed-Ds*. The first batch of MiG-21MFs was airlifted to Angola by Soviet Air Force Antonov An-22 Antey (*Cock*) transports on 8th January 1976 and reassembled on 13th

Angolan Air Force (FAPA) insignia as used in the 1980s.

Three FAPA MiG-21*bis* fighters, including C310 and C314, share a hardstand with a Cuban Air Force MiG-29UB. Note the different camouflage schemes and the early-style black/yellow/red striped rudders.

Angolan Air Force (FAPA) MiG-21MF C41 in early insignia armed with R-3S AAMs; this aircraft was flown by the Cuban General Rafael del Pino

FAPA MiG-21*bis Fishbed-L* C326 shot down on 24th September 1984; note the roundel on the tail

FAPA MiG-21*bis Fishbed-L* C314 in non-standard three-tone camouflage carrying four UB-16-57 FFAR pods and a MiG-23 style drop tank

FAPA MiG-21*bis Fishbed-N* C340 which crash-landed in Namibia in 1989

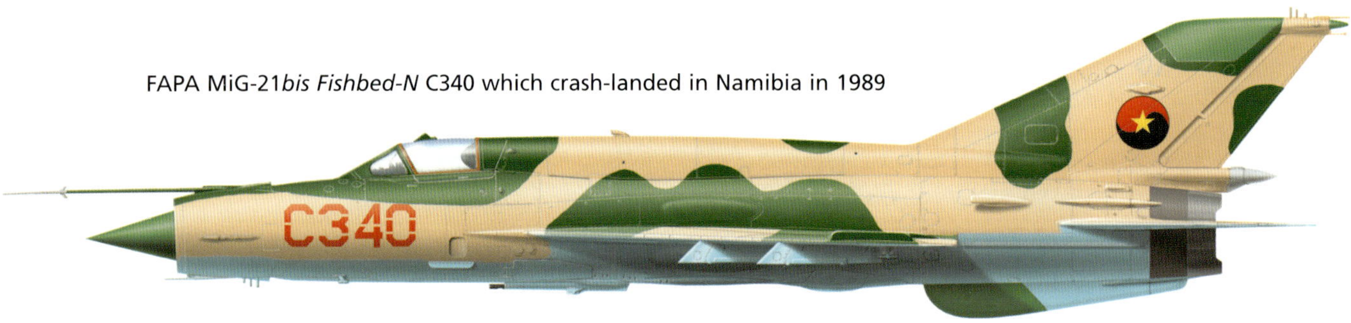

FAPA MiG-21*bis Fishbed-N* C350

Angolan Air Force MiG-21*bis* C340 was captured by the South Africans after force-landing in Namibia in 1989. It is now on display in the South African Air Force Museum.

March 1976; these aircraft had the Cuban serials '510 Red' through '521 Red', becoming C41 through C52 upon delivery. A few months later the Soviet Union supplied eight MiG-21F-13 (*izdeliye* 74) *Fishbed-C* fighters. 12 more MiG-21MFs arrived from the Soviet Union in 1980, followed in 1983 by 12 MiG-21*bis* (*izdeliye* 75B) *Fishbed-L* fighters; further supplies of up to 35 MiG-21*bis* came in 1985, followed by another tranche in 1986. The MiG-21MFs and *Fishbed-Ls* wore a sand/olive drab or sand/brown/green/dark green camouflage. It appears that the ones flown by Cuban pilots seconded to FAPA had the rudder painted red, yellow and black (top to bottom), like the MiG-17Fs, though these colours disappeared later. In contrast, the MiG-21s flown by Angolan pilots sported the FAPA roundels (red/black with an S-shaped colour division line and the star in the middle) on the fin and on the wings (port/upper side and starboard/underside, US style).

As regards trainer versions, the Soviet Union supplied two MiG-21US (*izdeliye* 68A) *Mongol-Bs* in 1976 with the *Fishbed-Cs*; at least two MiG-21USs or MiG-21U (*izdeliye* 66-600) *Mongol-As* followed in 1980, a further two to four MiG-21USs in 1983 and almost certainly MiG-21UM *Mongol-Bs* in 1986. Unfortunately accurate figures of the number of MiG-21s procured by the FAPA-DAA are not available. In 2010 it was reported that some of the MiG-21s had been refurbished by Odesaviaremservis (Odessa Aircraft Repair Service) in the Ukraine and returned to active status.

The Angolan MiG-21s saw service with the following three regiments. The 24th Air Instructional Regiment included the 9th Fighter Training Squadron with some seven MiG-21USs and five MiG-21UMs. The 25th Air Combat Fighter Regiment comprised the 11th Fighter Squadron (with a mix of MiG-21F-13s, MiG-21MFs and MiG-21*bis* at various times) for reconnaissance and close air support (CAS), the 12th FS and 13th FS for interception; the latter two squadrons flew the MiG-21MF and/or MiG-21*bis* until replaced by the MiG-23ML in 1987. The 26th ACFR (comprising the 14th FS, 15th FS and 16th FS) was a CAS unit and had no MiG-21s after 1988. No operational MiG-21s remained by 2007.

The Angolan Air Force received several variants of the Mikoyan MiG-23. Initial deliveries reportedly included about 20 of the MiG-23MF *Flogger-B* fighter (or the simplified MiG-23MS *Flogger-E*). They were based at Cahama, Catumbela, Cuito Cuanavale, Lubongo, Luena, Menongue, Moçamedes and Saurimo. These were superseded in the interceptor role by 20 MiG-23ML *Flogger-G* fighters (sometimes reported as MiG-23MLAs) delivered in 1985. In 1987-88 the FAPA received two squadrons of MiG-23MLs from the Soviet Air Force – specifically, the 20th Fighter Division/821st Fighter Regiment based at Spassk-Dal'niy in the Far East; the fighters were suitably modified at an aircraft repair plant prior to delivery. The total number of

Here, C340 is seen being worked on in South Africa.

Angolan Air Force (FANA) MiG-23ML C-474 and MiG-23UB I-28 await redelivery after being refurbished by Odesaviaremservis.

Flogger-Gs delivered is estimated as at least 80. The fighter variants were operated by the 12th and 13th Fighter Squadrons of the 25th ACFR. Conversion and proficiency training was undertaken on initially six MiG-23UB *Flogger-Cs*. More MiG-23s were reportedly delivered in 1998 (six from Russia and 14 from Belarus); two aircraft were reportedly resold to Namibia by April 2000. The MiG-23s wore a variety of sand/olive drab or sand/brown/olive drab camouflage schemes with the roundel on the fin, and it seems that each aircraft had an individual pattern. In 2005 the 18 surviving MiG-23MLs were refurbished and upgraded to MiG-23MLD *Flogger-K* standard by Odesaviaremservice, receiving a Sapfeer-21 radar and a modified SUV-23 weapons control system allowing R-73 and R-77 air-to-air missiles to be carried. In so doing they received a new camouflage scheme and new FANA roundels lacking the star.

In January 1999 the FANA purchased a single Sukhoi Su-27S *Flanker-B* fighter and a single Su-27UB *Flanker-C* combat trainer in the Ukraine (despite initial reports of 14 single-seaters and four trainers being delivered, these were the only two Angolan Su-27s). Unfortunately operational details are hard to come by. The *Flankers* operated from Catumbela airport, Lubango, in the Southern Military Region and wore a sand/green/olive drab camouflage known as the Nile Valley scheme because a similar camouflage was worn by Egyptian Air Force MiG-21s. Oddly, they wore no serials most of the time. The single-seater eventually crashed on 20th July 2006 while practicing for a display flight, killing pilot Samuel Chipalavela.

(Note: An Angolan Su-27 allegedly crashed on 16th November 2000. Some reports say the aircraft was shot down by UNITA rebels on 19th November, using a 9K38 Igla man-portable air defence system (MANPADS), and flown by Ukrainian mercenary pilot Igor' Valenchenko. However, the serial of this aircraft was reported as 272; considering that FAPA/FANA serials always have a letter prefix indicating the aircraft's role and are assigned in

Another Angolan MiG-23UB, I-31, shares the hardstand at Odessa with an (ex-) Egyptian MiG-21.

batches, this makes the whole story somewhat doubtful.)

In 1982 Angola ordered its first specialised strike aircraft, receiving ten Sukhoi Su-22 *Fitter-F* fighter-bombers (occasionally misidentified as the 'Su-20M' – a variant which never existed). Later, in the spring of 1990, Angola received 14 brand-new Su-22M4 *Fitter-Ks* and Su-22UM3K *Fitter-G* trainers from the Soviet Union (eventually, more than 30 were delivered). The fighter-bombers were based at Catumbela, Luanda (Belas airport), Saurimo and Namibe (the latter base was also known as Yuriy Gagarin AB); they wore a similar three-tone camouflage to the MiG-23s but lacked national insignia. The 26 RACB (*Regimento Aéreo de Caças-Bombardeiros* – 26th Fighter-Bomber Regiment) comprising three squadrons was stationed at the latter location since 1986; the second squadron, 15 ECB (*Esquadrão de Caças-Bombardeiros* – 15th Fighter-Bomber Squadron), operated the *Fitter-F/Ks*. The Su-22UM3Ks, on the other hand, were operated by the 10 EI (*Esquadrão de Instrução* – 10th Training Squadron) of the 24 ENAM (*Ecole Nacional d'Aviação Militar* – 24th National Military Aviation College) at Negage.

In 2000 the original deliveries from the Soviet Union were augmented by 13 former Slovak Air Force aircraft – eleven Su-22M4s and two Su-22UM3Ks; according to some reports, ten *Fitter-Ks* and two *Fitter-Gs* were bought in 1996, followed by four and one respectively in 2001. These were delivered wearing their original Slovak serials and reserialled on site; they were repainted for the new owner but not refurbished then. Additional aircraft were later procured from Belarus. According to press reports, the FANA had 36 *Fitters* in 2002; by 2004 this number had risen to 42. In 2007 the surviving Angolan Su-22s were refurbished by the ARZ No.558 aircraft repair plant in Baranovichi, Belorussia.

In 1987 Cuba sent a squadron of 20 Mikoyan MiG-23BN *Flogger-H* fighter-bombers to Angola; these were seconded to the FAPA, retaining their Cuban serials (unfortunately none have been identified). The MiG-23BNs saw no action in the First Angolan Civil War, serving only as a deterrent for the South African interventionists, as they could strike at any base in South Africa.

There have been reports that eight Sukhoi Su-24M *Fencer-D* tactical bombers delivered in August 1999 and several more in 2000-01, and that the bombers were operated by the rebel UNITA forces in Angola. However, the only evidence of this to date is a very poor-

Top: Another aspect of I-29 at Odessa in company with several Ukrainian Air Force MiG-23MLDs and MiG-23UBs (plus a MiG-29).

Above: The tails of these FANA MiG-23s seen at Odessa show the current Angolan roundels with no star in the middle. Note the radar warning receiver aerials.

Angolan Air Force (FAPA) MiG-23MF C455 carrying four UB-32A FFAR pods

FAPA MiG-23ML C477 armed with R-23 and R-60M AAMs

Angolan Air Force (FANA) MiG-23UB I-31 armed with R-3S AAMs

quality photo of one such aircraft – reportedly supplied by Belarus – taken from a long way off. Russia expressly denied having delivered any *Fencers* to Angola.

In early 1988 Angola placed an order for 12 single-seat Sukhoi Su-25K *Frogfoot-A* attack aircraft and two Su-25UBK *Frogfoot-B* trainers, enough to equip a squadron. Later, these were augmented by at least a further three two-seaters. By March 1988 all the Su-25s had been crated for delivery by sea from Novorossiysk to Luanda. The export variant for Angola differed from the domestic version in having certain 'tropicalised' items of

Freshly overhauled Angolan Air Force (FANA) Su-22UM3K I-34. Surprisingly, the new-style Angolan roundels were removed after delivery for unknown reasons.

Angolan Air Force Su-22M4 C-509 on display with an array of weapons, including napalm tanks. The lack of national insignia was typical for the Second Angolan Civil War period.

avionics and equipment, a different identification friend-or-foe (IFF) system and the exclusion of guided missiles or guided bombs from the weapons range.

A hand-picked 14-man group of Soviet specialists was formed to assist the Angolans in mastering the Su-25; it was headed by Maj. Rafis Gumerov, with Maj. Leonid Chernov as Flight Planning Navigator and Capt. Vladimir Kornev as Flight Commander. The group included three pilots and a team of tech staff from two of the first Soviet Su-25 units – the 80th OShAP (Independent Ground Attack Air Regiment) at Sital-Chai, Azerbaijan, and the 90th OShAP at Artsyz in the Ukraine. All members of the training support group were seasoned Afghan War veterans.

On 26th November 1988 the group arrived at Luanda. The operating base selected for the Su-25 was situated near the port of Namibe, some 170km (105 miles) from the border with the then South African-administered territory of Namibia. The base already hosted the abovementioned 26 RACB equipped with MiG-21MFs, Su-22M4Ks and Su-22UM3Ks; the Su-25 was to form the third squadron of this composite regiment. (Later, the *Fitters* moved to Lubango, whereupon the 26th Regiment became a purely ground-attack unit.) The base also hosted a squadron of Cuban MiG-23ML *Flogger-G* interceptors deployed to provide air cover for the southern regions of Angola. Conversion training courses were organised for the FAPA ground personnel and, as the Angolans generally had a fairly good command of the Russian language, they mastered the main maintenance tasks and some simple repair work for the Su-25 by February 1989. By this time, the Su-25s had been finally assembled. Assembly and test flying of the aircraft was carried out by factory teams from Tbilisi (for the Su-25Ks) and Ulan-Ude (for the Su-25UBKs).

After familiarisation flights in the two Su-25UBKs, the Angolan students went on to general aircraft handling, navigation training,

One more FANA *Fitter*, Su-22M4 C510, at Baranovichi.

Angolan Air Force (FAPA) Su-22M4 C501 with two RBK-250 cluster bombs and two PTB-800 drop tanks

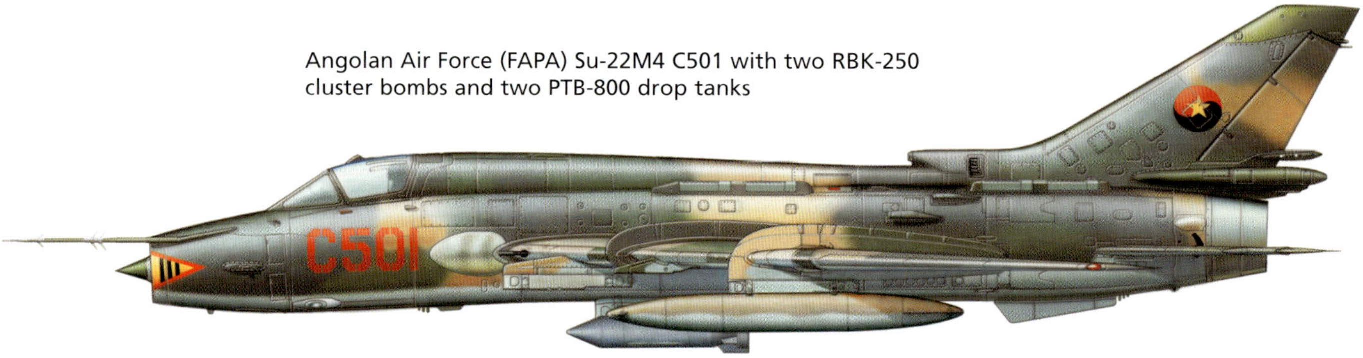

Angolan Air Force (FANA) Su-22UM3K I-34

formation and group flying and weapons application. By August 1989, two flights of the new squadron were ready for solo flying, but basic training continued up to the autumn of 1990. Three of the Su-25s were lost in accidents caused by poor airmanship. Generally the Su-25s were little used and were grounded after March 1991, left unprotected to corrode in the humid climate. However, a little later Angola acquired an additional batch of Su-25s, both to augment its inventory of combat aircraft and for the training of FAPA pilots. These were possibly based at Saurimo.

The assets of the FAPA/FANA's Airlift Command (*Comando da Aviação de Transporte*) are concentrated at Luanda International airport. Over the years the Angolan Air Force operated at least 21 second-hand Antonov An-12A, An-12B, An-12BP and An-12BK *Cub* transports. As a rule, they (and the other FANA transports) wear a medium grey colour scheme with light grey undersurfaces. In addition to the three-digit serials, they have registrations in the D2-M** range (M for military) allocated as ATC callsigns. These are occasionally applied in lieu of serials, in which case the 'Força Aérea Nacional de Angola' titles are painted over; in fact, some have only been seen with such quasi-civil registrations. In addition, the Angolan government also operated at least four An-12s registered in other blocks (mostly D2-E**). One of these, D2-EAD, reportedly crashed on 19th September 1984. Take out the digit from the registration, and the result is 'DEAD'. *Nomen est omen*?

The principal tactical transport of the FAPA was the Antonov An-26 *Curl* turboprop twin, of which close to 50 were operated in military markings; at least two such aircraft were shot down by ground fire during the civil war. Eight *Curls* had quasi-civil registrations in the D2-M** series (although not all of them necessarily wore overt military markings before that). Additionally, at least 14 An-26s operated by the Angolan government were civil-registered (mostly in the D2-E** block). A real oddity was a single Antonov An-24RT *Coke* transport (the An-26's immediate precursor) acquired from a Moldovan airline. Starting in 1998, the FANA took delivery of eight Antonov An-32A *Cline* tactical transports; at least seven of them were second-hand examples bought in Russia and Moldova.

The FANA fleet reportedly included two examples of the Antonov An-28 *Cash* STOL turboprop twin acquired secondhand in the CIS in 2005. Another oddball type was a single Antonov An-30A-100 *Clank* – an executive conversion of the An-30A photo survey aircraft acquired from the Ukraine in 2007.

The Angolan Air Force acquired nine Antonov An-72/An-74 *Coaler-A/B* STOL transports on the second-hand aircraft market, starting in 2001. Interestingly, while most of them have regular FANA serials or registrations

Angolan Air Force (FANA) Su-22M4 '3619 Yellow' still wearing its former Slovak Air Force serial

FANA Su-22M4 C510 No.2 with fuselage pylons fitted; note the differently oriented roundel

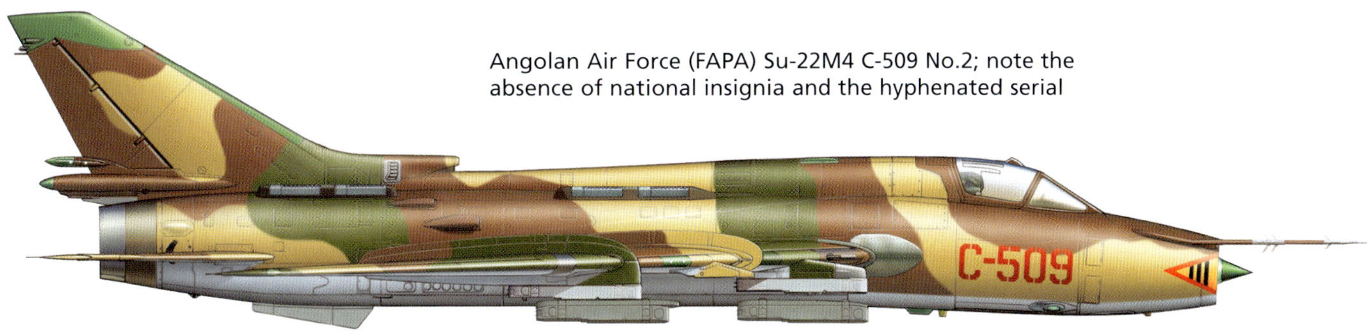

Angolan Air Force (FAPA) Su-22M4 C-509 No.2; note the absence of national insignia and the hyphenated serial

in the D2-F** or D2-M** series, two aircraft are registered outside Angola (in Equatorial Guinea and São Tomé & Principe) for unknown reasons. All the Antonov types listed above are operated by the 5th Light Transport Squadron of the 23rd Air Transport Regiment (a composite unit) at Luanda, though the An-12s are arguably not 'light' transports.

The FANA's most valuable air transport asset is undoubtedly the Il'yushin IL-76TD *Candid-A* four-turbofan heavy transport, which entered service in 2002. Almost all of the 13 aircraft seen to date wear grey Air Force colours (except one, which has a blue/white civil style colour scheme) and 'IL-76TD' nose titles. In fact, however, most of them are former Ukrainian Air Force IL-76MD *Candid-Bs*, and one machine is actually a converted IL-78 *Midas-A* refuelling tanker. They serve with the 6th Transport Squadron of the 23rd ATR.

Presidential VIP transport duties were performed by a single Tupolev Tu-134AK *Crusty* airliner (in its radar-nosed export version with a 'solid' nose housing a Groza-M134 weather radar). The aircraft was delivered in 1977 with 'Força Aérea Angolana No.1' titles but these were changed to 'Republica Popular de Angola' by November 1981. For travels over shorter distances there was a pair of Yakovlev Yak-40 *Codling* trijets – a civil-registered Yak-40 'Salon 2nd Class' VIP aircraft and a convertible Yak-40K that initially wore a military serial but later assumed a civil identity. These aircraft were operated by the 7th Transport Squadron of the 23rd ATR and have since been replaced by Western business jets.

As for the FAPA/FANA's rotary-wing component, the Mil' Mi-8 made up the backbone of the helicopter fleet. Starting around 1978, the Angolan Air Force was a major operator of both the 'first-generation' models – the Mi-8T/Mi-8TV *Hip-C* utility/assault helicopter, the Mi-8PS VIP helicopter – and the 'second-generation' Mi-8MTV/Mi-17 *Hip-H*. A few Mi-8s

Angolan Air Force (FAPA) Su-25K B-18 carrying FAB-250M62 low-drag bombs

were civil-registered and operated by the government flight, but most had overt military markings. Some of the Mi-8TVs were retrofitted with those vortex-type dust/debris filters on the engine air intakes introduced on the Mi-8MT; other helicopters were equipped with imported intake filters designed by the US-based Pall Corporation. The Mi-8s were operated by the 1st and 2nd Helicopter Squadrons of the 21st Transport Helicopter Regiment.

Attrition among the Angolan Mi-8s was consistently high. Much of it was due to combat losses sustained in the two civil wars; other losses, however, were caused by poor airmanship and the frequent dust storms in desert areas causing 'brownouts' on landing.

A number of Mil' Mi-6 *Hook* heavy-lift helicopters were also delivered, serving with the 21st THR/1st HS, but their reported serials are doubtful, as they coincide with those of certain Mi-17s. Three of the Mi-6s were reportedly destroyed in October-November 1985 during the First Angolan Civil War.

The Mil' Mi-24 attack helicopter was introduced by the FAPA in the mid-1980s. Initial deliveries were of the Mi-24D (Mi-25) *Hind-D* version; a fresh batch of Mi-24V (Mi-35) *Hind-Es* was delivered in 1986. The Angolan Air Force took delivery of more than 30 Mi-25s and Mi-35s. 12 Mi-25s and 14 Mi-35s were reportedly on strength in November 2000, and more helicopters, including Mi-24P (Mi-35P) *Hind-Fs*, followed in 2001 as attrition replacements. They saw service with the 3rd and 4th Helicopter Squadrons of the 22nd Transport/Assault Helicopter Regiment.

The Angolan 'crocodiles' wore a tan and chocolate brown camouflage with pale grey undersurfaces and warning inscriptions in Portuguese. The Mi-24Vs delivered in 1986 had a late-model ESM suite featuring a Beryoza radar homing and warning system, an L-166V-11E infrared countermeasures (IRCM) jammer and faired triple ASO-2V chaff/flare dispensers on the rear fuselage sides. Strangely enough, though, they were not fitted with air/exhaust mixers.

The enemy – the UNITA rebels – also had six Mi-24Ds bought in early 1999 from Ukrainian surplus stocks during the Second Angolan Civil War. The helicopters were based at Luca and flown by mercenary pilots.

Below: Men at work on a FAPA Su-25K (B-605) carrying drop tanks and UB-32A FFAR pods; note the lack of insignia.

Bottom: Two FAPA Su-25UBKs parked in an earthen revetment.

Angolan civil wars

Immediately after Angola gained independence, the three insurgent groups that had fought against Portugal fell out on the issue of how the country should be run or, more probably, who should be head of state. These were the Popular Movement for the Liberation of Angola (MPLA – *Movimento Popular por la Libertação de Angola*) led by José Eduardo dos Santos and supported by the USSR, who sup-

This anonymous Angolan Su-25K carries a mix of six UB-32A FFAR pods and two PTB-800 drop tanks.

plied them with arms via Cuba; the National Front for the Liberation of Angola (FNLA – *Frente Nacional por la Libertação de Angola*) backed by South Africa, Zaïre and the CIA; and the National Union for the Total Independence of Angola (UNITA – *União Nacional por la Independencia Total de Angola*). UNITA led by Dr. Jonas Savimbi had split off FNLA in March 1966. Soon this confrontation turned into open warfare between the Marxist FNLA government (supported by the USSR and its allies, notably Cuba) and UNITA (backed by South Africa and the Western powers) – the First Angolan Civil War. Finding itself in an embattled position, the MPLA turned to Cuba for help. The first task of the Cuban troops, which had landed in Angola in February 1975 (before the Portuguese handed over power), was to convert the MPLA guerrillas into a professional army. Thousands of Cuban volunteers joined the army, and massive arms deliveries were made.

In September 1975 South Africa resorted to armed intervention in Angola; the South African Defence Force (SADF) linked up with FNLA guerrillas to march on the capital, Luanda, approaching within 20 km (12.4 miles) of the city by 18th December 1975. By this time, however, enough Cuban soldiers and equipment had arrived to halt the advance; the South African government took the view that the bad publicity it was already receiving over the policy of apartheid was bad enough and it would be unwise to exacerbate the bad feeling any further, so in January 1976 the SADF began withdrawing from Angola. FNLA and the Portuguese troops took over SADF positions. The USA, not wanting to become embroiled again in a foreign war, backed down as well, instructing the CIA to cease all operations in Angola.

In January 1976 the Cuban government sent enough MiG-17F and MiG-21MF fighters to allow the MPLA to field a squadron of each type that became operational on 21st January 1976. This date is acknowledged as the day on which the FAPA-DAA was created. Shortly afterwards the USSR reinforced it by supplying MiG-21F-13s. Their first mission was to attack FNLA bases in Redondo and Nea Lisboa on 25th January 1976 and within a few weeks MiG-21MF fighter-bombers had strafed airfields at Nea Lisboa, destroying or damaging a number of light aircraft used to bring in supplies.

At first the MiG-21s (and later the more modern fighters) were flown not by Angolans but by Soviet, Cuban and East European volunteers. For example, the MiG-21F-13s were Soviet-manned while the MiG-21MF and MiG-21*bis* squadrons had Cuban pilots, and several of these units operated autonomously outside the direct control of the FAPA-DAA. Slowly but surely Angolan pilots were trained and received sufficient experience to man the MiG-21 squadrons after the departure of the Soviets and Cubans in 1989.

The MPLA, now reinforced by 15,000 Cuban troops and air support, overpowered the FNLA, which had ceased to exist by the end of March 1976, its supporters fleeing to Zaïre or allying themselves to UNITA. The latter could now count on South Africa, Rhodesia and covert CIA help. In the meantime, MPLA, which controlled Luanda, assumed the mantle of government of Angola from the Portuguese.

This poor-quality picture taken from a long way off is just about the only photo evidence of Su-24Ms being delivered to Angola.

The sole Angolan single-seat Su-27 with the controversial serial C605

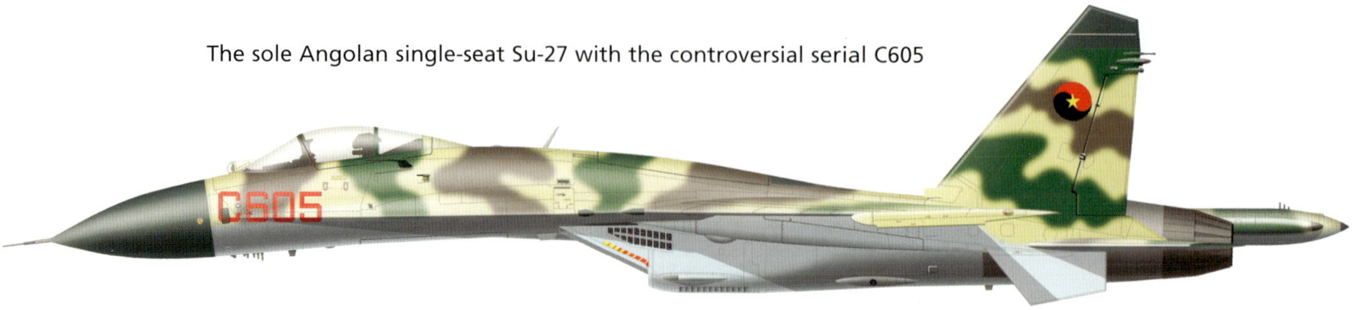

The sole Angolan Su-27UB with the equally controversial serial I-51

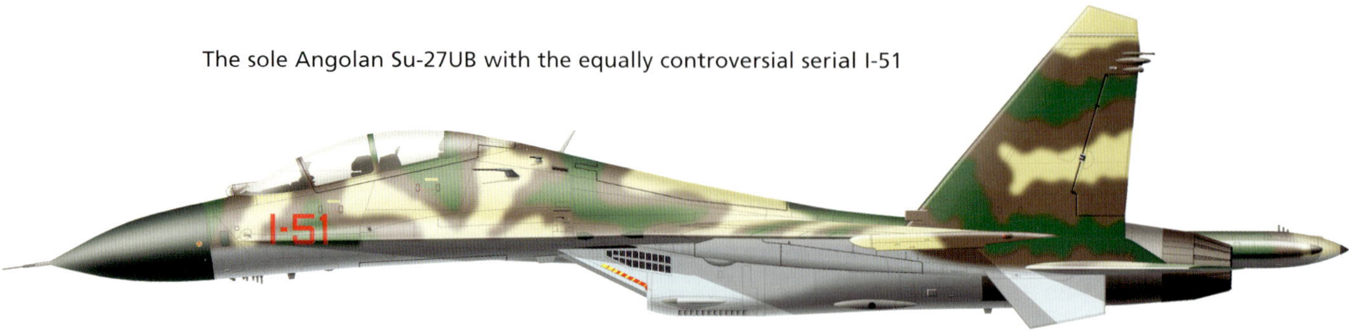

In March, having convincingly defeated the FNLA, the MPLA approached the common borders of Angola, Zaïre and Zambia and, turning south, steadily advanced on UNITA-held territory. FAPA MiG-21MFs attacked Gago Coutinho airfield with unguided rockets, destroying an Air Congo Fokker F.27 Friendship airliner hired by the CIA to bring in food from Rhodesia (some sources, however, say the cargo was arms, not food!). The President of Zaïre, Mobutu Sese Seko, whose country owned Air Congo, demanded $2,000,000 compensation from the CIA, threatening to switch sides if not paid, but was eventually persuaded to settle for $600,000. UNITA retreated to the haven of the undeveloped South-East of Angola, which, having few roads, was ideally suited for guerrilla warfare. People's Armed Forces for the Liberation of Angola (FAPLA – *Forças Armadas Populares de Libertação de Angola*), continued the struggle against UNITA, which was rapidly becoming stronger with US help. Of paramount importance to UNITA was the supply of MANPADS

Angolan and Ukrainian personnel pose with the FANA Su-27UB, which wears no insignia whatever.

and anti-aircraft guns, as it had no air force of its own. However, it did have support from the South African Air Force (SAAF), which defended the border against any accidental incursions by FAPA aircraft and did not hesitate to venture over the border to attack MPLA targets. On the night of 4th April 1986 an Angolan MiG-23ML piloted by Lt.-Col. Gonzalez attacked a pair of SAAF Lockheed C-130 Hercules transports supposedly carrying arms for UNITA but the R-24 missile failed to detonate as it passed the target.

Matters came to a head in 1987 when FAPLA decided to attack the UNITA-held cities of Jamba and Mavinga, and a fierce battle began at the city of Cuito Cuanavale. Even though FAPLA suffered enormous casualties, UNITA and SADF could not capture the city. Either way, both sides were now prepared to negotiate and agreed to the withdrawal of foreign troops.

The Cuban and Soviet military aid stopped in 1988-89; but then, so did the involvement of the SADF, leaving the MPLA and UNITA to slug it out between themselves until a ceasefire was agreed in 1991. However, peace did not reign for long. In 1992, dissatisfied with the outcome of the national elections which the MPLA convincingly won, UNITA took up arms again, starting the Second Angolan Civil War. The government forces were on the retreat, suffering such heavy losses that the Angolan government had to contract the proactive security (read: mercenary) company Executive Outcomes Ltd. (EO) based in Pretoria, South Africa, in order to train the troops in counter-insurgency (COIN) warfare and ultimately recapture the strategically important oil- and diamond-rich areas. Though mockingly dubbed the Diamond Dogs of War by some observers, EO have on occasion managed to secure peace where others had failed dismally. After that, the bitter civil war continued with sporadic ceasefires until 22nd February 2002 when Savimbi was killed in an ambush, and a final and lasting ceasefire was finally agreed. In 2003, UNITA took the political path, giving up the armed struggle.

Here the same aircraft is caught by the camera as it becomes airborne.

At the early stages of the fighting, MiG-21s played a leading role until the arrival of the MiG-23MF swing-wing fighters in the 1980s, but attrition was very high owing to conditions of terrain and weather – quite apart from enemy action. 19 MiG-21MFs are known to have been lost; three (possibly four) were claimed to have been shot down by SAAF Dassault Mirage F.1CZ pilots between 1981 and 1985. The following cases are known (in all cases the fighters were flown by Cuban pilots). On 15th March 1981 a MiG-21MF piloted by Maj. Leonel Ponce Torriente fell victim to a shoulder-launched anti-aircraft missile near Cabinda; the pilot ejected and was rescued. On 2nd May a MiG-21MF flown by Capt. Carlos Benitez Estacio was hit by anti-aircraft artillery, crash-landing at Lubango. On 6th November 1981 SAAF pilot Maj. Johann Rankin shot down a MiG-21MF flown by Lt. Donacio Valdez Espinoza near Mulondo; this was the first SAAF 'kill' after the Korean War.

Starboard side view of the Angolan Su-27UB sandwiched between an Aero L-39 Albatros and a MiG-23.

Angolan Air Force (FANA) An-12BK D2-MBN; note that the aircraft has been demilitarised by deleting the tail barbette and overpainting the gunner's station windows

On 5th October 1982 a second clash between SAAF Mirage F.1CZs and Angolan MiG-21MFs took place. (Some sources state the version of the *Fishbeds* as MiG-21*bis*; however, one of the MiGs was serialled C47, which makes it impossible. Well, even more serious misidentification is known to have occurred in the heat of the battle.) That day at 1028 hrs the FAPA-DAA air defence radar at Cahama detected several SAAF aircraft approaching between Virey and Tchibemba – an English Electric Canberra PR. Mk 2 photo reconnaissance aircraft from 12 Squadron tasked with reconnoitring Cahama, plus a pair of escorting Mirage F.1CZs from 3 Squadron at Ondangwa flown by Maj. Johan Rankin and his wingman Capt. Cobus Toerien. A pair of MiG-21MFs piloted by Lt. Raciel Marrero Rodriguez and Lt. Gilberto Ortiz Perez (serial C47) took off from Lubango at approximately 1042 hrs and were vectored towards the Canberra by ground controlled intercept (GCI) controller Lt. Barbaro Perez Duran. The pair of Mirages had fallen behind, taking off late due to the wingman having engine starting problems, and this may have led the Angolans to believe that the Canberra was making a foolhardy unescorted run.

It has to be said that the radar at Cahama was experiencing difficulties, and most of the MiG-21s' scrambles from Lubango turned out to be false alarms; hence the more experienced pilots tended to bully the younger pilots into doing these alerts. Lt. Rodriguez was a case in point – a Pilot Third Class with only 320 flying hours. This time, however, it was no false alarm – this was the real thing.

In turn, the MiGs were detected by the SADF radar picket at Dayton, and the South African controller Capt. Les Lomberg instructed the Canberra to head south whilst vectoring the Mirages north towards the MiGs, telling them to climb to 9,150 m (30,000 ft). Maj. Rankin visually detected the MiG-21s 9.25 km (5.74 miles) away to his right at his own flight level. The Mirages then jettisoned their drop tanks and went into afterburner whilst making a hard right hand turn. At this very moment Lt. Perez also spotted the enemy; the MiGs also did a hard right and a dogfight commenced. The R-3 missiles fired by the MiGs missed their targets. As the Mirages began manoeuvring the Cuban pilots lost sight of them and began a new turn, trying to re-establish visual contact; in so doing Capt. Toerien almost collided with Lt. Rodriguez. Two minutes later, looking through his rear view periscope, Lt. Perez and saw Maj. Rankin's Mirage 800-1000 m (2,640-3,280 ft) behind him. Informing Rodriguez that he was under attack, he did an abrupt semi-reversal and levelled out. Rankin fired two Matra R.550 Magic IR-homing missiles at Perez' aircraft, one at an altitude of 3,000 m (9,840 ft) and the other at 1,500 m (4,920 ft); the first, being fired at the limit of the 'kill' zone, lost lock-on and went ballistic, but the other one exploded immediately behind Perez's MiG-21. The latter

Angolan and Cuban servicemen pose for a group photo beside FAPA An-26 T-233. Note the optional bomb racks on the fuselage.

Despite its very military colour scheme, this An-26 wears the civil registration D2-END.

disengaged and headed for Lubango, trailing smoke. Perez had not felt the impact and therefore did not realise that he had been hit. MiG-21MF C47 landed safely at Lubango, but witnesses said afterwards the aircraft was so full of holes that it looked like a sieve.

Meanwhile, Lt. Rodriguez saw the Mirages right in front of him; he also noticed a bright flash of an explosion to his left (obviously one of Rankin's missiles). The Mirages split, one going up and the other turning left, and Rodriguez lost sight of them. At this point he kept turning right at maximum speed, whilst trying to communicate with the GCI controller. When he finally got through, Lt. Duran ordered him descend to 2,000 m (6,560 ft) on a heading of 270° and search that area. Here it deserves mention that Maj. Rankin's Mirage F.1CZ serialled 203 had the new blue/grey air superiority camouflage, giving rise to the nickname 'Le Spectre', whilst the other Mirage still had the old green/sand tactical camouflage. This is where the low-visibility scheme paid off; through his periscope Lt. Rodriguez could see Capt Toerien's aircraft some 1,800 m (5,900 ft) behind him but he was too late to realise that Rankin's aircraft was only some 500 m (1,640 ft) away. Entering the target's turn radius, Rankin fired a volley of 30-mm cannon fire at some 300 m (990 ft) range, scoring hits on the MiG-21's port wing.

The combatants' accounts of what happened next are rather contradictory. The South Africans reported that the MiG started streaming fuel, which then exploded, and the explosion was clearly captured by Maj. Rankin's gun camera. Rankin's aircraft went smack into the resulting fireball and the engine surged, causing him to cut the engine and restart it before heading back to base. Capt. Toerien pursued the rapidly descending MiG, which had turned northwards, trailing black smoke from the port side of the rear fuselage, but gave up the chase when his leader reported his compressor stall. At this point he had lost sight of Maj. Rankin and had been alerted that another pair of MiG-21s had scrambled from Lubango (these eventually failed to locate the enemy, returning to base after running low on fuel). Both Mirages returned to Ondangwa.

Lt. Rodriguez, however, claims that he still had control of the aircraft after being hit,

Non-airworthy and minus propellers, An-32 T-256 languishes beside two Angolan Air Force Embraer EMB-312 Tucanos used as COIN aircraft.

The Angolan Air Force (FANA) An-30A-100 in quasi-civil guise as D2-MBO; note the overpainted serial underneath the registration

although the aircraft was indeed on fire; hydraulic and oil pressures were OK, the only untoward thing being the aircraft's tendency to bank right. When about 45 km (28 miles) away from Lubango he advised the tower that the damage to his port wing might prevent the landing gear from extending normally (this radio message was picked up by SADF listening posts but attributed to Lt. Perez). Lt. Rodriguez claims to have landed safely; yet, at least one of the MiGs was written off. Due to this disastrous attempt at intercepting the South Africans, Col. Bilardel (the CO of the unit based at Lubango) was removed from his post and demoted.

Another MiG-21MF piloted by Capt. Raúl Fernandez Sanchez was lost at Lubango on 23rd April 1983. On 20th October (or November?) 1983 MiG-21MF C50 flown by Lt.-Col. Henry Perez was damaged by AA fire over Luena but made it back to base. There were non-combat incidents as well. On 16th March 1976 MiG-21MF C41 piloted by Gen. Rafael del Pino ran out of fuel and force-landed near the very village where Jonas Savimbi had been born; the aircraft was recovered and apparently returned to service. Its abovementioned sister ship C50 was less lucky – it was subsequently written off in a crash landing caused by hydraulics failure.

28 MiG-21*bis* were lost between 1984 and 1994 – presumably mostly to UNITA ground fire, excluding two aircraft: C340 piloted by Lt. Vinez was damaged in an off-field forced landing caused by navigational error and captured by the South Africans, the other one was wrecked at Luanda. A MiG-21*bis* piloted by Lt. Fidel Perez crashed at Lubango on 29th December 1983. A sister ship flown by Capt. Pausides Hechevaria was lost at Malange on 9th August 1984. Two more examples piloted by Lt. Sergio Herrera and 1st Lt. Raúl Quiala crashed at Lubango on 31st July 1985 and 7th July 1986 respectively, both pilots losing their lives.

Other reported Angolan losses included a MiG-21R flown by Capt. Chacón which crashed at Luena on 21st September 1983. Capt. Andres Valle, Capt. Roberto Hernandez Alvarez and Capt. Guillermo Meriño flying MiG-21s of an unspecified version were shot down and killed at Lubango on 2nd October 1983, 10th November 1983 and 21st September 1984 respectively. Additionally, a

An-32 T-250 illustrates the livery worn by Angolan Air Force transports

MiG-21U flown by pilots Pastrana and Meriño crashed at Lubango on 8th December 1982. Another two-seater was shot down by UNITA forces near Luvuei on 28th October 1987; the pilots, Lt. Col. Manuel Rojas and Capt. Ramón Quesada Aguilar, ejected and were captured.

The first MiG-23MSs delivered to the FAPA were used to defend Luanda, but as more MiG-23MLs were received, many were sent to support the army and transferred to bases further south, such as Luena, Mavinga, Mocamedes and Xangongo. Although there were interceptions of South African Air Force aircraft, the *Floggers*' operations were mostly concerned with ground attack. Angolan losses were high, as UNITA rebels were able to field Western-sourced anti-aircraft artillery and General Dynamics FIM-92A Stinger MANPADS; 40 FAPA aircraft of various types were lost in the battle for Mavinga in 1987.

Until the arrival of the MiG-23 fighters, the SAAF's Mirage F.1CZ reigned supreme and in an encounter with MiG-21MFs the latter usually pulled the short straw. The advent of the *Flogger* changed that; up to then the FAPA pilots had chosen to avoid the SAAF, but now the Cubans were determined to challenge the SAAF's air superiority over the battlefield. Shortly after the MiG-23's arrival on the Angolan TO two Mirage F.1CZs were shot down in two separate incidents.

The MiG-23ML pilots preferred to use their superior speed for 'slash and dash' attacks, whilst the South Africans preferred to 'mix' it with the enemy. On 10th September 1987 a Mirage F.1CZ fired a Matra Magic missile at a MiG-23ML but missed.

Another incident that took place on the afternoon of 27th September 1988 was a lesson in formation flying. Four Mirage F.1CZs scrambled from Rundu AB to intercept a pair of MiG-23MLs flown by Maj. Alberto Ley Rivas and Lt. Juan Carlos Chávez Godoy who were flying top cover for some assault helicopters. The Cuban pilots made a head-on attack against one pair of Mirages flown by Cmdt. Carlo Gagiano (leader) and Capt. Arthur Piercy (wingman). Two R-23R missiles were fired; one of them passed harmlessly over Cmdt. Gagiano's aircraft but the other exploded alongside Capt. Piercy's Mirage serialled 206, inflicting damage to the tail section. The Mirage plummeted earthwards but Piercy was able to recover it and head back to Rundu at ultra-low level when the starboard fuel pump and No.2 hydraulic system knocked out. As the fighter landed, it turned out that the brake parachute was also damaged and would not deploy; as a result, the jet overran the 2000-m (6,560-ft) runway and crashed through the perimeter fence. As it careered over rough ground the ejection seat fired spontaneously; Piercy's parachute had no time to open fully and he suffered serious injuries on impact. Mirage F.1CZ '206' was declared a write-off and cannibalised for spares. Bottom line: MiG vs Mirage, 1-0.

FAPA MiGs actively patrolled the airspace over southern Angola to prevent South Africa from supplying arms to UNITA. On 7th August 1988 this led to a major international incident that could have easily been interpreted as an assassination attempt against a foreign head of state. The Botswana Defence Force Air

Angolan Air Force An-72 D2-FGF keeps company with an An-28 and an L-410UVP at Odessa, where it came for heavy maintenance.

Wing British Aerospace 125-800 business jet serialled OK-1 (c/n 258112) was carrying President Quett Ketumile Joni Masire and eight other government officials from Gaborone to Luanda. As the aircraft cruised at 10,670 m (35,000 ft) it entered the area covered by the surface-to-air missile (SAM) site at Cuito Cuanavale; since the BAe 125 did not respond to IFF interrogation, it was coded hostile and designated as a legitimate target. Four missiles were fired at it but missed because the radar guidance system had not been properly tuned. Next, however, the target information was forwarded to Menongue AB, and MiG-23ML C479 piloted by Capt. Albert Olivarez Horta scrambled to intercept, closing in on the target at Mach 1.9. Acting against his instructions, Horta launched an attack without visually identifying the target first, firing two medium-range missiles. The first, an IR-homing R-24T, took the biz-jet's starboard engine right off, the aircraft pitching up violently and stalling in so doing; the explosion also caused a total decompression and knocked out all the navigation equipment. The fuel tanks in the starboard wing were punctured by shrapnel but luckily there was no fire. The occupants of the BAe 125 were definitely lucky: the other missile, a semi-active radar-homing R-24R, went for the detached engine falling earthwards, not for the aircraft itself. The pilots managed to recover from the stall and executed an emergency descent, making a safe emergency landing at Kuito Bie, some 40 km (25 miles) behind the point where the aircraft had been hit. Several passengers, including President Masire, and the aircraft captain (Col. Albert Scheffers, CO of the Air Wing) received various injuries and were treated at the local hospital (the President later underwent further treatment in the UK); the aircraft was declared a write-off. The co-pilot, Arthur Ricketts (a training captain seconded from BAe), later received the Diploma of Outstanding Airmanship for saving the aircraft and everyone inside it.

The Angolan government acknowledged its fault, presenting an official apology and paying for the damage done. On the other hand, Lt.-Col. Eduardo Gonzalez (the commander of one of the Cuban units deployed in Angola at the time) rightly pointed out that violations of the war zone that was officially closed to air traffic occurred almost on a daily basis, meaning something like this was bound to happen eventually, and that it was highly irresponsible for the biz-jet's crew to fly across what was universally known as a war zone.

On another occasion, a MiG-23 shot down a Lockheed C-130 Hercules bringing supplies to UNITA forces with cannon fire. The SAAF's Atlas (Aermacchi) MB.326M Impala Mk.1 ground attack aircraft were considered easy targets for the MiG-23s.

As a result of several ceasefires breaking down, in 1994 the Angolan government hired the abovementioned company Executive Outcomes to provide air support. Accordingly EO began using at least three Angolan MiG-23MLs. In June 1994, flown by South African pilots, they flew successful strike missions against UNITA bases near Cafunfo. This, and the delivery of supplies paradropped by Russian-flown IL-76 transports, demoralised

Another quasi-civil Angolan Air Force transport – this time an IL-76TD.

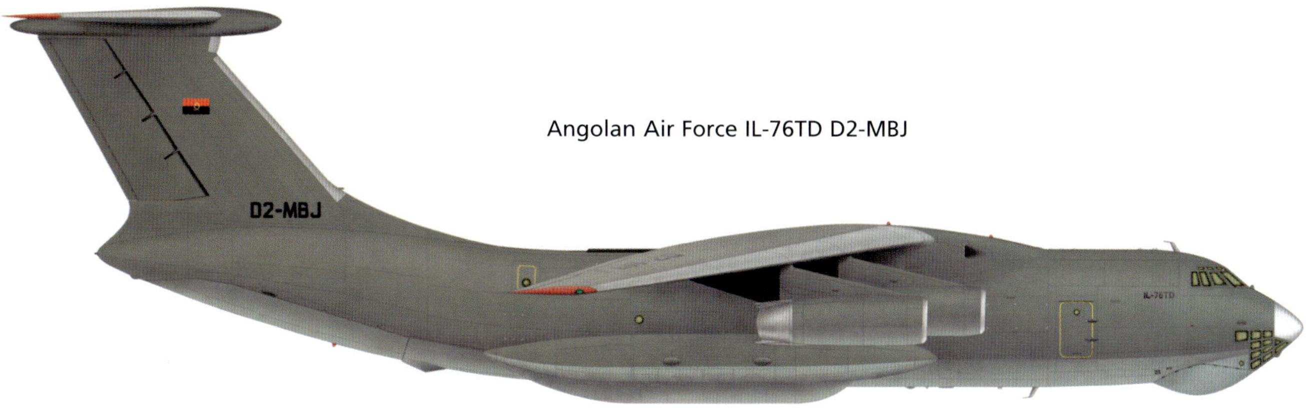

Angolan Air Force IL-76TD D2-MBJ

D2-FCN has IL-76TD nose titles but is a 'Falsie' – that is, an IL-76MD (albeit demilitarised); note the flare dispensers in the main gear fairings revealing the aircraft had participated in the Afghan War when it was CCCP-76651 with the Soviet Air Force

the rebels and allowed the government forces to launch a successful offensive, taking Cafunfo on 15th July. Incidentally, airmanship is one thing but ground support is another, and both are equally important, as the EO pilots found out – the hard way. On 27th July 1994 a MiG-23ML C434 had just taken off on a sortie from Cabo Ledo when the 'low fuel' light came on – something that seemed utterly impossible at this phase of the flight, since the tanks were supposed to be full. Minutes later, the engine began to miss and flamed out as the pilot turned the fighter around and headed back to base. The landing gear was extended but failed to lock down due to low hydraulic pressure and collapsed as the fighter touched down, slithering on its belly. It turned out that the aircraft had not been refuelled before the sortie. There's discipline for you. Although technically repairable, the MiG was abandoned.

In the mid-1980s Angola received a number of newish Su-22s from the USSR to bolster the FAPA, followed by Su-22M4s. They took the place of the MiG-21MF fighters which were transferred to Lubango AB, after which the 26th Regiment at Namibe became a fighter-bomber unit. The Soviets sent Lt.-Col Nikolai Voloshanovskiy, later replaced by Aleksey Shevtsov, on attachment to the commander of the Angolan squadron. Now brought up to strength, the regiment, which by this time was commanded by Capt. Chávez, was the most operationally efficient in the FAPA, a force to be reckoned with.

The regiment first saw action in the summer campaign of 1985 when eight Su-22s took part in Operation *Second Congress* alongside 30 MiG-17Fs, 50 MiG-21, 30 MiG-23MLs and more than 100 helicopters. The aim of the operation was to capture the strategically important town of Mavinga in Cuando Cubango province and destroy UNITA assets in the area, ideally capturing the UNITA headquarters at Jamba.

As a rule, large-scale operations in Angola were undertaken in the summer when the heat in the southern hemisphere had somewhat abated, permitting both people and equipment to function more or less normally. At that time the UNITA had undertaken a number of successful incursions in the central provinces and in the vicinity of the capital, so a new major offensive against the rebels was launched in July 1987. It was decided to transfer a flight of Su-22M4s there to destroy enemy positions, concentrations and bases with bombing and ground attacks and so pave

H344, one of the Angolan Air Force (FAPA) Mi-17s.

the way for an offensive by government troops. At the same time the Su-22M4s were to operate from the forward airfield at Menongue together with MiG-21 fighters in support of operations on the eastern front.

The FAPA enjoyed aerial supremacy at this stage but the UNITA air defences had become stronger, bolstered by South African-supplied Stingers and weaponry captured from the MPLA. At this stage the South African Defence Force (SADF) and UNITA claimed no fewer than 25 'kills' against MiGs and Su-22s. Among other Angolan jets, Su-22 C510 (the first to have this serial) was shot down by MANPADS in July 1986; two more were lost between August 1987 and April 1988. The presence of MANPADS forced the FAPA to launch ground attack sorties from altitudes of 2,000 m (6,560 ft) and up – sometimes as high as 4,000 m (13,120 ft).

A large batch of new Su-22M4s was purchased from the Soviet Union in 1988-89. These aircraft took part in the large-scale operations against the UNITA in the Bie, Cunene, Cuando Cubango and Moxico provinces. However, the enemy was still receiving fresh supplies of Stingers, augmented by captured 9K34 Strela-3 (NATO SA-14 *Gremlin*) MANPADS of Soviet provenance. As a result, losses continued; for example, Su-22M4 C512 was shot down by MANPADS near Cahama on 10th April 1991.

H-620 was one of several Angolan Air Force Mi-17s to wear all-white colours. Note the non-standard common intake filter for the two engines (manufactured by the Pall Corp.).

During the Second Angolan Civil War, Executive Outcomes supplied four pilots who had at one time flown with the South African Air Force. Having been told which aircraft were available to the Angolans, they immediately elected to fly the Su-22M4. The Angolans refused, however, and required the pilots to spend several months on the Pilatus PC-7 trainers used by FANA as COIN aircraft. During this period the head of EO, Eeben Barlow (a veteran of army intelligence and of the SADF's elite 32nd Battalion), used every means at his disposal at every level for his pilots' right to fly the Su-22.

The Angolans themselves liked the Su-22 very much. The South Africans said this machine was the pride of FAPA, and the Angolans spared no effort on their servicing so that the aircraft, at least outwardly, were in very good condition. After much prevarication, the South Africans were given permission to fly the Su-22M4 operationally, but at the last minute the head of the Angolan Air Force operations section, who had flown these aircraft himself, said no. Thus, the South Africans had to make do with the PC-7 on which they provided cover for Mil' Mi-17 helicopters as well as carrying out reconnaissance and artillery spotting.

In June 1993 all available strike aircraft, including the Su-22M4s, participated in an offensive against UNITA forces near Huambo. However, the offensive collapsed 120 km (74.5 miles) short of the objective. In 1994, however, the government forces' offensive (supported by EO) was far more successful; huge areas were regained, including Cafunfo,

Angolan Air Force Mi-17 H-624 in an unusual all-white colour scheme; note the Russian titles

Cuito Cuanavale, and Huambo. However, the government forces were in need of a respite and reorganisation. Intensive negotiations with the UNITA went on throughout 1995-96, while sporadic fighting continued. Eventually, a ceasefire was achieved in 1997 and a Government of Unity and National Reconciliation was formed. However, hostilities resumed in 1998, UNITA forming two mechanised brigades that began an offensive in south-eastern Angola, seizing several cities. Again, the Angolan government went shopping for additional weapons (among other things, in Russia and Belarus) and employ mercenaries in order to check the rebels' advance and stop them from getting supplies. At this stage the newly acquired Su-27 and Su-27UB fighters were used as forward air controllers (FACs) for the strike aircraft – primarily the Su-22M4s.

The Su-25K arrived on the Angolan TO in 1989 when the civil war had been raging for some 14 years. The government forces relied heavily on the use of air assets, hoping to achieve a decisive success in the long-standing confrontation. In the opinion of Lt.-Gen. Pyotr Goosev, the chief advisor to the Soviet Military Mission in Angola, 1989 was to be marked as a turning point in the war. The Su-25 was to be brought into the conflict as a 'weapon of victory', along with armoured vehicles, tanks and artillery, plus the Mi-24 helicopter that had acquired such a fearsome reputation in COIN operations in other parts of the world.

Basic training continued up to the autumn of 1990, since the Angolans did not show any particular desire to get involved in combat. The low skill level of the Angolan pilots was soon to show itself and three of the Su-25s fell victim to the actions of their own pilots. In February 1990 one Angolan pilot retracted the undercarriage before he had reached take-off speed; the nosewheel retracted first and, after scraping its nose along the full length of the concrete strip, the aircraft ran off the end and exploded, the pilot managing to eject just in time. Three months later, two other Angolan 'aces' making a formation take-off started a banking turn, wingtip to wingtip, having scarcely got airborne. The aircraft collided, breaking up in mid-air and crashing on the edge of the airfield. Both pilots ejected at very low level; the flight leader punched out as soon as he felt the jolt of the collision and the wingman followed suit when his aircraft, which had lost a wing, started tumbling uncontrollably.

By October 1990, the Su-25 squadron at Namibe had flown a total of around 1,500 hours. Only one flight had completely mastered the weapons application phase of the training, including the squadron commander, known as 'Captain Jaime'. Bombing and air-to-ground gunnery were practised at a range situated about 7 km (just over 4 miles) north of Namibe airfield, where targets were made out of old car tyres and burned-out vehicles dragged off the highway.

It was decided to involve the unit in the operations planned for the summer of 1990. By then, as noted earlier, the UNITA had undertaken a number of successful raids in the central provinces of Angola and in the vicinity of the capital Luanda, and the Angolan military leadership decided it was now time to fling the Su-25s into the fray. The Su-25 were tasked with destroying rebel positions by delivering heavy strikes against deployed UNITA groups and bases, to clear the way for the advance of the ground forces. A flight of Su-25s was deployed to a base near Luanda, with everything needed for their operations, including spares, weapons and ammunition, being flown in from Namibe by chartered Lockheed C-130s or FAPA An-26 transports as required.

Mi-35 H357 in low flight, showing the camouflage scheme typical of Angolan Air Force (FAPA) *Hinds*.

The Angolan pilots prudently maintained a safe altitude when flying combat sorties, not descending below 5,000-7,000 m (16,500-23,000 ft), and avoided diving in order not to be picked off by anti-aircraft fire. Bombing raids involved the use of OFAB-100-120 and OFAB-250-270 high-explosive/fragmentation bombs and RBK-500 cluster bombs loaded with fragmentation bomblets. When conducting attacks on road vehicle convoys, UNITA rebel camps and settlements occupied by guerrillas, the Su-25s sometimes used 57-mm S-5 unguided rockets launched from UB-32-57 pods. The FAPA's use of the Su-25 was contrary to the recommended procedures defined in the Soviet combat manual and it turned out that the aircraft's manoeuvrability, accuracy of weapons delivery and firepower, which were essential in this particular combat situation, were not actually exploited. Worse, in the opinion of the Angolan pilots the Su-25 was inferior to the fast MiGs, which were capable of 'unloading' their bombs and disappearing from danger in a matter of seconds. It was not so much a question of the aircraft's combat effectiveness, but rather of the Angolan pilots' willingness to fly and fight. After having flown 25 combat sorties in a month, the Angolans refused to take part in any further action, flew the Su-25s back to Luanda and stayed there.

In October 1990, some of the Soviet pilots and the majority of the technical personnel were rotated, but their actual work level was also reduced. After the unsuccessful 'trial' of the Su-25 in the abortive 'Luanda operation', the Angolan Air Force practically stopped using the type operationally. The Angolan pilots flew very little and, as a result, quickly lost even the so-so level of flying competence they had 'accumulated' over the past months.

A gradual reduction in the size of the Soviet advisory group in Angola began from around December 1990. By March 1991, the flying operations of the Su-25 squadron had ceased and the aircraft were left unprotected, to corrode in the humid climate. Also by this time, the Cuban contingent had left and President Eduardo dos Santos, who had been unable to deal with his opponents by military means, also declared his 'disillusionment with, and condemnation of Marxism'. The Su-25 did not take any further part in military operations and did not even fly after this. However, a little later, Angola acquired an additional batch of Su-25s, both to augment its inventory of combat aircraft and for the training of FAPA pilots. These were possibly based at Saurimo.

Originally flown by Cuban and East German crews, the Angolan Mi-25s saw action against the UNITA guerrillas and SADF. In addition to their usual missions they always escorted major supply convoys, which were a vulnerable and lucrative target. The *Hind-Ds* were much used against the UNITA near the towns of Mavinga and Cassinga which were the scene of fierce fighting. Western reports that all Angolan Air Force Mi-25s had been destroyed or rendered unserviceable by late 1985 gives some indication of the intensity of the fighting; for example, a Mi-25 was shot down on 3rd June 1985 and another on 23rd June.

The opposing forces admitted the high survivability of the *Hind*, noting that weapons with a calibre less than 23 mm were useless against it. Hence both UNITA and the SADF made wide use of Soviet-built air defence systems captured from the FAPLA – ZU-23-2 double-barrel 23-mm AA guns, ZSU-23-4 Shilka quadruple self-propelled AA guns and 9K32 Strela-2/-2M and 9K38 Igla-1 MANPADS. Apart from that, SADF units were also armed with Bofors AA guns and Cactus SAM batteries. FAPA helicopter pilots responded to this threat by inventing new tactics. Flying at treetop level was one; the dense jungle muffled engine noise and rotor slap quite effectively, and the enemy could not detect the chopper until it was on them. Not to be outdone, the enemy soon began posting 'crocodile hunters' armed with MANPADS on the treetops to strip the Angolans of this advantage.

A fresh batch of Mi-35s was delivered to Angola in 1986 as attrition replacements. Between October 1987 and early 1988 they were in action near Cuito Cuanavale, inflicting heavy losses in both personnel and materiel on the SADF. After the South African withdrawal from the conflict the UNITA had to deal with the omnipresent 'crocodiles' on its own. Three Mi-35s were shot down on 27th September 1988, 22nd August 1989 and 25th February 1990. Another *Hind* was lost on 28th January 1990 when it crashed during a sand storm, killing two of the seven occupants and injuring the others.

When Soviet, Cuban and East German military advisors left, the Angolan Air Force immediately faced huge problems maintaining equipment serviceability and personnel skill. Today, only a handful of Angolan *Hinds* remain operational because new equipment and spares supplies from Russia have been reduced to a trickle or dried up completely. Yet these few are still in action against UNITA rebels – ironically, supported by some South Africans who were once the enemy. On the other hand, UNITA now has its own Mi-24s bought from the Ukraine in early 1999. At least one FAPLA Mi-25 was reportedly shot down by UNITA rebels over the mountains of central Angola in the first week of January 1999.

In addition to waging war on home ground, Angola sent six MiG-21*bis* fighters to Democratic Republic of Congo, extending aid to President Laurent Desiré Kabila during the Second Congo War which began in August 1998. The MiGs operated from Kinshasa IAP, delivering bomb and rocket strikes against the Rally for Congolese Democracy (RCD) rebels who were advancing on Kinshasa. Later, more aircraft arrived from Angola – Su-25s, MiG-23MLs and combat helicopters. The rebels claimed to have shot down 18 'Zimbabwean MiGs' by ground fire; however, no Zimbabwean jets were present there, and while some aircraft (apparently Angolan ones) were destroyed the number cannot be 18. While troops loyal to Kabila succeeded in saving Kinshasa, they never totally defeated the rebels. A United Nations-brokered ceasefire was signed in Lusaka, Zambia, on 1st August 1999 by the Democratic Republic of Congo, Angola, Namibia, Zimbabwe, Rwanda and Uganda, but regrettably the RCD refused to sign. Most signatories broke the ceasefire at one time or another, even though over 5,000 UN peacekeepers were there to monitor it. It was not until 17th December 2002 that the Global and All-Inclusive Agreement brought an end to the hostilities.

Known Angolan Air Force (FAPA/FANA) aircraft of Soviet/ Russian origin are listed below. The serials of aircraft wearing overt military markings have prefixes denoting the aircraft's role: C for [*avion de*] *caça* (fighter), I for [*avion de*] *instrução* (trainer), B for *bombardeiro* (bomber), T for [*avion de*] *transporte* (transport), H for *helicoptero* and R for [*avion de*] *reconhecimento* (reconnaissance aircraft). Fighter, trainer and strike aircraft serials are red, while the transports' serials are black. The serials are usually applied without hyphens (with or without a space), although some are hyphenated. Several Mi-8MTV-1s have serials prefixed MAT (*Ministerio de Administracion Territorial* – Ministry of Territorial Administration).

Below: This Mi-35 serialled H-384 has a rather non-standard four-tone camouflage. The helicopter carries two drop tanks and two B-8V20 FFAR pods.

Bottom: The remains of a FAPA Mi-25 destroyed by UNITA action. Surprisingly, the engines have not been salvaged from the wreck.

Known Angolan Alr Force aircraft of Soviet/Russian origin				
Type	Serial/ registration	C/n	F/n	Notes
An-2	R-34	?		Crashed 23-4-1987
An-2	R-102	?		Destroyed by enemy action 19-1-1987
An-2	R-104	?		
An-2	R-108	?		Destroyed by enemy action 15-1-1987
An-2	R-109	?		Destroyed by enemy action 10-1-1987
An-12	T-300	?		Basic Aeroflot Soviet Airlines c/s, in service 2001. Damaged 7-8-2005
An-12B (+)	T-301	5343405		Ex-Tiramavia ER-ADM, in service 2001. Became, see next line
	D2-MAZ			
An-12	T-302	?		Basic Aeroflot c/s, in service 2001
An-12B (+)	T-303	3340903		All-white c/s, in service 2001. Became, see next line
	D2-MAW			
An-12	T-304	?		In service 2001. Crashed on landing Kanyengue 27-1-2002
An-12	T-305	?		In service 2003
An-12TB	T-306	402007		Ex-Tiramavia ER-ADB; in service 2003. Became, see next line
	D2-MBE			
An-12	T-307	?		Ex-Tiramavia ER-ACH; in service 2003. Became, see next line
	D2-MBH			Back to T-307. Crashed on take-off Menongue 16-5-2003
An-12	T-308	?		In service 2003
An-12B (+)	T-309	4342209		In service 2003
An-12	T-310	?		In service 2003
An-12BK	T-311	7345410		Ex-Adala Airways 9Q-CEN; in service 2005. Became, see next line
	D2-MBI			Damaged 29-9-2008, derelict Luanda
An-12A (+)	D2-MBD	2340608		Ex-Tiramavia ER-ACB, bought 2004; military identity unknown
An-12	T-314	?		Seen 2009
An-12	T-315	?		Seen 2011
An-12	T-316	?		Seen 2008
An-12A (+)	T-3**	2340801		Ex-EX-085, seen 2006
An-12BK	D2-MBN	9346809		Ex-Ukrainian Air Force 87 Red, seen 2008; military identity unknown
An-12	D2-MBR	?		Seen 2007
An-12	D2-MBU	?		Seen 2010
An-12BP (+)	D2-MBV	5343208		Ex-ER-AXH
An-12	D2-EAC	?		Government operated
An-12	D2-EAD	?		Government operated. Crashed 19-9-1984
An-12	D2-EAE	?		Government operated
An-12BP (+)	D2-FRI	7345210		Ex-Belorussian Air Force 10 Yellow, seen 2005
An-24RT	T-240	0911307		Ex-ER-AEU, callsign D2-MAV
An-26	T-52	3805		Ex/to Cuban Air Force, Cuban serial unknown
An-26	T-201	?		
An-26	T-202	?		Shot down 27-9-1983
An-26	T-203	?		
An-26	T-204	?		
An-26	T-205	?		
An-26	T-206	?		
An-26	T-207	?		
An-26	T-208	?		
An-26	T-209	?		
An-26	T-210	?		
An-26	T-211	?		
An-26	T-212	?		

An-26	T-213	?		
An-26	T-214/ D2-MHI	?		
An-26	T-215	?		
An-26	T-216 D2-MAB	11403		Became, see next line
An-26	T-217	?		
An-26	T-218	?		
An-26	T-219	?		
An-26	T-220	?		
An-26	T-221	?		
An-26	T-222	?		
An-26	T-223 D2-MPQ	13704		Became, see next line
An-26	T-224	?		Crashed 30-1-2001
An-26	T-225	?		Damaged 19-2-2001
An-26	T-226	?		
An-26	T-227	4902		Ex-Aeroflot Soviet Airlines CCCP-26611
An-26	T-228	?		Derelict Luanda
An-26	T-229	?		
An-26	T-230	?		Derelict Luanda
An-26	T-231	?		
An-26	T-232	7306		Ex/to Cuban Air Force CU-T1432
An-26	T-233	?		
An-26	T-234	?		
An-26	T-235	?		
An-26	T-236	?		
An-26	T-237	7704		Ex-Cuban Air Force CU-T1237. Shot down 27-4-1988
An-26	T-238			
An-26	T-239	7907		Ex/to Cuban Air Force CU-T1239
An-26	T-241	?		
An-26	T-242	?		
An-26	T-243	?		
An-26	T-244	?		
An-26	T-701	?		
An-26	T-702	?		
An-26	T-704	?		
An-26	T-705	?		
An-26	D2-EFN	?		Government operated. Derelict Luanda
An-26	D2-END	?		Government operated; green/grey camouflage
An-26	D2-ENI	?		Government operated
An-26	D2-EPN	7804		Government operated
An-26	D2-EPO	?		Government operated
An-26	D2-EPP	?		Government operated
An-26	D2-EPQ	10808		Government operated. Retired, preserved as tourist attraction Luanda
An-26	D2-EPR	?		Government operated
An-26	D2-EPU	?		Government operated
An-26	D2-EZG	?		Government operated. Seen 1989
An-26	D2-EZH	?		Government operated. Seen 1989
An-26	D2-EZI	?		Government operated. Seen 1989
An-26	D2-FAB	?		Sometimes reported in error as an An-32
An-26	D2-FAC	?		Sometimes reported in error as an An-32
An-26	D2-FAD	?		Sometimes reported in error as an An-32
An-26	D2-FAE	?		Sometimes reported in error as an An-32
An-26	D2-MAD	?		
An-26	D2-MIR	?		Ex/to Cuban Air Force CU-T1423
An-26	D2-MOP	13607		

An-26	D2-TAD	4609		
An-26	D2-TAE	?		
An-30A-100	D2-MBO	1401		Ex-ARP 410 Airlines UR-30000, bought 2004. Became, see next line
	T-240	?		Seen 2006. Back to D2-MBO
An-32A	T-250	2202		Ex-RA-69329
An-32A	T-251	2104		Ex-ER-AEH
An-32A	T-252	2102		Ex-Ufa Engine Production Association (IMPO) RA-48098 (ER-AFO not taken up). Became, see next line
	D2-MAX			Seen 2003. Back to T-252 by 3-2009
An-32A	T-253	?		
An-32A	T-254	2106		Ex-ER-AWT
An-32A	T-255	1409		Ex-Komsomol'sk-on-Amur Aircraft Production Association (KnAAPO) RA-48979 (ER-AWU not taken up)
An-32A	T256	2108		Ex-ER-AEA; all-white c/s, no titles or national insignia. Sold as S9-BOH
An-32A	D2-FEN	1410		Ex-ER-AWI. Became, see next line
	T-257			
An-72	T-700	365.720.60.640	0606	Ex-Russian Air Force RA-72928, purchased by 2001; sold to Georgia as 4L-AAL
An-72V	T-701	365.470.95.907	1408	Ex-S9-BOU, formerly operated by Angola as such
An-72	T-702	365.720.93.876	1210	Ex-Tiramavia ER-AWS. Became, see next line
	D2-MAQ			Blue/white civil-style c/s, later overall grey c/s (after overhaul in 2009)
An-72-100	T-703	365.720.70.688	0704	Ex-Tiramavia ER-AWQ. Became, see next line
	D2-FEP			
An-72	T-704	365.720.**.***	?	Seen 2009
An-72	T-705	365.720.**.***	?	Seen 2010
An-72	T-750	365.720.**.***	?	Seen 2011
An-72	D2-FGF	365.720.**.***	?	
An-74	D2-MBF	365.470.60.649	0609	Ex-Tiramavia ER-AEO; seen 2007
An-72	D2-MBP	365.720.20.385	0305	Ex/to EK-72903; blue/white civil-style c/s
An-72	D2-MBT	365.720.70.695	0707	Ex-Enimex ES-NOB, seen 2-2008. Became, see next line
	D2-FGE			
An-72-100	D2-MBW	365.720.80.780	0804	Ex-Enimex ES-NOK
An-72-100	D2-MBX	365.720.95.909	1410	Ex-Enimex ES-NOH
An-72-100D	3C-QQO	365.720.96.914	1503	Ex-Russian Air Force RA-72982. To Enimex as ES-NOI
IL-76TD	T-900	0053462872	4708	Ex-Ukrainian Air Force IL-76MD UR-76651. Became, see next line
	D2-FCN			Tail gunner's station retained/demilitarised
IL-76TD	T-901	0063470107	5307	Ex-Ukrainian Air Force IL-76MD UR-76694. Became, see next line
	D2-FCM			Tail gunner's station retained/demilitarised?
IL-76TD	T-902	1013409280	8210	Ex-Atlant Hungary HA-TCK. Became, see next line
	D2-MBJ			Registered 2006; back to T-902 same year
IL-76TD	T-905	0063469062	5206	Ex-Ukrainian Air Force IL-76MD UR-76688. Became, see next line
	D2-FEM			Tail gunner's station retained/demilitarised? Named 'Rei-Eraikui'; back to T-905
IL-76TD	T-906	?	?	Damaged 27-8-2009
IL-76TD	T-907	?	?	
IL-76TD	T-909	?	?	
IL-76TD	D2-FCO	0043454615?	4104?	Ex-ER-IBE? Seen 2006; blue/white c/s, no titles
IL-76 (+)	D2-FDX	063407170	0503	Ex-Russian Air Force RA-86747
'IL-76TD'	D2-FEW	0073475239	5610	Ex-Ukrainian Air Force IL-78 UR-76721; possibly T-904 allocated
IL-76TD	D2-FGG	?	?	Ex-Servisair, transferred 2011
IL-76TD	D2-MAA	1013408246	8202	Ex-Uzbekistan Airways UK 76813
IL-76TD	D2-MBQ	1013407223	8106	Ex-Uzbekistan Airways UK 76811
UTI-MiG-15	C-01?	?		Strange serial for a trainer

MiG-17F?	C21	?		Also reported as a UTI-MiG-15!
MiG-17F	C22	?		Gate guard at Luanda airport
MiG-17F	C23	?		Natural metal finish
MiG-17F	C24	?		Sand/brown camouflage, red serial
MiG-17F	C25	?		Natural metal finish
MiG-17F	C26	?		Natural metal finish
MiG-17F	C27	?		Natural metal finish
MiG-17F	C28	?		Natural metal finish
MiG-17F	C29	?		
MiG-17F	C-721	?		Possibly ex-C21. Destroyed 5-12-1985
MiG-17F	C-724	?		Possibly ex-C24. Destroyed 5-12-1985
MiG-21MF?	C10	?		
MiG-21MF?	C11	?		
MiG-21MF?	C12	?		
MiG-21MF?	C15	?		
MiG-21MF?	C16	?		Shot down by MANPADS 19-12-1986
MiG-21MF?	C17	?		
MiG-21MF?	C20	?		
MiG-21MF?	C30	?		Shot down by MANPADS 22-9-1986
MiG-21MF?	C32	?		
MiG-21MF?	C35	?		
MiG-21MF	C40	?		Sand/olive drab camouflage. Damaged in dogfight with SAAF Mirage F.1CZ 5-10-1982
MiG-21MF	C41	?		D/D 8-1-1976. Sand/olive drab camouflage. Damaged 16-3-1976
MiG-21MF	C42	?		D/D 8-1-1976. Shot down by MANPADS 12-12-1987
MiG-21MF	C43	?		D/D 8-1-1976.
MiG-21MF	C44	?		D/D 8-1-1976.
MiG-21MF	C45	?		D/D 8-1-1976.
MiG-21MF	C46	?		D/D 8-1-1976. Damaged beyond repair 25-11-1987
MiG-21MF	C47	?		D/D 8-1-1976. Also reported as MiG-21*bis*. Damaged near Lubango 5-10-1982
MiG-21MF	C48	?		D/D 8-1-1976. Sand/olive drab camouflage. Damaged on landing 1980 but repaired
MiG-21MF	C49	?		D/D 8-1-1976.
MiG-21MF	C50	?		D/D 8-1-1976. Sand/brown/green/olive drab camouflage. Reported damaged Lubango 20-11-1983 (some sources say Luena 20-10-1983); gate guard Lubango AB
MiG-21MF	C51	?		D/D 8-1-1976. DBR Negage 29-12-1986, derelict
MiG-21MF	C52	?		D/D 8-1-1976. Crashed 13-8-1988
MiG-21MF	C53	?		Sand/brown/olive drab camouflage. Noted as a wreck Luanda Airport 2001
MiG-21MF	C54	?		Shot down by Stinger MANPADS near Cazongo 20-8-1986
MiG-21MF	C55	?		
MiG-21MF	C56	?		
MiG-21MF	C61	?		
MiG-21MF	C62	?		
MiG-21MF	C63	?		Shot down by MANPADS 21-11-1990
MiG-21MF	C64	?		Shot down by MANPADS near Cuemba 13-9-1984
MiG-21MF	C88	?		
MiG-21MF	C118	?		Also reported as MiG-21*bis*
MiG-21MF	C130	?		Also reported as MiG-21*bis*
MiG-21MF	C205	?		Also reported as MiG-21*bis*
MiG-21MF	C206	?		Also reported as MiG-21*bis*
MiG-21MF	C207	?		Also reported as MiG-21*bis*
MiG-21MF	C208	?		Also reported as MiG-21*bis*. Shot down by MANPADS 19-8-1986
MiG-21MF	C209	?		Also reported as MiG-21*bis*. Shot down by MANPADS 29-5-1989
MiG-21MF	C211	?		Also reported as MiG-21*bis*. Crashed 4-8-1989
MiG-21MF	C212	?		Also reported as MiG-21*bis*. Shot down by MANPADS 21-8-1991
MiG-21MF	C213	?		Also reported as MiG-21*bis*
MiG-21MF	C214	?		Also reported as MiG-21*bis*
MiG-21MF	C215	?		Also reported as MiG-21*bis*

MiG-21MF	C216	?		Also reported as MiG-21*bis*
MiG-21MF	C217	?		Also reported as MiG-21*bis*
MiG-21MF	C218	?		Also reported as MiG-21*bis*. Crashed 5-7-1988
MiG-21MF	C219	?		Also reported as MiG-21*bis*. Shot down by MANPADS 24-7-1991
MiG-21MF	C224	?		Also reported as MiG-21*bis*
MiG-21MF	C228	?		Also reported as MiG-21*bis*
MiG-21MF	C232	?		Also reported as MiG-21*bis*
MiG-21MF	C234	?		Also reported as MiG-21*bis*
MiG-21MF	C239	?		Also reported as MiG-21*bis*
MiG-21MF	C241	?		Also reported as MiG-21*bis*
MiG-21MF	C250	?		Also reported as MiG-21*bis*
MiG-21MF	C255	?		Also reported as MiG-21*bis*
MiG-21MF	C268	?		Also reported as MiG-21*bis*
MiG-21MF	C272	?		Also reported as MiG-21*bis*. Shot down 10-11-2002
MiG-21MF?	C281	?		Also reported as MiG-21*bis*
MiG-21MF?	C285	?		Also reported as MiG-21*bis*
MiG-21MF?	C289	?		Also reported as MiG-21*bis*
MiG-21*bis*	C300	N750*****	?	
MiG-21*bis*	C301	N750*****	?	
MiG-21*bis*	C302	N750*****	?	Shot down 5-2-1990
MiG-21*bis*	C303	N750*****	?	
MiG-21*bis*	C304	N750*****	?	
MiG-21*bis*	C305	N750*****	?	Crashed 19-8-1987
MiG-21*bis*	C306	N750*****	?	
MiG-21*bis*	C307	N750*****	?	Crashed 7-4-1988
MiG-21*bis*	C308	N750*****	?	
MiG-21*bis*	C309	N750*****	?	
MiG-21*bis*	C310	N750*****	?	Crashed 22-6-1987
MiG-21*bis*	C311	N750*****	?	Shot down by MANPADS 18-3-1991
MiG-21*bis*	C312	N750*****	?	Shot down 11-11-1987
MiG-21*bis*	C313	N750*****	?	
MiG-21*bis*	C314	N750*****	?	Chocolate brown/dark green/yellow camouflage
MiG-21*bis*	C315	N750*****	?	
MiG-21*bis*	C316	N750*****	?	Crashed 26-12-1987
MiG-21*bis*	C317	N750*****	?	
MiG-21*bis*	C318	N750*****	?	
MiG-21*bis*	C319	N750*****	?	
MiG-21*bis*	C320	N750*****	?	
MiG-21*bis*	C321	N750*****	?	Variously claimed shot down 9-/12-/14-8-1985 near Lucusse or 6-3-1991!
MiG-21*bis*	C322	N750*****	?	
MiG-21*bis*	C323	N750*****	?	Reported written off – date not recorded
MiG-21*bis*	C324	N750*****	?	
MiG-21*bis*	C325	N750*****	?	Shot down by MANPADS 19-9-1989
MiG-21*bis*	C326	N750*****	?	Shot down near Luso 24-9-1984
MiG-21*bis*	C328	N750*****	?	
MiG-21*bis*	C328	N750*****	?	
MiG-21*bis*	C329	N750*****	?	Crashed 15-3-1987
MiG-21*bis*	C330	N750*****	?	Crashed 2-9-1987
MiG-21*bis*	C331	N750*****	?	Crashed 26-12-1987
MiG-21*bis*	C332	N750*****	?	
MiG-21*bis*	C333	N750*****	?	
MiG-21*bis*	C334	N750*****	?	
MiG-21*bis*	C337	N750*****	?	
MiG-21*bis*	C338	N750*****	?	
MiG-21*bis*	C339	N750*****	?	
MiG-21*bis*	C340	N75096900	75-69-**?	Tan/green camouflage. Captured by SADF after off-field emergency landing 14-12-1989; preserved SAAF Museum, Swartkop

MiG-21*bis*	C341	N750*****	?	Crashed 27-12-1987
MiG-21*bis*	C342	N750*****	?	
MiG-21*bis*	C343	N750*****	?	Crashed 27-12-1987
MiG-21*bis*	C344	N750*****	?	
MiG-21*bis*	C345	N750*****	?	
MiG-21*bis*	C346	N750*****	?	
MiG-21*bis*	C347	N750*****	?	Destroyed on ground at Huambo 22-6-1985
MiG-21*bis*	C348	N750*****	?	
MiG-21*bis*	C349	N750*****	?	
MiG-21*bis*	C350	N750*****	?	Tan/green camouflage. DBR in forced landing in Zaïre in the mid-1990s
MiG-21*bis*	C351	N750*****	?	
MiG-21*bis*	C352	N750*****	?	
MiG-21*bis*	C353	N750*****	?	
MiG-21*bis*	C354	N750*****	?	
MiG-21*bis*	C355	N750*****	?	
MiG-21*bis*	C356	N750*****	?	
MiG-21*bis*	C357	N750*****	?	
MiG-21*bis*	C358	N750*****	?	
MiG-21*bis*	C359	N750*****	?	Shot down by MANPADS 8-1-1989
MiG-21*bis*	C360	N750*****	?	
MiG-21*bis*	C361	N750*****	?	Crashed 14-3-1988
MiG-21*bis*	C362	N750*****	?	
MiG-21*bis*	C363	N750*****	?	Crashed 31-1-1987
MiG-21*bis*	C364	N750*****	?	
MiG-21*bis*	C365	N750*****	?	Shot down by MANPADS 30-1-1989
MiG-21*bis*	C366	N750*****	?	Reported written off – date not recorded
MiG-21*bis*	C367	N750*****	?	
MiG-21*bis*	C368	N750*****	?	Shot down by Stinger MANPADS 10-9-1989 or 13-9-1989
MiG-21*bis*	C369	N750*****	?	
MiG-21*bis*	C370	N750*****	?	Shot down by Stinger MANPADS 4-3-1991
MiG-21*bis*	C371	N750*****	?	
MiG-21*bis*	C372	N750*****	?	
MiG-21*bis*	C373	N750*****	?	
MiG-21*bis*	C374	N750*****	?	Shot down by Stinger MANPADS 16-2-1991
MiG-21*bis*	C375	N750*****	?	Destroyed on ground at Cuito Cuanavale 7-3-1988
MiG-21*bis*	C376	N750*****	?	
MiG-21*bis*	C377	N750*****	?	
MiG-21*bis*	C378	N750*****	?	Crashed near Lubango AB 26-6-1994
MiG-21*bis*	C379	N750*****	?	
MiG-21*bis*	C380	N750*****	?	Crashed 9-7-1988
MiG-21*bis*	C381	N750*****	?	
MiG-21*bis*	C382	N750*****	?	Possibly DBR in forced landing at Tembo 8-1988
MiG-21*bis*	C384	N750*****	?	
MiG-21*bis*	C385	N750*****	?	
MiG-21*bis*	C386	N750*****	?	
MiG-21*bis*	C387	N750*****	?	
MiG-21*bis*	C388	N750*****	?	
MiG-21*bis*	C389	N750*****	?	
MiG-21*bis*	C390	N750*****	?	
MiG-21*bis*	C391	N750*****	?	
MiG-21*bis*	C392	N750*****	?	Crashed 13-12-1986
MiG-21*bis*	C393	N750*****	?	
MiG-21*bis*	C394	N750*****	?	
MiG-21*bis*	C395	N750*****	?	
MiG-21*bis*	C396	N750*****	?	
MiG-21UM	I10	?	?	Crashed near Menongue 30-7-1987
MiG-21UM	I11	?	?	

MiG-21UM	I12	?	?	
MiG-21UM	I13	?	?	Two-tone green camouflage
MiG-21UM	I14	?	?	Shot down near Menongue 19-8-1985
MiG-21UM	I15	?	?	
MiG-21UM	I16	?	?	
MiG-21UM	I17	?	?	
MiG-21UM	I18	?	?	
MiG-21UM	I19	?	?	
MiG-21UM	I45	?	?	Light grey overall, red/yellow/back rudder
MiG-23MLA	C400	03903*****	1****	
MiG-23MLA	C401	03903*****	1****	
MiG-23MLA	C402	03903*****	1****	
MiG-23MLA	C403	03903*****	1****	
MiG-23MLA	C404	03903*****	1****	Destroyed on the ground 2-5-1991
MiG-23MLA	C405	03903*****	1****	
MiG-23MLA	C406	03903*****	1****	
MiG-23MLA	C407	03903*****	1****	Destroyed on the ground 2-5-1991
MiG-23MLA	C408	03903*****	1****	Shot down by MANPADS 5-1-1990
MiG-23MLA	C409	03903*****	1****	Destroyed on the ground 2-5-1991
MiG-23MLA	C410	03903*****	1****	
MiG-23MLA	C411	03903*****	1****	Destroyed on the ground 2-5-1991
MiG-23MLA	C412	03903*****	1****	Shot down 31-12-1987
MiG-23MLA	C413	03903*****	1****	
MiG-23MLA	C414	03903*****	1****	
MiG-23MLA	C415	03903*****	1****	
MiG-23MLA	C416	03903*****	1****	
MiG-23MLA	C417	03903*****	1****	
MiG-23MLA	C418	03903*****	1****	
MiG-23MLA	C419	03903*****	1****	
MiG-23MLA	C420	03903*****	1****	
MiG-23MLA	C421	03903*****	1****	Shot down by MANPADS 16-10-1989
MiG-23MLA	C422	03903*****	1****	
MiG-23MLA	C423	03903*****	1****	Donated to Cuba in 1989
MiG-23MLA?	C424	03903*****	1****	Converted to MiG-23MLD in 2005? Crashed at Lubango 21-7-2005
MiG-23MLA	C425	03903*****	1****	
MiG-23MLA	C426	03903*****	1****	
MiG-23MLA	C427	03903*****	1****	
MiG-23MLA	C-428	03903*****	1****	Green/brown camouflage
MiG-23MLA	C429	03903*****	1****	
MiG-23MLA	C430	03903*****	1****	
MiG-23MLA	C431	03903*****	1****	Shot down 6-12-1985
MiG-23MLA	C432	03903*****	1****	Shot down 15-6-1990
MiG-23MLA	C433	03903*****	1****	Shot down 5-10-1990
MiG-23MLA	C434	03903*****	1****	Operated by Executive Outcomes; derelict Cabo Ledo after wheels-up landing 27-7-1994
MiG-23MLA	C435	03903*****	1****	
MiG-23MLA	C436	03903*****	1****	Operated by Executive Outcomes; derelict Saurimo AB mid-1990s with no serial
MiG-23MLA	C437	03903*****	1****	
MiG-23MLA	C438	03903*****	1****	
MiG-23MLA	C439	03903*****	1****	
MiG-23MLA	C440	03903*****	1****	Donated to Cuba in 1989
MiG-23MLA	C441	03903*****	1****	Abandoned at Cuito Cuanavale
MiG-23MLA	C442	03903*****	1****	Donated to Cuba in 1989
MiG-23MLA	C443	03903*****	1****	Donated to Cuba in 1989
MiG-23MLA	C444	03903*****	1****	
MiG-23MLA	C445	03903*****	1****	Donated to Cuba in 1989
MiG-23MLA	C446	03903*****	1****	

MiG-23MLA	C447	03903*****	1****	
MiG-23MLA	C448	03903*****	1****	
MiG-23MLA	C449	03903*****	1****	
MiG-23MLA	C450	03903*****	1****	
MiG-23MLA	C451	03903*****	1****	
MiG-23MLA	C452	03903*****	1****	
MiG-23MLA	C453	03903*****	1****	
MiG-23MLA	C454	03903*****	1****	Operated by Executive Outcomes; red serial later changed to black!
MiG-23MF	C455	03902*****	?	Light green/dark green/brown camouflage
MiG-23MLA	C456	03903*****	1****	
MiG-23MLA	C457	03903*****	1****	Reported shot down by MANPADS 10-4-1991 but seen intact in 1997!
MiG-23MLA	C458	03903*****	1****	
MiG-23MLA	C459	03903*****	1****	
MiG-23MLA	C460	03903*****	1****	
MiG-23MLA	C461	03903*****	1****	Shot down in Bie Province 29-4-1988
MiG-23MLA	C462	03903*****	1****	
MiG-23MLA	C464	03903*****	1****	
MiG-23MLA	C465	03903*****	1****	
MiG-23MLA	C466	03903*****	1****	
MiG-23MLA	C467	03903*****	1****	Shot down 28-1-1991?
MiG-23MLA	C468	03903*****	1****	
MiG-23MLA	C469	03903*****	1****	
MiG-23MLA?	C-470	03903*****	1****	Converted to MiG-23MLD
MiG-23MLA?	C-471	03903*****	1****	Converted to MiG-23MLD
MiG-23MLA?	C474	03903*****	1****	Converted to MiG-23MLD
MiG-23MLA	C475	03903*****	1****	
MiG-23MLA	C477	03903*****	1****	Reported shot down by MANPADS?-9-1987 but this is not confirmed
MiG-23MLA	C478	03903*****	1****	
MiG-23MLA	C479	03903*****	1****	
MiG-23MLA	C482	03903*****	1****	Crashed after 1988 due to engine failure, pilot OK
MiG-23UB	I20	?	?	Crashed near Humpata, Huila Province, 11-11-2008
MiG-23UB	I21	?	?	
MiG-23UB	I22	?	?	
MiG-23UB	I23	?	?	
MiG-23UB	I24	?	?	
MiG-23UB	I25	?	?	
MiG-23UB	I26	?	?	
MiG-23UB	I27	?	?	
MiG-23UB	I-28	?	?	
MiG-23UB	I-29	39040072	?	Ex-Soviet (Ukrainian?) Air Force. Green/dark green/brown camouflage, green radome
MiG-23UB	I-30	29020407	?	Ex-Soviet (Ukrainian?) Air Force. Green/dark green/brown camouflage
MiG-23UB	I-31	39040097	?	Ex-Soviet (Ukrainian?) Air Force. Green/dark green/brown camouflage
Mi-8T	D2-EUA	?	?	Basic Aeroflot c/s. Seen 2003 wfu
Mi-8...	D2-EUF	?	?	
Mi-8...	D2-EUH	?	?	
Mi-8...	D2-FMH	?	?	
Mi-8...	D2-FMI	?	?	
Mi-8...	D2-TAD	?	?	
Mi-8T	H-01	?	?	Crashed
Mi-8T	H-02	31202	?	Donated to Cuba; preserved Havana Victory Park
Mi-8T	H-03	?	?	Crashed
Mi-8T	H-04	?	?	Crashed
Mi-8...	H-5	?	?	
Mi-8T	H-06	?	?	Crashed 17-5-1978
Mi-8...	H-6	?	?	Crashed 20-9-1986
Mi-8...	H-7	?	?	

Mi-8...	H-8	?	?	
Mi-8...	H-9	?	?	
Mi-8...	H10	?	?	
Mi-8...	H11	?	?	
Mi-8...	H12	?	?	
Mi-8PS	H14	?	?	Overall matt grey
Mi-8TV	H16	?	?	Overall matt grey
Mi-8...	H18	?	?	
Mi-8...	H22	?	?	
Mi-8...	H27	?	?	
Mi-8...	H30	?	?	Crashed 28-3-1988
Mi-8...	H32	?	?	
Mi-8...	H33	?	?	Crashed 29-10-1985
Mi-8...	H38	?	?	
Mi-8...	H39	?	?	
Mi-8...	H40	?	?	
Mi-8...	H42	?	?	Crashed 6-12-1987
Mi-8...	H43	?	?	
Mi-8...	H44	?	?	
Mi-8...	H45	?	?	Crashed 6-12-1987
Mi-8...	H46	?	?	Crashed 6-4-1987
Mi-8...	H48	?	?	
Mi-8...	H50	?	?	
Mi-8...	H51	?	?	
Mi-8...	H52	?	?	
Mi-8...	H53	?	?	
Mi-8...	H54	?	?	Crashed 31-7-1990
Mi-8...	H55	?	?	
Mi-8...	H56	?	?	
Mi-8...	H57	?	?	Derelict Luanda
Mi-8...	H58	?	?	Crashed 20-11-1987
Mi-8...	H59	?	?	
Mi-8T	H60	?	?	Crashed; wreck dumped Huambo
Mi-8T	H61	?	?	Equipped with Mi-8MT-type intake filters
Mi-8...	H64	?	?	
Mi-8...	H65	?	?	
Mi-8...	H73	?	?	
Mi-8...	H75	?	?	Crashed 21-4-2001
Mi-8...	H82	?	?	Crashed 25-2-1987
Mi-8...	H84	?	?	Crashed 8-8-1989
Mi-8...	H87	?	?	
Mi-8...	H92	?	?	
Mi-8...	H94	?	?	Crashed 29-12-1987
Mi-8...	H95	?	?	
Mi-8...	H96	?	?	
Mi-8...	H98	?	?	Crashed 29-6-1985
Mi-8...	H99	?	?	
Mi-8T	H344	?	?	Sand/brown/green camo
Mi-17	H-500	?	?	
Mi-17	H-501	?	?	Crashed 19-9-1987 or 29-9-1985
Mi-17	H-502	?	?	Derelict Huambo by 1985
Mi-17	H-503	?	?	Crashed 13-9-1989
Mi-17	H-505	?	?	Crashed 29-9-1985
Mi-17	H-507	?	?	
Mi-17	H-508	?	?	Crashed 9-8-1987
Mi-17	H-509	?	?	Crashed 13-4-1988
Mi-17	H-513	?	?	Crashed 26-9-1987

Mi-17	H-514	?	?	Crashed 18-4-1991
Mi-17	H-515	?	?	
Mi-17	H-516	?	?	Crashed, date unknown
Mi-17	H-518	?	?	Crashed 19-7-1987
Mi-17	H-519	?	?	
Mi-17	H-521	?	?	
Mi-17	H-522	?	?	
Mi-17	H-523	?	?	
Mi-17	H-524	?	?	
Mi-17	H-525	?	?	Crashed, date unknown
Mi-17	H-527	?	?	
Mi-17	H-528	?	?	
Mi-17	H-529	?	?	
Mi-17	H-530	?	?	Crashed 26-5-1989
Mi-17	H-531	?	?	
Mi-17	H-532	?	?	
Mi-17	H-533	?	?	Crashed 6-1-1989
Mi-17	H-534	?	?	Crashed 26-10-1994
Mi-17	H-535	?	?	
Mi-17	H-536	?	?	
Mi-17	H-537	?	?	
Mi-17	H-538	?	?	Derelict Huambo
Mi-17	H-539	?	?	
Mi-17	H-540	?	?	
Mi-17	H-541	?	?	
Mi-17	H-542	?	?	
Mi-17	H-543	?	?	
Mi-17	H-544	?	?	Crashed 26-10-1995, derelict Luanda
Mi-17	H-546	?	?	
Mi-17	H-547	?	?	
Mi-17	H-548	?	?	
Mi-17	H-549	?	?	
Mi-17	H-550	?	?	
Mi-17	H-551	?	?	
Mi-17	H-552	?	?	
Mi-17	H-553	?	?	Crashed 6-1-1989
Mi-17	H-554	?	?	
Mi-17	H-555	?	?	
Mi-17	H-556	?	?	
Mi-17	H-557	?	?	
Mi-17	H-558	?	?	
Mi-17	H-559	?	?	
Mi-17	H-560	?	?	
Mi-17	H-561	?	?	
Mi-17	H-562	?	?	
Mi-17	H-563	?	?	
Mi-17	H-565	?	?	
Mi-17	H-566	?	?	
Mi-17	H-569	?	?	
Mi-17	H-570	?	?	
Mi-17	H-573	?	?	
Mi-17	H-575	?	?	
Mi-17	H-577	?	?	
Mi-8MTV-1	H-579	95997	?	
Mi-17	H-584	?	?	Crashed 26-10-1994
Mi-17	H-587	?	?	
Mi-17	H-588	?	?	

Mi-17	H-589	?	?	
Mi-17	H-591	?	?	Crashed 5-2-2000
Mi-17	H-593	?	?	
Mi-17	H-594	?	?	
Mi-17	H-595	?	?	Overall matt green c/s, extra tanks, no insignia
Mi-17	H-606	?	?	
Mi-17	H-610	?	?	
Mi-17	H-613	?	?	
Mi-8MTV-1	H-616	?	?	Green c/s
Mi-17	H-618	?	?	
Mi-17	H-620	?	?	All-white, Pall air intake filters
Mi-17	H-622	?	?	
Mi-17	H-623	?	?	
Mi-17	H-624	?	?	All-white, Pall air intake filters
Mi-8MTV-1	MAT.7	95733	?	Ex-Aeroflot CCCP-25117, D/D 1-1993
Mi-8MTV-1	MAT.8	95736	?	Ex-Aeroflot CCCP-25120, D/D 1-1993
Mi-8MTV-1	MAT.9	95593	?	Ex-Aeroflot CCCP-25453, D/D 1-1993
Mi-8MTV-1	MAT.10	95735	?	Ex-Aeroflot CCCP-25119, D/D 1-1993
Mi-8MTV-1	MAT.15	95653	?	Ex-Aeroflot CCCP-25505, D/D 1-1993
Mi-8MTV-1	MAT.16	95654	?	Ex-Aeroflot CCCP-25506, D/D 1-1993
Mi-8MTV-1	MAT.17	95655	?	Ex-Aeroflot CCCP-25507, D/D 1-1993
Mi-8MTV-1	MAT.18	95734	?	Ex-Aeroflot CCCP-25118, D/D 1-1993
Mi-24D	H-301	?	?	Shot down 3-1-1991
Mi-24D	H-302	?	?	Crashed ?-6-1985; derelict Huambo
Mi-24D	H-303	?	?	
Mi-24D	H-304	?	?	
Mi-24D	H-305	?	?	
Mi-24D	H-306	?	?	
Mi-24D	H-307	?	?	
Mi-24D	H-308	?	?	Crashed (date unknown); wreck dumped Huambo
Mi-24D	H-309	?	?	Shot down 1-9-1988
Mi-24D	H-310	?	?	
Mi-24D	H-312	?	?	
Mi-24D	H-313	?	?	
Mi-24D	H-314	?	?	Shot down 29-9-1985
Mi-24D	H-316	?	?	Shot down 27-9-1985
Mi-24D	H-317	?	?	
Mi-24D	H-318	?	?	
Mi-24D	H-319	?	?	
Mi-24D	H-320	?	?	
Mi-24D	H-322	?	?	Shot down 27-9-1985
Mi-24D	H-323	?	?	Shot down ?-9-1985
Mi-24D	H-325	?	?	
Mi-24D	H-327	?	?	
Mi-24D	H-330	?	?	
Mi-24D	H-331	?	?	
Mi-24D	H-332	?	?	Derelict Luanda
Mi-24D	H-333	?	?	Derelict Luanda
Mi-24D	H-334	?	?	
Mi-24D	H-335	?	?	
Mi-24D	H-336	?	?	
Mi-24D	H-337	?	?	
Mi-24D	H-338	?	?	Shot down 4-9-1987
Mi-24D	H-339	?	?	
Mi-24V	H-350	?	?	
Mi-24V	H-351	?	?	Crashed (date unknown)

Mi-24V	H-352	?	?	
Mi-24V	H-353	?	?	
Mi-24V	H-354	?	?	Shot down 21-3-1989
Mi-24V	H-355	?	?	
Mi-24V	H-356	?	?	
Mi-24V	H-357	?	?	Derelict Huambo
Mi-24V	H-358	?	?	
Mi-24V	H-360	?	?	Derelict Huambo
Mi-24V	H-361	?	?	
Mi-24V	H-362	?	?	
Mi-24V	H-363	?	?	
Mi-24V	H-365	?	?	WFU and cannibalised for spares
Mi-24V	H-366	?	?	
Mi-24V	H-367	?	?	
Mi-24V	H-368	?	?	
Mi-24V	H-369	?	?	
Mi-24V	H-370	?	?	
Mi-24V	H-372	?	?	
Mi-24V	H-373	?	?	
Mi-24V	H-374	?	?	
Mi-24V	H-375	?	?	
Mi-24P	H-376	?	?	
Mi-24V	H-379	?	?	
Mi-24P	H-380	?	?	
Mi-24P	H-381	?	?	
Mi-24V	H-384	?	?	Green/brown/dark brown camouflage. Wfu Luanda
Mi-24V	H-385	?	?	
Su-22	C501 (1)	?		
Su-22	C502 (1)	?		
Su-22	C503 (1)	?		
Su-22	C504 (1)	?		
Su-22	C505 (1)	?		
Su-22	C506 (1)	?		
Su-22	C507 (1)	?		
Su-22	C508 (1)	?		
Su-22	C509 (1)	?		
Su-22	C510 (1)	?		Shot down by MANPADS ?-7-86
Su-22M4	C501 (2)	?		Sand/chocolate/medium green/dark green camouflage
Su-22M4	C502 (2)	?		
Su-22M4	C503 (2)	?		Refurbished by ARZ No.558 and operational 2008
Su-22M4	C504 (2)	?		
Su-22M4	C505 (2)	?		
Su-22M4	C506 (2)	?		
Su-22M4	C507 (2)	41409		Tan/dark brown/bright green camouflage. Refurbished by ARZ No.558 and operational 2008
Su-22M4	C508 (2)	41510?		
Su-22M4	C-509 (2)	41511?		Sand/dark earth/green camouflage. Refurbished by ARZ No.558 and operational 2008
Su-22M4	C-510 (2)	41612		Sand/chocolate/medium green/dark green camouflage
Su-22M4	C511			
Su-22M4	C512			Shot down by MANPADS near Cahama 10-4-1991
Su-22M4	C513			
Su-22M4	C514			Sand/chocolate/medium green/dark green camouflage
Su-22M4	C515			Crashed Catumbela 21-3-1998, pilot killed
Su-22M4	C516			Grey/green/dark brown camouflage
Su-22M4	C 517			Sand/green/brown camouflage
Su-22M4	C 518			Damaged in belly landing at Luanda sometime in the mid-1990s but repaired. Derelict Luanda
Su-22M4	C519			Ex-Belorussian AF (Su-17M4)?

Su-22M4	C520?	?		Ex-Belorussian AF (Su-17M4)? Existence not confirmed but probable
Su-22M4	C521?	?		Ex-Belorussian AF (Su-17M4)? Existence not confirmed but probable
Su-22M4	C-522	36920		Ex-Slovak Air Force 3620 Yellow/'L'
Su-22M4	C550	?		Reported as ex-Slovak Air Force, see lines below!
Su-22M4	C551	?		Reported as ex-Slovak Air Force, see lines below!
Su-22M4	C-5...	36715		Ex-Slovak Air Force 3615 Yellow/'G'
Su-22M4	C-5...	36716		Ex-Slovak Air Force 3616 Yellow/'H'
Su-22M4	C-5...	36817		Ex-Slovak Air Force 3617 Yellow/'I'
Su-22M4	C-5...	36818		Ex-Slovak Air Force 3618 Yellow/'J'
Su-22M4	C-5...	36919		Ex-Slovak Air Force 3619 Yellow/'K'
Su-22M4	C-5...	40512		Ex-Slovak Air Force 4012 Yellow/'M'
Su-22M4	C-5...	40614		Ex-Slovak Air Force 4014 Yellow/'N'
Su-22M4	C-5...	40716		Ex-Slovak Air Force 4016 Yellow/'O'
Su-22M4	C-5...	40717		Ex-Slovak Air Force 4017 Yellow/'P'
Su-22M4	C-5...	40818		Ex-Slovak Air Force 4018 Yellow/'R'
Su-22UM3K	I31	175323*****		
Su-22UM3K	I32?	175323*****		Existence not confirmed but probable
Su-22UM3K	I33	175323*****		
Su-22UM3K	I-34	17532374609		Sand/chocolate/medium green/dark green camouflage; operational 2009
Su-22UM3K	I35	17532372307		Ex-Slovak Air Force 7207 Yellow/'T'; refurbished by ARZ No.558 and operational 2008
Su-22UM3K	I-36	17532372308		Ex-Slovak Air Force 7208 Yellow/'U'; refurbished by ARZ No.558 and operational 2008
Su-25K	B-18	255081*****	?	Probably reserialled to one of the following
Su-25K	B-600	255081*****	?	
Su-25K	B-601	255081*****	?	
Su-25K	B-602?	255081*****	?	Probably lost in accident
Su-25K	B-603	255081*****	?	
Su-25K	B-604	255081*****	?	Crashed 3-7-1994
Su-25K	B-605	255081*****	?	
Su-25K	B-606	255081*****	?	
Su-25K	B-607?	255081*****	?	Probably lost in accident
Su-25K	B-608	255081*****	?	
Su-25K	B-609	255081*****	?	
Su-25K	B-610	255081*****	?	
Su-25UBK	I-40	382201*****	?	
Su-25UBK	I-41	382201*****	?	
Su-25UBK	I-44	382201*****	?	
Su-25UBK	I-45	382201*****	?	
Su-25UBK	I-48	382201*****	?	
Su-27	C 110	369110*****		Reported as C605 but no photo proof of serial! Ex-Ukrainian Air Force. Crashed 20-7-2006
Su-27UB	I100	963104*****	?	Reported as I-51 but this is an Aero L-39C! Ex-Ukrainian Air Force
Tu-134AK	D2-ECC	(73)49830	3903	Possibly reserialled SG-104 in 1983 for a while. Wfu Luanda
Yak-40K	T-450 D2-MAS	9830558		Ex-Novosibirsk Aircraft Production Association RA-87962, seen 2001. Became, see next line
Yak-40 'Salon 2nd Class'	D2-EAG	9230122		Ex-Aeroflot CCCP-87819. Was leased to Aertirrena as I-JAKO. Wfu Luanda

Benin

The only known Soviet aircraft operated by the Benin Air Force (*Force Aérienne du Benin*) were three Polish-built examples of the Antonov An-2T *Colt* utility biplane. Delivered in 1979 and based at Cadjehoun AB near the capital of Cotonou, the aircraft were placed on the civil register as TY-ADP (c/n 1G174-27), TY-ATP (c/n 1G174-28) and TY-ALY (c/n 1G174-29). In 1982 all three An-2s were reportedly sold to the Soviet Union, serving on with the Soviet Air Force.

BF 8401, the sole MiG-17F delivered to Burkina Faso. It is seen here preserved at Ouagadougou. Note the derelict Mi-4 in the background.

Burkina Faso

On 5th August 1960 the French colony of Upper Volta gained independence, becoming the Republic of Burkina Faso. With French assistance, an embryonic air arm known as the Escadrille de Burkina Faso was formed in 1964 and later developed into a fully-fledged Burkina Faso Air Force (BFAF, or *Force Aérienne de Burkina Faso*; also referred to as *Armée de l'Air Burkinabé*, or AdAB).

Starting in the mid-1970s, the Soviet Union extended military aid to Burkina Faso, including deliveries of aircraft to the BFAF. Among other things, a single second-hand Mikoyan/ Gurevich MiG-17F *Fresco-C* fighter was delivered to the unnumbered Fighter Squadron (*Escadrille de Chasse*) at Ouagadougou airbase. Later, the squadron received eight Mikoyan/ Gurevich MiG-21*bis* (*izdeliye* 75A) *Fishbed-L* fighters and two MiG-21U trainers, probably the MiG-21UM (*izdeliye* 69A) *Mongol-B* version. The MiG-21s came from Libya with no qualified personnel to operate them; they were soon withdrawn. By 1993, the surviving MiGs were all non-operational and stored at the base.

Helicopters were also supplied. The first Soviet choppers in BFAF service were two Mil' Mi-4 *Hound* utility helicopters; one has been preserved at Ouagadougou AB. Starting in 1990, the Burkina Faso Air Force acquired at least one Mil' Mi-8S *Hip-C* VIP helicopter for government use and four Mil' Mi-17 *Hip-H* utility helicopters in unarmed utility configuration; some of these were quasi-civil, being registered in the XT-M** block

XT-MAU, the Mi-8S VIP helicopter operated by the Burkina Faso government flight. Note the addition of Mi-8MT/Mi-17-style intake filters.

Burkina Faso Air Force Mi-17 BF-9001

Known Burkina Faso AIr Force aircraft of Soviet/Russian origin				
Type	**Serial/ registration**	**C/n**	**F/n**	**Notes**
MiG-17F	BF8401	?		Brown camouflage. Stored Ouagadougou
MiG-21*bis*	75 Red	?	?	Stored Ouagadougou 1997
Mi-8S	XT-MAU	37001	?	Mi-8MT-style intake filters. Seen 1998
Mi-17	XT-MCS	371M02	?	Seen 1998. Repainted in military markings by 2009 as, see next line
	BF-9001			Sand/brown camouflage
Mi-17	XT-MCV	227M204	?	Seen 1998
Mi-17	BF-9202	?	?	Seen 9-2010, two-tone green camouflage; later sand/brown camouflage
Mi-35	BF-0504	?	?	Nose machine-gun not fitted

Burkina Faso Air Force Mi-35 BF-0504

Burundi AIr Force Mi-25 BAF-1214 illustrates its unusual 'water-melon' camouflage scheme.

Burundi

Burundi, another former French colony in central Africa, is a military operator of Mil' helicopters. The Burundi National Army (*Armée Nationale de Burundi*) operated a couple of quasi-civil ex-Russian Mi-8T *Hip-C* utility helicopters bought in September 1998, an ex-Bulgarian Mi-8MTV *Hip-H* and two or three Mi-24V *Hind-E* attack helicopters purchased from Ukrainian surplus stocks by January 2002.

Additionally, several Antonov An-12 *Cub* freighters have been on the Burundi register, and an Il'yushin IL-76TD *Candid-A* freighter was leased from Russia. It is probable that they were used in the interests of the military, since the Burundi National Army has no military transports of its own.

Known Burundi Air Force aircraft of Soviet/Russian origin				
Type	Serial/ registration	C/n	F/n	Notes
An-12B (+)	9U-BHN	3341506		Second An-12 with this registration (see next line!); ex D2-FRC, opb Air Vitoria, seen in 2002. Sold to Mango Airlines 2005 as 9Q-CVT
An-12BP (–)	9U-BHO	8345503		Originally 9U-BHN No.1; ex-Renan ER-ACK. Sold to Mango Airlines 2005 as 9Q-CVM
An-12B	9U-NHN	?		Operator and fate unknown
IL-76TD	RA-76419	1023414470	8708	Operated by Burundi BCR Charter, leased from Arkhangel'sk Airlines 7-1993
Mi-8T	9U-BFJ	99357573	9843?	Ex-Aviaenergo RA-22150. Wfu Bujumbura by 9-2007
Mi-8T	9U-BFK	99357581	9844?	Ex-RA-22151. Wfu Bujumbura by 9-2007
Mi-8MTV	BAF-411	103M12	?	Ex-Bulgarian Air Force '412 White'. Sold to Georgia as 4L-TBS by 2009
Mi-24V	BAF-1214	?	?	Seen 8-2005; two-tone green camouflage
Mi-24V	BAF-1250	?	?	Seen 11-2007

Cameroun

The Cameroun Air Force (*Armée de l'Air du Cameroun*) received a single VIP-configured Mil' Mi-4S *Hound* helicopter. It was registered TJ-XAP, although Camerounian quasi-civil aircraft usually have registrations in the TJX-** series instead of the normal TJ-*** (and the correct rendering would thus be TJX-AP). The helicopter was written off in a crash before 1975.

Looking forlorn with no rotors and rotted cockpit covers, a Cape Verde Air Force Mi-17 languishes at Praia airport.

Cape Verde

The air arm of this former Portuguese colony, Cape Verde Air Force (*Força Aérea Caboverdaine*), operated at least three Antonov An-26 *Curl* tactical transports, one of which is serialled FA-C01.

Also, at least three Mil' Mi-8 *Hip* helicopters (two Mi-8T *Hip-Cs* and a Mi-8MT *Hip-H*) were delivered; they wore FACV roundels but no serials, the Mi-8MT bearing the legend 'Lobo do Mar' (Sea Wolf). All three ended up derelict at Praia International Airport.

Chad

The Chad Air Force (*Force Aérienne Tchadienne*) obtained at least four second-hand Antonov An-26 transports from places as far apart as Russia and Libya. While the ex-Russian aircraft had been bought, the ex-Libyan Arab Air Force example was war booty captured at the LAAF base in the Aouzou Strip. Like all Chad Air Force aircraft, the *Curls* were quasi-civil – which, in the case of the transports, facilitated flights abroad.

The only Soviet combat aircraft in Chad Air Force service are six Sukhoi Su-25 attack aircraft – four Su-25 *Frogfoot-A* single-seaters and two Su-25UB *Frogfoot-B* combat trainers. All six were purchased from the Ukraine, two each being delivered in 2008, 2009 and 2010 by Antonov Airlines An-124-100 Ruslan (*Condor*) transports. The jets wear a sand/tan camouflage. The Su-25s saw action against rebels operating from bases abroad (in Darfur, Sudan); their first known sortie was on 14th May 2009 when a pair of Su-25s attacked a rebel camp near Geneina, Sudan, followed by two more raids in the following days to pave the way for an operation by the ground forces. Similar operations were undertaken in December 2009.

In March 2000 the Chad Air Force bought its first two Mi-8MTV-1 *Hip-H* helicopters. These were second-hand machines originally believed to be from Ukrainian Army stocks (later they turned out to be ex-Russian civil examples). Wearing a tan/brown camouflage scheme and military serials along with civil reg-

Below: Chad Air Force Su-25 TT-QAI undergoes maintenance at N'djamena. The 'land rocket' sitting alongside is a turbine-powered ground power unit converted from a PTB-800 drop tank.

Bottom: Chad Air Force Su-25UB TT-QAH in early guise with the 'Q' mispainted as the Arabic numeral 9.

istrations, the helicopters were based at Faya Largeau AB and flown by Ukrainian mercenary pilots against rebels in northern Chad. On 3rd October 2002 both machines were damaged on the ground (not destroyed, as sometimes reported) by rebels who overran Faya Largeau and have been grounded since. After that, the Chad Air Force acquired a further four Mi-8MTV/Mi-17s, one of which was lost to enemy action or in an accident.

The first two Chadian Mi-24 attack helicopters were bought from the Ukraine in the summer of 2000; these were Mi-24D *Hind-Ds*. The helicopters wore a tan/green camouflage quite different from the standard Soviet patterns. Unlike most Chad Air Force aircraft, they

Another Chadian Su-25, TT-QAL, taxies out with two drop tanks and two UB-32A FFAR pods.

Chad Air Force An-26 TT-LAN

Chad Air Force An-26 TT-LAO

All Chad Air Force An-26s are quasi-civil and painted differently. TT-LAM seen here is all-white and wears full AIr Force markings with an unusually narrow yellow band on the fin flash.

An-26 TT-LAN wears this green/ grey camouflage. Note the former Russian registration RA-26234 painted out above the Chadian one.

An-26 TT-LAO is stored at the maintenance facility at Tripoli-Mitiga in this civil-style colour scheme with a cheatline in national flag colours.

Chad Air Force MI-24V TT-OAM/ 03028 wears a fin flash augmented by 'Forces Aeriennes Tchadiennes' titles.

Another view of the same helicopter. A non-airworthy Mi-8 is visible in the hangar beyond.

Mi-24V TT-OAP is seen here at the scene of its crash-landing on 12th June 2008; the helicopter broke its tailboom and lost the port wing and is certainly a write-off.

Chad Air Force Mi-17-1V TT-OAL

The first two Chad Air Force Mi-17s, TT-OAJ/4320 and TT-OAK/4321, dumped at Faya Largeau.

This Mi-171Sh with a stepped nose and a full-width cargo ramp is a recent addition to the Chad Air Force. Note the 250-kg bombs on the centre pylons.

TT-OAO/7415, another Chadian 'flatass' Mi-171Sh. Note the artwork on the starboard door and on the external armour.

Known Chad Air Force aircraft of Soviet/Russian origin				
Type	Registration/ serial	C/n	F/n	Notes
An-26	TT-LAM	?		Seen 1-1995; white with grey belly, Chad Air Force insignia
An-26	TT-LAN/14308	14308		Ex-Komiinteravia RA-26234, ex-RA-49274, ex-Luftwaffe 52+12, green/grey camouflage; active 12-2006
An-26	TT-LAO	?		First noted 3-2008, white with Chadian flag colours cheatline
An-26	TT-LAP	13110		Ex-Libyan Arab Air Force 8313. Stored N'djamena 2009
Mi-8MTV-1	TT-OAJ/4320	95826	?	Ex-RA-70899? Tan/brown camouflage. Damaged 3-10-2002, wfu Faya Largeau
Mi-8MTV-1	TT-OAK/4321	95831	?	Ex-RA-70900; tan/brown camouflage. Damaged 3-10-2002, wfu Faya Largeau
Mi-17-1V	TT-OAL (2)	148M01	?	Written off 27-4-2006; see Mi-24 below
Mi-17-1V	TT-OAM (2)	148M02	?	Seen 2-2005; see Mi-24 below
Mi-171Sh	TT-OAN/7344	59489617347?	?	Ex-17344 (Ulan-Ude factory demonstrator)? Stepped nose and cargo ramp/door; tan c/s. Seen 8-2008
Mi-171Sh	TT-OAO/7415	59489617415?	?	Ex-17415 (Ulan-Ude factory demonstrator)? Stepped nose and cargo ramp/door; tan c/s. Seen 8-2008
Mi-24D	TT-OAL/03010	?	?	First use of registration, see Mi-8 above, d/d 2000. Written off 26-4-2006
Mi-24D	TT-OAM/03028	?	?	First use of registration, see Mi-8 above
Mi-24V	TT-OAP	?	?	Crashed 12-6-2008
Mi-24V	TT-OAQ	?	?	Seen 2008
Mi-24V	TT-OAR	?	?	Seen 2008
Mi-24V	TT-OAS	?	?	Seen 2008
Su-25	TT-QAI	255081*****	?	Registration originally mispainted with an Arabic '9' instead of the 'Q'.
Su-25	TT-QAL	255081*****	?	Registration previously used by SIAI-Marchetti SF.260W Warrior shot down 26-11-2006; applied as 'TT-QA L'!
Su-25	TT-QAN	255081*****	?	
Su-25	TT-QAO	255081*****	?	
Su-25	TT-QA...	25508105011	?	
Su-25	TT-QAH	382201*****	?	Registration originally mispainted with an Arabic '9' instead of the 'Q'.
Su-25	TT-QAM	382201*****	?	

had 'Forces Aériennes Tchadiennes' titles (instead of 'Force Aérienne Tchadienne', that is). By July 2001 both *Hinds* had run out of service life and were due to go back to the Ukraine for refurbishment. Four examples of the newer Mi-24V *Hind-E* followed later. Again, they were flown by Ukrainian mercenary pilots and saw action against rebels in the north.

Congo (Congo-Brazzaville)

Confusingly there are two countries called Congo. The one dealt with first – the Republic of Congo (occasionally referred to as 'the

Early-style Congo Air Force roundel

Congo Air Force CS-102 '101' sits derelict at Pointe Noire in company with two MiG-17s, displaying faded old FAC insignia.

Congo Air Force UTI-MiG-15 (Aero CS-102) '101'

Republic of *the* Congo'), popularly known as Congo-Brazzaville after its capital, was formed from the French colony of Central-West Africa (or French Congo) on 15th August 1960. The air arm of Congo-Brazzaville is known as the *Force Aérienne Congolaise* (FAC) and was formed in the early 1960s. For the first ten years of its existence it relied on French equipment before switching to Soviet aircraft; that was when Dénis Sassou Nguesso was President and Congo was Soviet-aligned.

Congolese military aircraft usually wear alphanumeric serials with the same TN- nationality prefix as the Congolese civil registrations. The fighters were an exception, with number-only serials. The first Soviet fighter type operated by the Congo Air Force was probably the Mikoyan/Gurevich MiG-17, and enough were reportedly delivered sometime between 1975 and 1977 to equip a single squadron. Often reported in error as MiG-17F *Fresco-Cs*, they were in fact late-pro-

Two views of Congo Air Force MiG-17 '118 Black' derelict at Pointe Noire, also with old-style roundels.

duction MiG-17s *sans suffixe* (*Fresco-As*). By 1991 all of them had been retired at Pointe Noire (Agostinho Neto airport).

Jet training was initially accomplished with Czechoslovak-built Mikoyan/Gurevich UTI-MiG-15 *Midget* trainers, of which at least two were supplied. Again, the aircraft were retired at Pointe Noire.

The first supersonic fighter to see service with the Congo Air Force was reportedly the Mikoyan/Gurevich MiG-19S *Farmer-C*. Unfortunately, no details are known as to the number delivered and in-service dates.

What *is* certain is that the Congo Air Force operated the Mikoyan MiG-21. The first MiG-21*bis* (*izdeliye* 75B) *Fishbed-N* fighter and two MiG-21UM *Mongol-B* trainers arrived in 1988 (some sources say 1986), and within two years the total was reported as 16 and five respectively. Some of them had old-style red roundels with a yellow crest in the middle, as applied to the MiG-17s; others sported new red/yellow/black roundels.

At least one squadron was commissioned in 1991, but in 1997 many of the MiGs were out of use stored in a hangar at Pointe-Noire. They saw next to no action in the civil war that began in June that year when the Congolese army attempted to wipe out the 'Cobras' militia recruited by ex-President Dénis Sassou Nguesso, who had lost the presidential election to Pascal Lissouba five years earlier. Surprisingly, the Force Aérienne Congolaise did not intervene, letting mercenaries on both sides (and the Angolan army backing Sassou Nguesso) do the fighting; either the aircraft

Current Congo Air Force roundel

Congolese Air Force MiG-21*bis* '511 Black' and '525 Black' hangared at Brazzaville. These aircraft saw action in the Congolese civil war.

Two more FAC MiG-21*bis* ('509 Black' and '513 Black') withdrawn from use at Pointe-Noire.

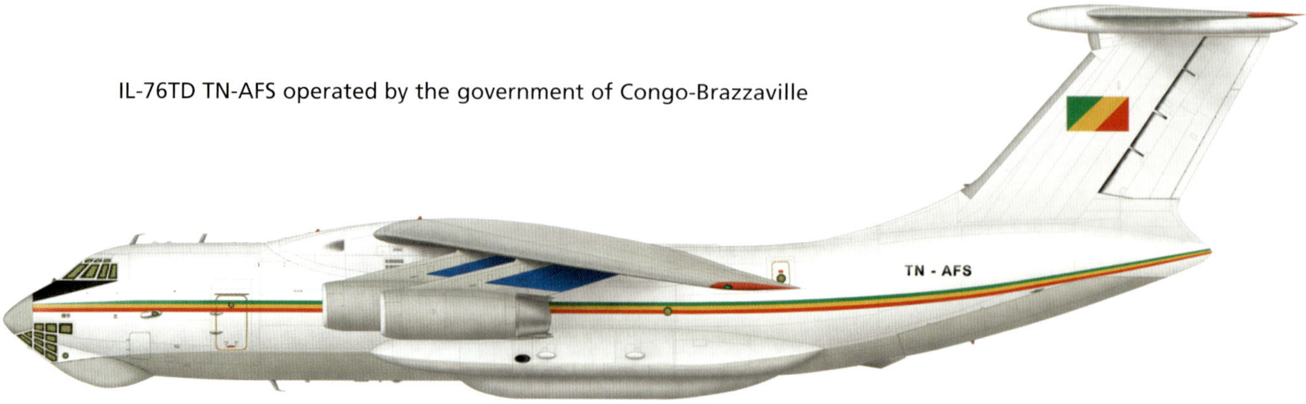

IL-76TD TN-AFS operated by the government of Congo-Brazzaville

were all unserviceable or the Congolese pilots were waiting to see which side was likely to gain the upper hand. Eventually, supported by another militia – the 'Ninjas' headed by Bernard Kolelas, – the 'Cobras' launched a decisive attack on Brazzaville; on 13th October 1997 two MiG-21s bombed the building of the Ministry of the Interior where President Lissouba had holed up. Two days later Sassou-Nguesso assumed presidential power again.

The latest fighter type delivered to Congo-Brazzaville is reportedly the Mikoyan MiG-29, the Congo Air Force ordering five MiG-29SE (*izdeliye* 9.13SE) *Fulcrum-C* single-seaters and one MiG-29UB *Fulcrum-B* combat trainer in 2001.

The transport component of the Congo Air Force initially included at least eight Il'yushin IL-14 *Crate* piston-engined transports, one of which was an IL-14S VIP aircraft used as the presidential machine. Later, two Antonov An-24RV *Coke* twin-turboprop short-haul airliners were transferred from Lina Congo, the flag carrier. Additionally, the Congo Air Force took delivery of two factory-fresh An-24T transports – a straightforward adaptation of the An-24B airliner with a ventral loading/paradrop door; these were later augmented by two Antonov An-32 *Cline* transports with a cargo ramp and bigger engines giving a higher payload and better 'hot-and-high' capabilities. At least one Il'yushin IL-76TD *Candid-A* trans-

Three Congo Air Force Mi-8Ts withdrawn from use at Brazzaville. The one on the left has new roundels, the other two have old ones.

This Congolese 'Evil Monster' Mi-35 wears a rather unusual two-tone green camouflage. Note the absence of intake filters. The dumped GAZ-66 lorries in the background are probably ground support vehicles at the base.

Gleaming with fresh paint, Congo Air Force Mi-35P TN-359 sports impressive 'flying eagle' nose art. Note that the 'danger' stencils on the tail are in Russian ('opasno').

port is also operated, wearing a cheatline in national flag colours and a large Congolese flag on the fin. VIP transport duties were performed by a single Il'yushin IL-18 *Coot* four-turboprop medium-haul airliner.

As regards rotary-wing aircraft, Congo bought at least three Mil' Mi-2 *Hoplite* light helicopters from the German Police after 2004; one of these continued to be used in a police capacity by its new owner. On the heavier end of the scale, at least five Mil' Mi-8 *Hip-C* medium helicopters, including an ex-Lithuanian example, have been delivered; at least three of these were withdrawn from use by 1991. They were augmented by at least two second-hand Mi-17 *Hip-Hs* bought in 1993. Two Mil' Mi-35P *Hind-F* attack helicopters (the export version of the Mi-24P) were delivered in 1997 and saw action in the civil war that began that year. However, commenting on this, a Russian Foreign Ministry spokesman denied that Russia had delivered any Mi-24s to the Republic of Congo 'lately'; this means the helicopters in question were not new-build Mi-35Ps but ex-Soviet Army Mi-24Ps from Ukrainian or Belorussian surplus stocks. The Mi-8s and Mi-35s are known to have been used by Lissouba's 'Zulu' militia against the 'Cobras' and 'Ninjas', with which Lissouba was at odds; two of them were shot down by

Known Congo Air Force aircraft of Soviet origin

Type	Serial/ registration	C/n	F/n	Notes
An-24B 'Salon'	TNK-AL	87304505		Government flight. Became, see next line
	TN-101			Reserialled by 1973. Became, see next line
	TN-220			
An-24RV 'Salon'	TN-111	17306606		Ex-TN-ABX. Became, see next line
	TN-222			
An-24RV 'Salon'	TN-224 (1)	17306607		Ex-TN-ABY; first use of serial, see An-32 below
An-24T	TNK-AN	1021803?	?	Probably became, see next line

	TN-102			Probably became, see next line
	TN-221			
An-24T	TNK-AO	1021804?	?	Probably became, see next line
	TN-103			Probably became, see next line
	TN-223			
An-32	TN-224 (2)	?		Second use of serial; seen 8-2004
An-32	TN-227	?		seen 8-2004
IL-14	TN-104	?		Became, see next line
	TN-225			
IL-14	TN-109	?		Became, see next line
	TN-226			
IL-14	TN-112	?		Became, see next line
	TN-227			
IL-14	TN-117	?		Became, see next line
	TN-228			
IL-14	TN-120	?		Became, see next line
	TN-229			
IL-14S	203	?		Possibly TN-203; presidential aircraft
IL-14	TN-215	?		
IL-14	TN-240	?		
IL-18 'Salon'	TN-105	?		Presidential aircraft
IL-76TD	TN-AFS	1033415504	8806	Ex-EL-WTA; formerly operated by Centrafrican Airlines
UTI-MiG-15	101 Black	612769		Czech-built (Aero CS-102). Wfu Pointe Noire
UTI-MiG-15	102 Black	109948-19		Soviet-built (Ulan-Ude plant). Preserved at Air Force HQ by 1997
MiG-17	106 Black	N54211933		Wfu Pointe Noire, seen 5-1997
MiG-17	109 Black	N54210417		Wfu Pointe Noire, seen 5-1997
MiG-17	112 Black	N54210865		Wfu Pointe Noire, seen 5-1997
MiG-17	115 Black	N54211827		Wfu Pointe Noire, seen 5-1997
MiG-17	118 Black	N54211944		Wfu Pointe Noire, seen 5-1997
MiG-17	120 Black	?		
MiG-17	121 Black	N54211057		Wfu Pointe-Noire, seen 5-1997
MiG-17	124 Black	?		Wreck dumped Pointe Noire 1997
MiG-21*bis*	505 Black	N750*****	?	
MiG-21*bis*	507 Black	N750*****	?	Stored Maya-Maya 1997; camouflaged, old insignia
MiG-21*bis*	509 Black	N75087309	?	Grey/green/brown camouflage, new insignia. Noted stored Pointe Noire 1997
MiG-21*bis*	511 Black	N75087406	?	Grey/green/brown camouflage, new insignia. Noted stored Pointe Noire 1997
MiG-21*bis*	513 Black	N75087444	?	Grey/green/brown camouflage, new insignia. Noted stored Pointe Noire 1997
MiG-21*bis*	515 Black	N75087458	?	Noted stored Pointe Noire 1997
MiG-21*bis*	517 Black	N750*****	?	Stored Maya-Maya 1997; camouflaged, old insignia
MiG-21*bis*	519 Black	N750*****	?	
MiG-21*bis*	521 Black	N750*****	?	Stored Maya-Maya 1997
MiG-21*bis*	523 Black	N750*****	?	Stored Maya-Maya 1997; camouflaged, new insignia
MiG-21*bis*	525 Black	N75096461	75-69-...?	Grey overall. Noted stored Pointe Noire 1997; grey overall, new insignia
MiG-21*bis*	527 Black	N750*****	?	Stored 1997 Maya-Maya; natural metal finish, old insignia
MiG-21UM	501 Black	?	?	
MiG-21UM	503 Black	?	?	Stored Maya-Maya 1997; sand/green camouflage, old insignia
MiG-21UM	529 Black	?	?	Crashed 11-10-1989
MiG-21UM	530 Black	?	?	Stored 1997 Maya-Maya; natural metal finish, new insignia
MiG-21UM	550 Black	?	?	

Mi-2	TN-657	538117033		Police. Ex-German Police D-HZPA, ex-East German Police DDR-VPE
Mi-2	TN-658	538840114		Ex-German Police D-HZPD, ex-East German Police DDR-VPH
Mi-2	TN-659	569341085		Ex-German Police D-HZPK, ex-East German Police DDR-VGF, ex-EGAF/Border Troops '314 Black'
Mi-8	TN-043	?	?	Wfu Brazzaville by 6-2004
Mi-8	TN-232	?	?	Wfu Brazzaville by 6-2004
Mi-8	TN-241	?	?	Wfu Brazzaville by 6-2004
Mi-8	TN-243	?	?	Wfu Brazzaville by 6-2004
Mi-8T	TN-355	99050212	9605	Ex-Lithuanian AF. Wfu Brazzaville by 6-2004
Mi-8T	TN-???	10825	?	Ex-Air Transport Europe OM-DXO
Mi-17	TN-358	226M55	?	Ex-Costa Rican Police MSP-016
Mi-17	TN-???	550M22	?	
Mi-35P	TN-359	?	?	Probably ex-CIS Mi-24P, special colours
Mi-35	TN-???	?	?	Probably ex-CIS Mi-24P

Additionally, several Congolese airlines have operated Antonov An-12 *Cub* freighters. Though formally civil aircraft, they may be used to perform a military airlift role.

Type	Registration	C/n	F/n	Notes
An-12	TN-ACM	?		
An-12B (+)	TN-AFJ	401908		Ex-Von Haaf Air D2-FVG No.2
An-12BP (–)	TN-AFR	8345502		Ex-RA-12951. Shot down by UNITA guerrillas near Capenda, Angola, 30-6-1999
An-12BP (+)	TN-AFW	4342305		Ex-Soviet Air Force '14 Red'. Sold as, see next line
	TN-AGC			Air Atlantis, operated 1999
An-12BK	TN-AGE	4342409		Hellier International; ex-Polyot Russian Airlines RA-11320, bought 1999
An-12BP (–)	TN-AGK (1)	9346504		Ex-Mikma ER-ACG. Sold to Forner Airlines as ST-AQQ
An-12B (+)	TN-AGK (2)	402006		Ex-Aviostart LZ-ASY
An-12A (+)	TN-AGY	1340206		Aero Fret Business; ex-Tiramavia ER-ADC, bought 2004
An-12BK	TN-AGZ	8345403		Trans Air Congo; ex-Tiramavia ER-ACZ, bought 2004; later Aero Fret Business (2005-06). Sold to EX Aero Fret Business as EX-124 No.1 mid-2006
An-12A (+)	TN-AHA	0901306		Aero Fret Business; ex-Tiramavia ER-ACL, bought 2004
An-12BP (–)	TN-AHD	8345504		Natalco Airlines, operated since 1999 as ST-AQF; reregistered 2004
An-12B (+)	TN-AGH	4342404		Trans Air Congo; ex-Anton Air International 7P-ANB, bought 2002. Sold to Phoenix Aviation as EX-11760
An-12BP (–)	3C-AAL	00347003		Ex-KNG Transavia Cargo, bought 2004. To Groupe Rubuye
An-12B	3C-QQL	3341803		Ex-UN 11003, leased from Troika Leasing 2001. Sold to DRC as 9Q-CIH

MANPADS. Later three Mi-8s and at least one Mi-24 were captured intact at Pointe-Noire by the 'Cobras' and the Angolan army which supported them, crossing the border.

The Congolese Navy also operates Soviet aircraft, having ordered two Mil' Mi-14 *Haze* amphibious helicopters (exact version unknown), the first of which was delivered in September 2009. Since the type is no longer in production, these are again second-hand examples.

Democratic Republic of Congo (formerly Zaïre)

The Belgian Congo gained sovereignty on 30th June 1960 to become the Congo Republic (alias Congo-Kinshasa, after its capital), a name it held until 27th October 1971 when it was officially renamed Zaïre. In 1996 a civil war (subsequently known as the First Congo War) broke out in Zaïre; this ended a year later when President Mobutu Sésé Séko was overthrown by insurgents led by Laurent Desiré Kabila and assisted by Ugandan and Angolan troops. Kabila later became the new president, and the country's name was changed again to the Democratic Republic of Congo (DRC).

All of the Soviet military hardware operated by the Zaïre Air Force (*Force Aérienne Zaïroise*), and subsequently by the Congo Air Force (confusingly, the DRC's air arm is again known as the *Force Aérienne Congolaise*) appears to be second-hand. Thus, the fighter component

Democratic Republic of Congo Air Force MiG-23UB FG-2000
(ex-Libyan Arab Air Force '7920')

Above: DRC Air Force MiG-23UB FG-2000 hangared at Kinshasa.

Top right: DRC Air Force Su-25K FG-500 is towed by a Mercedes Unimog U1300.

Right and above right: Su-25Ks FG-500 and FG-502 hangared at Kinshasa. Note how the fin flash changed over time (the all-blue version is the old one).

included four Mikoyan MiG-21PFM *Fishbed-Fs* and a single MiG-21UM *Mongol-B* trainer purchased from Serbia in 1996 via the French private military company Geolink and based at Gbadolite. The MiG-21s did not fly a single sortie as Gbadolite was overrun by Rwanda-based insurgents before the fighters had even been assembled. Later (around 1998), the Congo Air Force obtained a pair of MiG-23MS *Flogger-E* fighters and a pair of MiG-23UB *Flogger-C* combat trainers from Libya. These were delivered to Kinshasa-N'djili airport by Jamahiria Air Transport Il'yushin IL-76s.

The strike component included ten Sukhoi Su-25K *Frogfoot-A* attack aircraft ordered in late 1999; the deal was reported to be worth US$6 million. For once, these were new aircraft supplied by the Tbilisi Aircraft Plant (Tbilaviamsheni, or TAM) in Georgia. Deliveries took place between November that year (the first four aircraft) and January 2000 (the remaining six), the *Frogfeet* being airlifted by Antonov An-124 Ruslan (*Condor*) heavy transports. Some sources, however, state that only eight were delivered. The Su-25Ks are operated by an unidentified unit at Kamina and sport the latest version of the Congolese insignia – the national flag. Flown by Ukrainian mercenary pilots, in 1999-2000 they took part in the Second Congo War (1998-2003). This

Known Zaïrean/Congo (DRC) Air Force aircraft of Soviet origin				
Type	**Serial/ registration**	**C/n**	**F/n**	**Notes**
An-12B (+)	9T-TCH (1)	5343103		Blue/white c/s. Ex-S9-GAW; seen 12-2009
An-12BP (+)	9T-TCH (2)	8345807		Red/white c/s, demilitarised. Ex-S9-PSK; seen 12-2009 (!) either this or previous aircraft must be a mispaint
An-12	9T-TCI	5343006		Ex-S9-PSM
An-12BP (+)	9T-TCM	5343507		Ex-3X-GEO; all-white, no titles
An-12A (+)	9Q-CZD (?)	2400702		Ex-XU-315
An-12BP (+)	EX-001	5343606		Kyrghyz registration used as a 'flag of convenience'? White with red tail, no titles. Operated by the Congolese Presidential Flight since mid-2006
An-26	9T-TAA	?		Grey c/s, 'stars and bars' insignia
An-26	9T-TAB	?		Grey c/s
An-26	9T-TAD	?		Crashed 29-11-2003
An-26	9T-TAE	?		
An-26	AT-922	?		
An-26	9Q-CBM	?		
An-26	?	57314301		Ex-RA-26210; seen 3-2010
MiG-23MS	FG-2002?	?	?	Ex-Libyan Arab Air Force, status unknown
MiG-23MS	FG-2003?	?	?	Ex-Libyan Arab Air Force, status unknown
MiG-23UB	FG-2000/7920	B1037920	?	Ex-Libyan Arab Air Force '7920'
MiG-23UB	FG-2001?/8301	B1038301?	?	Ex-Libyan Arab Air Force '8301'?
Mi-2T	9T-HT12	516625040		Ex-Air Transport Europe OM-KJN; bought 1999?
Mi-8	9T-HM5	?	?	Stored Lubumbashi 9-2005
Mi-17	9T-HM7	?	?	
Mi-8T	9T-HM 20	?	?	
Mi-8	9T-HM21	?	?	
Mi-8S	S9-HEA	?	?	Operated by the Congolese Presidential Flight
Mi-24V	9T-HM1	?	?	Delivered to Zaïre
Mi-24V	9T-HM2	?	?	Delivered to Zaïre?
Mi-24V	9T-HM3	?	?	Delivered to Zaïre? Derelict Gbadolite by 9-2006
Mi-24V	9T-HM4	?	?	Delivered to Zaïre?
Mi-24V	9T-HM11	?	?	
Mi-24V	9T-HM12	?	?	
Mi-24K	9T-HM14	353462*******	?	
Mi-26T	9T-MIM/15	34001212***	?	Possibly reserialled as, see next line
	9T-HM15			Grey/green camouflage
Mi-26T	RA-06021	34001212321	?	Sold to DRC but not delivered; scrapped Luik Bierset, Belgium
Su-25K	FG-500	25508110***	?	
Su-25K	FG-501	25508110***	?	
Su-25K	FG-502	25508110***	?	
Su-25K	FG-503?	25508110***	?	Crashed 12-2006?

began after the Kinshasa government unwisely asked the foreign military contingents to leave and was left with virtually no military of its own; this precipitated insurgency supported by Rwandan and Ugandan intervention, which was opposed by an invited Zimbabwean and Angolan military contingents. Thus, on 11th May 1999 they simultaneously attacked Rwandan troops at Goma and Uvira; on 2nd June 1999 they flew more

Democratic Republic of Congo Air Force An-12B 9T-TCH – the first to be thus registered

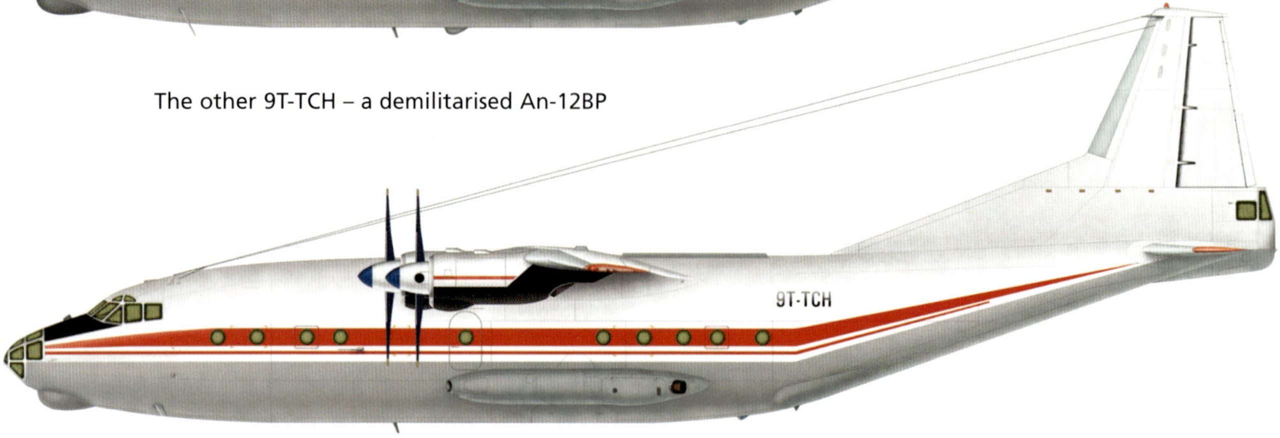

The other 9T-TCH – a demilitarised An-12BP

DRC Air Force An-26 9T-TAA wearing 'stars and bars' insignia along with the registration.

strike sorties against Uvira and Bukavu. In 2000 their primary targets were in the Gbadolite area; their last known combat use was during the battle for Pepa in December 2000.

There were no combat losses among the *Frogfeet*. After the war, however, one Su-25K went missing in December 2006 (no wreckage was ever found) and another crashed fatally at Kisangani on 30th June 2007 after taking part in an Independence Day parade.

The transport aircraft fleet of the Congo (DRC) Air Force is almost entirely quasi-civil

Wearing civil markings only, An-26 9T-TAB is seen in company with a close relative – an An-24RV (UR-ELW) operated for the UN.

and registered in the 9T-T** series (whereas the Zaïrean/Congolese civil nationality prefix is 9Q-). It includes at least six well-worn Antonov An-12 *Cubs* (some of which are reportedly registered in the civil range or even outside the DRC) and at least six Antonov An-26 *Curl* tactical transports.

The rotary-wing element of the Congo (DRC) Air Force is based at Gbadolite, Goma and Lubumbashi, and all the helicopters have alphanumeric serials prefixed 9T-. In addition to western models the fleet includes at least one Mil' Mi-2 *Hoplite* liaison helicopter, at least four Mil' Mi-8 *Hip-C* and Mi-17 *Hip-H* medium utility helicopters, a single Mi-26T *Halo* heavy-lift

Although registered in São Tomé & Principe, this Mi-8PS (S9-MEA) is operated for the government of the DRC.

The DRC Air Force's sole Mi-26, 9T-HM15.

A rather unlikely *Hind* in this part of the world is 9T-HM14, a Mi-24K reconnaissance/artillery spotting helicopter (note the low-set camera port).

Sporting a rather different camouflage scheme, Mi-24V 9T-HM2 is seen parked behind a barbed wire fence at Goma.

Additionally, several airlines in the DRC have operated Antonov An-12 *Cub* and Il'yushin IL-76 *Candid* freighters. Though formally civil aircraft, they may have been used in a military capacity to augment the aircraft listed above.

Type	Registration	C/n	F/n	Notes
An-12BK	9Q-CEN	7345410		Adala Airways; ex-Belorussian Air Force '11 Yellow', bought 10-2001. Sold to the Angolan Air Force by mid-2005 as T-311
An-12B (+)	9Q-CIH	4341803		Ex-Skylink 9L-LEC. Crashed in Bukalaza, Uganda, 5-1-2005
An-12B (+)	9Q-CVG	4342404		Mango Airlines? Ex-Phoenix Aviation EX-11760, operated 2005. Crashed 25-5-2005
An-12BP (–)	9Q-CVM	8345503		Mango Airlines; ex-Air Vitoria 9U-BHO, bought 2005, reregistered 20-3-2006
An-12B (+)	9Q-CVT	3341506		Mango Airlines; ex-Air Vitoria 9U-BHN No.2, bought 2005, reregistered 20-3-2006. Crashed en route from Goma to Kisangani 13-7-2006
An-12BP (–)	RA-11101	01347703		Air Transport Office; briefly leased from Russian State Civil Aviation Research Institute (GosNII GA) 12-1993
An-12BK	S9-PSK	8345807		Mango Airlines; ex-EX-048
IL-76TD	RA-76419	1023414470	8708	VAC Virunga Air Charter; leased from Arkhangel'sk Airlines 8-1993

helicopter and at least six Mil' Mi-24V *Hind-E* attack helicopters, plus a Mi-24K *Hip-G* reconnaissance/artillery spotting helicopter. The first lot of second-hand Mi-24s was bought by the Zaïre Air Force around 1996, probably from CIS surplus stocks. One of them was lost on 27th March 1997 when it hit a power line mast and crashed, killing seven. More *Hinds* were apparently obtained in 2000. The helicopters were used by the Kinshasa government in the spring of 2000 to unblock the town of Ikela besieged by opposition guerrillas.

Ivorian Air Force MiG-23MLA TU-VCH sits impounded at Lome, Togo.

Here, both Ivorian MiG-23MLAs are seen at Abidjan in company with a French Air Force CASA CN.235M.

Côte d'Ivoire

This former French colony, which gained sovereignty on 7th August 1960, started purchasing second-hand Soviet military equipment after 2000 when Laurent Gbagbo was elected President. This was a forced measure because all the original French and American aircraft of the Ivorian Air Force (or FACI – *Force Aérienne de Côte d'Ivoire*) were unserviceable by then.

Deliveries to the 'new' Ivorian Air Force included four ex-Belorussian Air Force Sukhoi Su-25 attack aircraft supplied in 2003-04 (two *Frogfoot-A* single-seaters and two Su-25UB *Frogfoot-B* trainers). They were originally operated by Ukrainian and Belorussian mercenary pilots, but these were later fired and replaced by local ones.

Two Mikoyan MiG-23MLD *Flogger-G* fighters were also bought from Bulgaria via the Sofia-based company Metalika in 2003. In April they arrived in Abidjan aboard Il'yushin IL-76TD transports (owned by the Ukrainian airline Atlant-Soyuz but chartered by the Bulgarian cargo carrier Vega Airlines). There the fighters were reassembled and made check flights at the hands of a Bulgarian pilot on 23rd June. However, in October 2004 for reasons unknown both MiG-23MLDs were flown to Lomé, Togo, where they were impounded by the local authorities – at the request of France, no doubt.

The Ivorian Air Force also had a pair of Antonov An-12 *Cub* medium transports (first seen in 2005) and a single Antonov An-32 *Cline* tactical transport delivered in 2003. The An-32 was reportedly used (together with

Wearing an awesome sharkmouth, Ivorian Air Force Su-25 '02 Red' is seen in its hangar at Abidjan.

This Ivorian shark has had its teeth pulled by visiting French dentists!

Another view of Su-25UB '21 Red' with sister ship '20 Red' behind it. Both aircraft were wrecked at Yamoussoukro in a punitive raid by French commandos on 4th November 2004.

Ivorian Air Force An-12BP TU-VMA

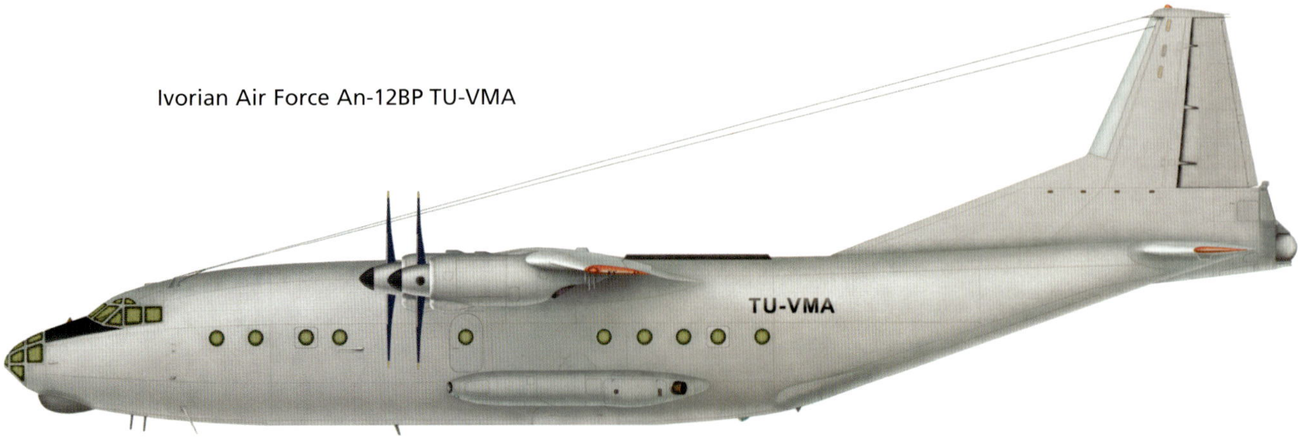

Mi-8 helicopters) to carry personnel and supplies to Yamoussoukro and Daloa located close to the frontlines, though the aircraft must have been outside the country in November 2004, thereby escaping destruction.

The rotary-wing element included an unknown number of Mil' Mi-8T/Mi-8PS *Hip-Cs* and Mi-17 *Hip-Hs* (two of which came from Bulgaria and retained Bulgarian civil registrations) and at least seven Mil' Mi-24V *Hind-E* and Mi-24P *Hind-F* attack helicopters, including at least two *Hind-Fs* delivered from Belorussia in 2002 and two from delivered Bulgaria in February 2003. Interestingly, the Mi-24Vs invariably carried UB-32-57 rocket pods with 57-mm (2.24-in) S-5 folding-fin aircraft rockets, while the Mi-24Ps were armed with the newer B-8V20 pods with 80-mm (3.15-in) S-8 FFARs.

However, we are jumping ahead of the story. On 19th September 2002 a failed coup d'état in Abidjan by several disgruntled Muslim officers, who had been fired by Gbagbo, sparked a civil war. The northern half of the country was controlled by three rebel factions – the Patriotic Movement of Côte d'Ivoire (MPCI – *Mouvement Patriotique de Côte d'Ivoire*), the Ivorian Popular Movement for the Great West (MPIGO – *Mouvement Populaire Ivorienne pour le Grand Ouest*) and the Movement for Justice and Peace (MJP), known collectively as the 'New Forces'. The north was predominantly Muslim, while the south was mainly Christian. France immediately launched Operation *Licorne* (Unicorn), sending an expeditionary force of more than 1,000 servicemen with transport aircraft and armed helicopters to Côte d'Ivoire to evacuate foreign nationals and act as peacekeepers; their presence temporarily halted the rebels' advance and a tentative ceasefire was reached in October 2002 – but not for long.

The Ivorian Mi-24s are among the few to sport huge sharkmouths.

Here, Ivorian Air Force Mi-24V TU-VHL is being refuelled. Note the slightly different sharkmouth.

The Mi-24Vs were flown by South African, Scandinavian and French mercenary pilots, while the ground crews were Ukrainian and Belorussian. The mercenaries operated as a stand-alone force, with little co-operation even with the government troops, to say nothing of the UN peacekeepers. The helicopters first saw action on 15th December 2002 when they attacked a rebel-held village in northern Côte d'Ivoire with unguided rockets. More sorties were flown on 27th December against a fuel/ammunition dump near Vavoua and on 31st December against a barge ferry near Menakro. Still, their operations had scant success because they crated a lot of collateral damage, many civilians being killed along with (or instead of) the guerrillas. The helicopters proved unable to stop the advance of the 'New Forces', who soon found themselves at war with the French troops.

A pair of Ivorian Mi-24Vs in flight.

A reconciliation process under international auspices started in January 2003. The Economic Community of West African States (ECOWAS) sent an additional 1,500-strong multi-national peacekeeping force. The

Known Ivorian Air Force aircraft of Soviet origin

Type	Serial/ registration	C/n	F/n	Notes
An-12	TU-VFA	?		Seen 2005
An-12BP (+)	TU-VMA	?		Seen 2008, grey overall; stored
MiG-23MLD	TU-VCH	2960327204	18421	Ex-Bulgarian Air Force '204 White'. Three-tone camouflage, registration applied as white outline on the fin and 'CH' repeated on the air intakes
MiG-23MLD	[TU-V]C...	2960327219	18427	Ex-Bulgarian Air Force '219 White'
Mi-8PS	LZ-CAE	10312	?	Ex-Heli Air Services; all-white c/s, no titles
Mi-8T	LZ-CAF	10314	?	Ex-Heli Air Services; all-white c/s, no titles
Mi-17?	TU-VHT	?	?	Also reported as a Mi-8T. All-white c/s, no titles. Destroyed by French troops at Presidential palace in Yamoussoukro 6-11-2004
Mi-24V	TU-VHL	?	?	Grey/green camouflage, large sharkmouth. Damaged at Yamoussoukro airport 6-11-2004; later repaired and moved to Abidjan
Mi-24V	TU-VHO	?	?	Grey/green camouflage, large sharkmouth. Seen Bouaké 8-2006
Mi-24V	TU-VHQ	?	?	Damaged at Abidjan airport 6-11-2004
Mi-24V	TU-VHR	?	?	Damaged at Abidjan airport 6-11-2004
Mi-24V	TU-VHS	?	?	Damaged at Yamoussoukro airport 6-11-2004; later repaired and moved to Abidjan
Mi-24V	TU-VH...	730207	?	Ex-Bulgarian Air Force '124 White'
Mi-24V	TU-VH...	750292	?	Ex-Bulgarian Air Force '126 White'
Mi-24P	H-385	?	?	
Mi-24V	H-388	?	?	
Su-25	02 Red (TU-VCL?)	255081*****	?	Large sharkmouth; FACI fin flash only but no roundels. Damaged by French troops at Abidjan airport 6-11-2004
Su-25	03 Red (TU-VCM?)	255081*****	?	Large sharkmouth; FACI fin flash only but no roundels. Damaged by French troops at Abidjan airport 6-11-2004
Su-25	23 Red (?)	255081*****	?	Reported but probably in error!
Su-25	24 Red (?)	255081*****	?	Reported but probably in error!
Su-25UB	21 Red (TU-VCI)	382201*****	?	Large sharkmouth. Damaged by French troops at Yamoussoukro airport 6-11-2004; later repaired and moved to Abidjan
Su-25UB	22 Red (TU-VCK)	382201*****	?	Damaged at Yamoussoukro airport 6-11-2004; later repaired and moved to Abidjan

French-brokered Linas-Marcoussis Accord saw the formation of a national reconciliation government; in February 2004 the United Nations Operation in Côte d'Ivoire (UNOCI) was established. On 4th July 2003 an 'End of War' agreement was signed. However, the belligerents would not give up the fight, again using the ceasefire to re-arm and regroup. In December 2003 the 'New Forces' declared their withdrawal from the disarmament process, demanding the pullout of the foreign military contingents; the national reconciliation government disintegrated, and a new round of the civil war became imminent. In June 2004 the rebels began renewed attacks on French citizens and UN peacekeepers along the ceasefire line. On 9th June 2004 the FACI Mi-24s were put into action again.

On 4th November 2004 Ivorian Air Force Su-25s began attacking rebel positions in Bouaké in northern Côte d'Ivoire (the unofficial 'capital' of the 'New Forces') and Korhogo, in contravention of the ceasefire agreement. The then French President Jacques Chirac warned President Laurent Gbagbo against trying to recapture the areas controlled by the 'New Forces', which Gbagbo regarded as insulting. Hence on 6th November two Su-25s attacked a camp of French peacekeepers near Bouaké with bombs and S-5 rockets, killing nine soldiers (plus an American working for the International Committee of the Red Cross) and wounding 31 others. The Ivorian government claimed the attack on the French was unintentional, presenting an official apology, but the French insisted that the attack had been deliberate. This was the last straw. Incensed by this attack, within hours Chirac ordered the destruction of the entire Ivorian Air Force and the seizure of the airport in the Ivorian capital of Yamoussoukro so that additional troops could be brought in. That same day French commandos raided Yamoussoukro airport, damaging two of the Su-25s (which had landed to refuel and re-arm) with Milan anti-tank guided missiles, while a French Army

Wearing only the last two letters of its registration, Djibouti Air Force Mi-8T J2-MAM has a machine-gun and its ammunition box suspended from the port inboard pylon.

Djibouti Air Force Mi-8T J2-MAM

Aérospatiale SA.342 Gazelle helicopter destroyed a Mi-24V, a Mi-24P and a Mi-8T parked in the Presidential palace compound with unguided rockets. The other two *Frogfeet* and two more Mi-24s were shot up at Abidjan airport in retaliation for another attack on French troops (this time by a Mi-24 and a gunman with a grenade launcher who damaged a French Air Force C.160F Transall transport at Abidjan airport).

Another ceasefire was reached in December 2004, but it was another two years before the end of hostilities was formalised. (On 4th March 2007 a peace accord was signed and Guillaume Soro, one of the rebel leaders, was appointed Prime Minister by President Gbagbo as one of the steps towards reconciliation.) Under special United Nations permission, the Ivorian Air Force resumed limited operations on 24th January 2005 when one of the surviving Mi-24s (and a BAC Strikemaster) made several check flights over Yamoussoukro. Both were subsequently flown out to Abidjan; so were the Su-25UBs damaged at Yamoussoukro, which the UN had permitted the FACI to repair but not to re-arm.

Its port engine missing, Mi-2 J2-MAO sits on the dump at Djibouti-Ambouli alongside an equally dead Mi-8T (J2-MAM).

Djibouti

The Djibouti Air Force (Force Aérienne Djiboutienne) operated a mix of Soviet and western types; the Soviet machines were all second-hand aircraft, mostly delivered in 1990. The entire fleet was concentrated at the sole airbase at Ambouli; all the aircraft wore an overall light grey colour scheme and were registered in the J2-M** quasi-civil block, usu-

Showing an altogether different camouflage, Mi-8T J2-MAS comes in to land past an IL-18D freighter.

Djibouti Air Force Mi-8T J2-MAS

Also wearing the tan/chocolate camouflage and full registration, Mi-35 J2-MBD is seen in a training sortie.

Another angle on the same helicopter; note the absence of the machine-gun in the nose.

J2-MBC, another Djibouti Air Force Mi-35, carrying B-8V20 FFAR pods.

Known Djibouti Air Force aircraft of Soviet origin				
Type	**Serial/ registration**	**C/n**	**F/n**	**Notes**
An-28	[J2-M]AT	1AJ009-14		Ex-Riga Air Club YL-KAA, ex-Aeroflot CCCP-28948. Derelict Ambouli AB by 3-2011
Mi-2	[J2-M]AK	?		Crashed
Mi-2	[J2-M]AO	?		Derelict Ambouli AB by 6-2006
Mi-2	[J2-M]AP	?		Crashed
Mi-2	[J2-M]AQ	?		Crashed
Mi-8T	[J2-M]AJ	?	?	Derelict Ambouli AB by 5-2001
Mi-8T	[J2-M]AL	?	?	Derelict Ambouli AB by 6-2003
Mi-8T	[J2-M]AM	3113	?	Ex-Aeroflot CCCP-25242. Sand/green camouflage; armed version (retrofit). Derelict Ambouli AB by 5-2001
Mi-8T	[J2-M]AR	13133?	?	Crashed
Mi-8T	J2-MAS	?	?	Sand/brown camouflage
Mi-8T	[J2-M]AU	4541	?	Ex-Aeroflot CCCP-25806. Sand/green camouflage; armed version (retrofit). Derelict Ambouli AB by 6-2003
Mi-17	[J2-M]AV	?	?	Crashed
Mi-17	[J2-M]AW	341M14	?	Sand/brown camouflage
Mi-35	J2-MBC	?	?	Sand/brown camouflage. Stored Ambouli AB by 3-2011
Mi-35	J2-MBD	?	?	Sand/brown camouflage. Derelict Ambouli AB by 3-2011
Mi-35	J2-MBE (1)	?	?	Reported 2007; fate unknown, registration transferred to a Let L-410UVP-E20 aircraft by 2010

ally only the last two letters of the registration being carried visibly.

The only fixed-wing aircraft of Soviet origin in the Djibouti Air Force was a single Antonov An-28 *Cash* twin-turboprop feederliner. On the other hand, there were quite a few helicopters – four Mil' Mi-2 *Hoplite* liaison helicopters, five Mil' Mi-8T *Hip-Cs* bought in unarmed utility configuration and refitted to assault configuration, two Mi-17 *Hip-Hs* and three Mil' Mi-35 *Hind-E* attack helicopters. The latter were not seen until March 2006 and so may be a later acquisition.

Equatorial Guinea

Equatorial Guinea (formerly Spanish Guinea) has been operating Soviet aircraft in a military capacity since 1978. The first operator, however, was not the air arm but the government (presidential) flight, which took delivery of two VIP-configured Yakovlev Yak-40 *Codling* trijets in 1978 and 1979. In addition to these, the government flight also operated a VIP-configured Antonov An-24RV *Coke* twin-turboprop airliner, a pair of Antonov An-32 *Cline* transports (one of which is a recent addition, having been delivered new in 2008), one or

010 Blue, an Equatorial Guinea Air Force Mi-24V armed with UB-32A FFAR pods.

Equatorial Guinea Government Mi-172 3C-LLE; inset shows the national coat of arms

This view of Mi-172 3C-LLE shows the airstair door; note the exhaust/air mixers for protection against heat-seeking SAMs. Sister ship 3C-LLD is identical.

Immaculate-looking Equatorial Guinea Air Force Su-25UB '026 White' at the 'last chance' checkpoint.

'027 White', the other Equatorial Guinean Su-25UB, at rest. Note the leopard's head artwork.

Known Equatorial Guinea Air Force aircraft of Soviet/Russian origin				
Type	**Serial/ registration**	**C/n**	**F/n**	**Notes**
An-24RV 'Salon'	3C-FMN	?		Last noted as such 8-1971. Became, see next line
	3C-1GE			
An-32B	3C-4GE (2)	3603		D/D 2008; green/brown camouflage, named 'Col. Cosme Ondo Nchama'. See Yak-40 below
An-32	3C-5GE	1609		D/D 7-7-1992; light grey c/s with national flag cheatline and coat-of-arms on tail, named 'Enrique Nvo Okenve'. Crashed on landing at San Antonio de Palé, Annobon Island, 16-4-2008, all 33 aboard killed
An-72	3C-CMN	365.720.92.858	1109	Delivered 8-2009; ex-Ukrainian Air Force '15 Blue', ex-[UR-]72959. Reregistered as, see next line
	3C-LLQ			Seen 1-2011, white with triple dark cheatline, no titles
An-72P	032 Black	365.760.96.926	1510	Delivered 10-2009; ex-Ukrainian Air Force/Border Guards '16 Blue', ex-[UR-]72984. White overall
IL-76TD	3C-LGF	0073479386	6007	Ex-Ecuatorial Cargo 3C-HAV, ex-Ukrainian Air Force IL-76MD UR-76748. Uzbekistan Airways livery style c/s, no titles. Possibly reregistered as, see next line
IL-76TD	3C-HHV	?	?	
Ka-29	019 Yellow	523502*******	?	Blue overall
Mi-172	3C-LLD	?	?	White c/s with national flag cheatline, exhaust/air mixers fitted for protection against MANPADS
Mi-172	3C-LLE	?	?	C/n reported as '246', which does not make sense. White c/s with national flag cheatline, exhaust/air mixers fitted for protection against MANPADS
Mi-24V	010 Red	353242*117620	?	Ex-Ukrainian Air Force; full c/n probably 3532424117620
Mi-24V	011	?	?	
Mi-24V	012	?	?	Ex-Ukrainian Air Force
Mi-24V	018	?	?	
Mi-24P	020 White	?	?	Ex-Ukrainian Air Force
Mi-24P	021 White	?	?	Ex-Ukrainian Air Force
Mi-24P	022 White	?	?	
Mi-26	30 Blue	34001212153	?	White overall; ex-Ukrainian Air Force 61 Yellow
Su-25	028 White	255081*****	?	Ex-Ukrainian Air Force
Su-25	029 White	255081*****	?	Ex-Ukrainian Air Force
Su-25UB	026 White	382201*****	?	Ex-Ukrainian Air Force; briefly serialled 031. Tan/green/dark green camouflage, leopard head nose art
Su-25UB	027 White	382201*****	?	Ex-Ukrainian Air Force; briefly serialled 032. Tan/green/dark green camouflage, leopard head nose art
Yak-40 'Salon 1st Class'	5C-MNB	9821557		Old registration prefix; delivered 1978. Became, see next line
	3C-CGE			
Yak-40 'Salon 2nd Class'	3C-4GE (1)	9940660		Delivered 1979. Retired, registration reallocated to An-32B

two An-72 *Coaler-A* twin-turbofan STOL transports, an Il'yushin IL-76TD *Candid-A* freighter and a pair of Mil' Mi-172 *Hip-H* VIP helicopters. All of them except the *Hips* (which were delivered new by Kazan' Helicopters) have been purchased from the Ukraine – even the IL-76TD was formerly a Ukrainian Air Force IL-76MD *Candid-B* that was converted to civil configuration. Most of the aircraft have non-standard registrations, the 3C- nationality prefix being followed by a number and the letters 'GE' (for Guinea Ecuatorial, in Spanish). The aircraft are based at Malabo-St. Isabel airport.

The Equatorial Guinea National Guard, on the other hand, began receiving Soviet hardware after the turn of the century, and again the aircraft were supplied by the Ukraine (from military surplus stocks). They include a single An-72P armed patrol aircraft delivered in late

2009, four Sukhoi Su-25 attack aircraft (two single-seat *Frogfoot-As* and two Su-25UB *Frogfoot-B* trainers) delivered in 2007, at least seven Mil' Mi-24 attack helicopters (including four Mi-24V *Hind-Es* and three Mi-24P *Hind-Fs*) delivered from 2002 onwards, a single Mil' Mi-26 *Halo* heavy-lift helicopter and at least one Kamov Ka-29 *Helix-B* shipboard transport/assault helicopter from Ukrainian Navy stocks. Again, most of the aircraft are based at Malabo, with a helicopter detachment at Bata.

The Eritrean Air Force roundel and fin flash.

Eritrea

Eritrea was predominantly Muslim and part of the Ottoman Empire until conquered and annexed by Italy in 1890. Ethiopia is a country with a large Christian population that had an independent existence until also conquered by Italy in 1936. In 1941, British Commonwealth Armies captured and occupied both countries and for a time Eritrea was a British protectorate. The UK gave Ethiopia full sovereignty under the Anglo-Ethiopian Agreement in December 1944.

On 2nd December 1950, by United Nations plebiscite, Eritrea was federated with Ethiopia; not only was Eritrea granted its own parliament but also given representation in the new Federal Ethiopian parliament. Ethiopia now had access to Red Sea ports, for without Eritrea it was landlocked.

In 1961, an Eritrean struggle for independence began and student protests swiftly turned into civil war. Haile Selassie I, Emperor of Ethiopia, repressed the revolt but played into the hands of his enemies by first dissolving the Federation in 1961, then closing down parliament and finally, in 1962, declaring Eritrea to be Ethiopia's 14th Province. The Eritrean People's Liberation Front (EPLF) soundly defeated the Ethiopian Army in the Battle of Asmara in January 1974; this victory gave the EPLF control of most its own country. The conflict continued after the revolution of September 1974 in which Haile Selassie I was deposed, and from a war of independence it

An Eritrean Air Force Mi-17 carrying UPK-23-250 cannon pods. Note the roundel facing the wrong way.

Eritrean personnel in civilian attire pose with Mi-17 ERAF 306. The paintwork appears to be rather faded.

Eritrean Air Force MiG-23BN ER-01 (named 'Lilo')

Despite receiving an Eritrean Air Force serial, this ex-Ethiopian MiG-23BN at Asmara was never made operational.

Eritrean Air Force MiG-29s ERAF 502 and ERAF 504 in line abreast formation.

Below: One of the Eritrean MiG-29UB trainers, ERAF 501.

Below left: An APA-5D ground power unit sits beside an ErAF MiG-29, suggesting an impending sortie.

Eritrean Air Force MiG-29 (*izdeliye* 9.12B) ERAF 508

Below: ErAF Su-27UB '609' in the Ukraine prior to delivery.

Bottom: Su-27 '608' illustrates the splinter camouflage worn by ErAF *Flankers*.

quickly escalated into an all-out war against the ruling Marxist regime (see Ethiopian entry). Two years after the latter had been toppled, Eritrea gained independence on 24th May 1993, after a bloody guerrilla war lasting more than 30 years.

In the aftermath of the civil war the nascent Eritrean Air Force (ErAF) appropriated a number of aircraft captured from Ethiopia. These included nine Mikoyan MiG-21MF *Fishbed-J* fighters (some sources suggest MiG-21*bis Fishbed-L*) and MiG-21R *Fishbed-H* reconnaissance aircraft at Asmara airport; they were reportedly not made airworthy and put into service, remaining in storage, though two aircraft have been reported with new Eritrean

'608' takes off past a Ukrainian IL-76 during a pre-delivery check flight. Note the false canopy on the underside of the nose.

Another Eritrean Su-27 before the application of insignia.

Freshly painted but still minus national insignia, this Su-27 awaits delivery to the Eritrean Air Force.

One of the Eritrean Su-27UBs parked at Asmara.

Su-27 '601' taxies at Asmara. This aircraft wears a slightly different style of camouflage.

Eritrean Air Force Su-27 '606', 6th Fighter Sqn, Asmara

Known Soviet aircraft in Eritrean Air Force service

Type	Serial	C/n	F/n	Notes
MiG-21MF	1058*	?	?	Captured at Asmara airport, taken over by ErAF, stored, present 1994
MiG-21MF	1065*	?	?	Captured at Asmara airport, taken over by ErAF, stored, present 1994
MiG-21MF	1082*	?	?	Captured at Asmara airport, taken over by ErAF, stored, present 1994
MiG-21R	1455*	?	?	Captured at Asmara airport, taken over by ErAF, stored, present 1994
MiG-21R	1461*	?	?	Captured at Asmara airport, taken over by ErAF, stored, present 1994
MiG-21R	1464*	?	?	Captured at Asmara airport, taken over by ErAF, stored, present 1994
MiG-21R	ER-101	94R*****	?	Sand/green camouflage, early insignia
MiG-21*bis*	ER-102	N750*****	?	
MiG-23BN	ER-01	03932*****	?	Green/light green/beige/brown camouflage, named 'Lilo'
MiG-29UB	ERAF 501	N509030*****	?	D/D 1998; green/grey camouflage, later standard two-tone grey camouflage
MiG-29	ERAF 502	29605*****	?	D/D 1998
MiG-29	ERAF 503	29605*****	?	D/D 1999. Shot down ?-1999?
MiG-29	ERAF 504	29605*****	?	D/D 1999
MiG-29	ERAF 505	29605*****	?	D/D 1999
MiG-29	ERAF 506	29605*****	?	D/D 2001
MiG-29	ERAF 507	29605*****	?	D/D 2001
MiG-29	ERAF 508	2960536535	51…?	D/D 2001
MiG-29	ERAF 509	29605*****	?	D/D 2001
MiG-29	ERAF 512	29605*****	?	
Mi-8…	2006	?	?	Version unknown. Ex-Ethiopian AF (old EtAF serial!), stored Asmara airport
Mi-8…	2008	?	?	Version unknown. Ex-Ethiopian AF (old EtAF serial!), stored Asmara airport
Mi-17	ERAF 306	?	?	Sand/brown camouflage
Mi-17	ERAF 308	?	?	
Mi-24A	1620	2201305		Ex-Ethiopian AF (old EtAF serial!), stored Asmara airport
Mi-35	?	?	?	Ex-Ethiopian AF '2108', captured 18-3-1999
Su-25	ERAF 702	255081*****	?	
Su-27	601	?		Reported in error as a Su-27UB
Su-27	603	?		No reports
Su-27	604	?		No reports
Su-27	605	?		No reports
Su-27	606	369110*****		
Su-27	607	?		No reports
Su-27	608	36911026310		Ex-Ukrainian AF/State Aviation Research Centre, tactical code unknown
Su-27UB	602	963104*****	?	
Su-27UB	609	96310424045	1407	Ex-Ukrainian AF/831st Fighter Regiment '72 Blue'

serials. Apart from these, no further MiG-21s were acquired by Eritrea.

Later, in the summer of 1998, the ErAF purchased a few MiG-29s (including MiG-29UB *Fulcrum-B* combat trainers) from Moldova (or, according to some reports, the Ukraine, which supplied four single-seaters and one trainer). Subsequently, in the summer of 2001 (immediately before the UN imposed an arms embargo), Eritrea bought a further four single-seat MiG-29s and two MiG-29UBs in Russia and at least another two in the MiG-29SE (*izdeliye* 9.12SE) *Fulcrum-A* version from the manufacturer, MAPO, in 2004. The fighters saw service with the No.5 Squadron stationed at Asmara. There were reports that the MiG-29s were procured with the assistance of Col. Vladimir Nefyodov, a former representative of the Russian arms export company Rosvo'oruzheniye. The contract included on-site maintenance support by a team of Russian specialists seconded to the ErAF. The pilots, on the other hand, were local ones – if the Eritrean government is to be believed.

The Eritrean MiG-29s were actively involved in the new war between Ethiopia and Eritrea which flared up in 1998; the apple of discord was the disputed Badme border region. The Soviet Union had supplied military aid to Ethiopia in the 1970s and 1980s. Both belligerents used Russian weapons; additionally, Russian (or, at any rate, Russian-speaking) military advisors and instructors were also present in both countries.

On 25th February 1999 Ethiopia launched an offensive called Operation *Sunset*. Ethiopian MiG-23BN fighter-bombers struck at key points in Eritrea, including the logistical centre at Harsele, water supply installations in Aseb and the airfields at Asmara and Aseb. Two ErAF MiG-29s scrambled to intercept the intruders, engaging them with R-27 medium-range AAMs, but missed. In turn, the Ethiopian pilots apparently called in reinforcements, and these arrived in the shape of Su-27s. The Eritreans attempted to disengage but the Su-27s gave chase and likewise attacked with R-27s. Both MiG-29s were damaged by the missiles exploding close behind, the Sukhois finishing off one of them with an R-73 short-range missile; the pilot ejected and was captured. The other MiG-29 (reportedly flown by none other than the Commander of the Eritrean Air Force, Brig. Gen. Habte Zion Hadgu) escaped but crashed about 6 km (3.72 miles) from Asmara; the fate of the pilot is unknown.

The following morning the Eritreans evened up the score: as the Ethiopian attacks continued, a solitary MiG-29UB intercepted two MiG-23BNs, destroying them with R-73 IR-homing AAMs. Shortly afterwards, however, two Su-27s arrived on the scene and attacked the *Fulcrum* with two R-27s, which missed. Some reports indicate that contact with the MiG-29 was lost shortly afterwards. However, Ethiopian sources say that although the first two missiles indeed missed, the Su-27 eventually destroyed the Eritrean fighter with either R-73s or cannon fire. While this kill remains unconfirmed, only one of the two Eritrean MiG-29UBs was ever seen again.

Soon afterwards Ethiopia launched Operation *Westwind*, delivering massive air strikes against Eritrean positions on the Mereb-Setit front several times a day. On 18th March 1999, as Eritrean MiG-29s tried to repel the MiG-21s and MiG-23s flying ground attack, they were pounced upon by the Ethiopian Su-27s flying top cover. As the first two *Fulcrums* approached, the *Flankers* fired at least two missiles against each of them.

Eritrean Air Force Su-25 ERAF 702

Ethiopian Air Force MiG-21MF '1087'

Again, one MiG-29 got away but the other was damaged by an R-27 exploding nearby and crashed, killing the pilot.

Interestingly, the border conflict between Ethiopia and Eritrea showed appallingly low performance of the R-27 missiles with which most of the 'kills' were scored; the 'kill' ratio was only about 16.6%. Out of sixteen R-27s on which details are known, only one scored a direct hit, three or four proximity fused, and the results of the other two remain unconfirmed, for four confirmed and one unconfirmed kills. At least ten (probably 18) others missed, or failed to launch altogether. Surprisingly, the far more experienced Russian and Ukrainian mercenary pilots fared no better than the local talent. Equally surprising was that not a single R-27 fired by the ErAF MiG-29s scored a hit. Reports from both air forces stress that rough handling during transport, improper storage conditions, and poorly trained armourers account for this poor performance.

Another relatively modern fighter operated by the ErAF is the Sukhoi Su-27, which was purchased after the Eritreans lost at least five MiG-29s in combat with Ethiopian Su-27s. According to press reports, four single-seat Su-27S *Flanker-Bs* and two Su-27UB *Flanker-C* combat trainers were in service with the No.6 Squadron at Asmara in June 2003; some sources suggest as many as nine may have been delivered. The single-seaters were listed as Su-27SKs, which is presumably an error because they were not new aircraft manufactured for export but acquired from Ukrainian Air Force surplus stocks. However, there have also been reports that no more than two ex-UAF *Flankers* – a Su-27S and a Su-27UB – were actually delivered in late 2002 after being refurbished by the Zaporozhye Aircraft Repair Plant (ZARZ) in the Ukraine; two further Su-27s were awaiting delivery but the Eritreans refused to accept them. They may have had a point – apparently the quality of the work at ZARZ turned out to be substandard, the fighters being grounded by malfunctions much of the time. Eventually, a couple of years later the ErAF had to send them off for overhaul again – this time at ARZ No.558 in Baranovichi, Belarus. The aircraft wear a light green/dark green/light grey splinter camouflage with a 'false canopy' painted on the underside of the nose.

The Eritrean Air Force's strike aircraft included an ex-Ethiopian Mikoyan MiG-23BN *Flogger-H* fighter-bomber which was not made operational. The ErAF reportedly bought eight Su-25 *Frogfoot-A* attack aircraft from Georgia in 2000; only one serial is known.

Soviet helicopter types in service with the Eritrean Air Force are the Mil' Mi-17 *Hip-H* utility helicopter (first noted in 2001) and the Mil' Mi-24V (Mi-35) *Hind-E* attack helicopter.

Ethiopian Air Force MiG-21MF '1127'

An Ethiopian Air Force MiG-21*bis* (1107) streams its brake parachute on landing.

Several Ethiopian Air Force Mi-24s were flown to Eritrea by defectors and subsequently included into the ErAF inventory. Additionally, at least one more example fell into Eritrean hands after being damaged by ground fire and landing in Eritrean territory on 18th March 1999. The helicopters are operated by the 3rd Squadron, likewise based at Asmara.

In December 2009 the United Nations imposed a weapons embargo on Eritrea because the latter was found guilty of supporting armed Muslim extremists. All members of the UN Security Council, including Russia, voted in favour of the embargo. This means further deliveries of Soviet/Russian aircraft to this part of Africa are unlikely in the near term… unless some unscrupulous arms dealer chooses to ignore the embargo.

A line-up of Ethiopian MiG-21s at Asmara airport (left to right: MiG-21*bis* 1127, MiG-21MF 1082 and MiG-21Rs 1461 and 1464).

Ethiopia

On 13th September 1974 Emperor Haile Selassie I of Ethiopia was overthrown by Marxist-minded officers led by Mengistu Haile Mariam and the country was declared a republic run by a military government known as the Derg (meaning 'council' in the local Amharic

The old Ethiopian Air Force roundel used in 1974-1991.

Ethiopian pilots pose with an EtAF MiG-23ML serialled 1009.

MiG-23BN '1265' with old Ethiopian Air Force (EtAF) insignia

MiG-23BN '1265' with current Federal Democratic Republic of Ethiopia Air Force (FDREAF) insignia

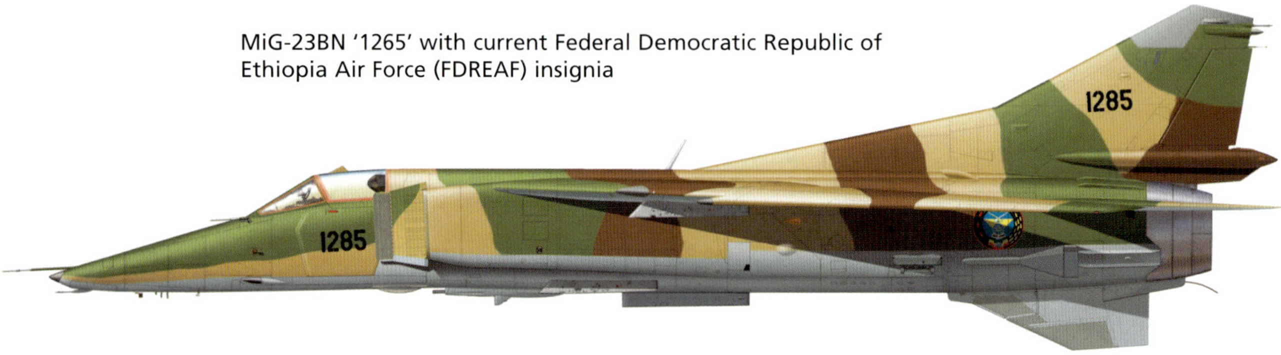

Ethiopian Air Force MiG-23BN '1265' on the hardstand of the L'vov Aircraft Repair Plant in company with Libyan MiG-23s – all of them undeliverable due to sanctions.

'1205', an EtAF MiG-23UB similarly stranded at L'vov.

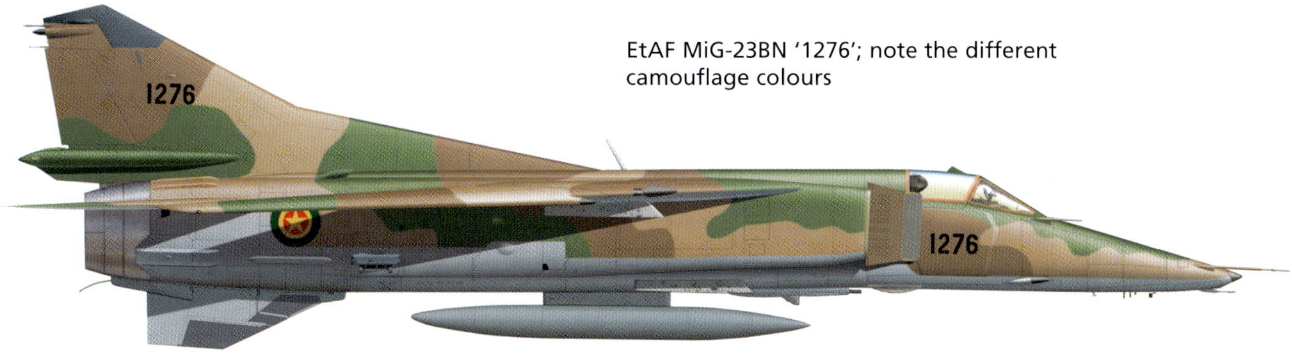

EtAF MiG-23BN '1276'; note the different camouflage colours

EtAF MiG-23UB '1205' carrying a PTB-800 drop tank

This MiG-23BN serialled 1270 was used in the Ethiopian War against Eritrean separatists.

MiG-23BN '1285' sporting new camouflage colours and FDREAF roundels. Note the non-standard IRCM flare dispenser installed locally by DAVEC below the port stabilator.

Ethiopian Air Force (EtAF) Su-25TK '1404' armed with Kh-29L laser-guided ASMs and R-73 AAMs

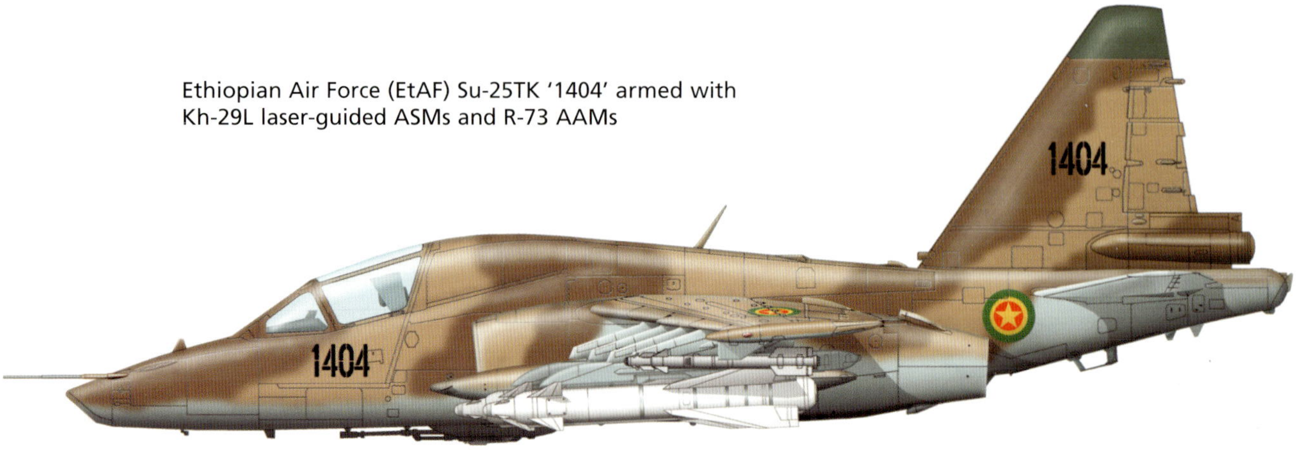

language). Accordingly the Imperial Ethiopian Air Force (IEAF) established in 1924 was renamed *Ye Ityopya Ayer Hayl* (simply 'Ethiopian Air Force' in Amharic, or EtAF). After the revolution the country was in chaos for several years while the Derg was trying to work out who would be Ethiopia's permanent leader; since 1961 there had been rebellion in the north-eastern province of Eritrea, which wanted to secede, and this continued and grew stronger. As if that weren't enough, Somalia decided to take advantage of Ethiopia's apparent weakness to unleash a war of aggression (more on this later).

The IEAF had relied on American hardware, but after the revolution the Carter administration cut off all military aid to Ethiopia. Hence Col. Mengistu Haile Mariam, who became President in 1977, turned to the Soviet Union for help in suppressing the mutiny in Eritrea and repelling the Somalian invasion. Soon Ethiopia began receiving a steady stream of Soviet weapons, though not necessarily new ones. The first fighters to arrive were 20 Mikoyan/Gurevich MiG-17F *Fresco-Cs* delivered in 1977 from South Yemen Air Force stocks, together with a few UTI-MiG-15 *Midget* trainers (though *Air International* reported in July 1978 that Ethiopia received at least 40 MiG-17s). Anyway, only 20 MiG-17s were reportedly in service by 1979; if the figure of 40 is true, this indicates high losses sustained in wars. 15 MiG-17s were still used as fighter-bombers in late 1991.

The MiG-17s were followed by 48 Mikoyan MiG-21MF (*izdeliye* 96F) *Fishbed-J* fighters in late 1977 and early 1978; the figure may be incorrect, as the total of 48 presumably included a few MiG-21UM (*izdeliye* 69) *Mongol-B* trainers. By December 1978, the 1st and 2nd Fighter Squadrons at Debre Zeit AB near Addis Ababa had re-equipped with MiG-21MFs; one of the squadrons was provided with aircraft from Cuba and manned by Cuban pilots seconded to the EtAF. Some sources suggest that as many as 36 of the MiG-21s were supplied by Cuba.

More deliveries of MiG-21MFs took place in 1982-83; the number was estimated to be at least 50, and possibly as high as 150. Again, this total included substantial numbers of MiG-21UMs. However, there are reports that only ten newer MiG-21MFs were actually delivered and operated by Cuban pilots on secondment to the EtAF, as Ethiopia wanted the more advanced MiG-21*bis* which the Soviet Union was not willing to supply just yet. At least eight MiG-21R (*izdeliye* 94R) *Fishbed-H* tactical

Ethiopian Air Force Su-27 '1804', Sqn 5, Debre Zeit AB

reconnaissance aircraft were also delivered. Attrition was very high due to inexperienced local pilots and to the subsequent wars against EPLF guerrillas and, from 1993 onwards, the Eritrean armed forces. 30 MiG-21*bis* (*izdeliye* 75A) *Fishbed-L* fighters were also delivered to the EtAF between 1986 and 1988.

After the fall of the Derg in 1991, when the EPLF guerrillas took Addis Ababa and Mengistu Haile Mariam fled the country, the EtAF became all but non-existent. Among other aircraft, there were about 20 intact MiG-21s but there was no one to fly them, as some of the pilots had fled abroad to escape persecution and the less lucky ones had been jailed for supporting the Mengistu regime. In 1995 – that is, two years after the formal secession of Eritrea – the new Ethiopian government started to revive the air arm, now known as the Federal Democratic Republic of Ethiopia Air Force (FDREAF). Ten MiG-21s were sent to Israel in 1997 for refurbishing and possible upgrading to MiG-21-2000 standard by Israel Aircraft Industries (IAI), but shortage of funds subsequently precluded the upgrade (though some sources do state that the EtAF operated MiG-21-2000s). These ten aircraft were an exception, as generally Ethiopian MiG-21s were overhauled in the Ukraine by the plants in Odessa and L'vov. By mid-2007 two units equipped with MiG-21s (about 18 fighters and six trainers) were reportedly still based at Dire Dawa (the 3rd Air Regiment) and Debre Zeit. In 2001 the EtAF began retiring the type and replacing it with the Su-27.

Another Mikoyan type used by the Ethiopian Air Force was the 'swing-wing' MiG-23. The first MiG-23BN *Flogger-H* fighter-bombers were delivered in 1977, followed by a further shipment in 1985 giving a total of 44. During the Derg regime there were reportedly 36 *Flogger-Hs* distributed between three regiments at Asmara (now in Eritrea), Dire Dawa (the 3rd Air Regiment) and Debre Zeit. In the 1990s the MiG-23BNs were concentrated at Bahir Dar AB. The MiG-23UB *Flogger-C* was used for conversion and proficiency training. Again, there were two shipments (in 1977 and 1985) giving a total of at least six (some sources say nine), and the trainers served with the same three units at Asmara, Dire Dawa and Debre Zeit. Additionally, Ethiopia purchased 18 MiG-23ML *Flogger-G* fighters (sometimes reported in error as '16 ex-Romanian MiG-23MF *Flogger-Bs*'). The total number of MiG-23s delivered to Ethiopia is close to 100 (the serials appear to support this

Female pilot Lt. Haimanot Gebremariam poses with her mount (Su-27 '1951'). In 2007 she and three other FDREAF pilots sought political asylum from the dictatorial Woyanne regime while on training in Israel and were sentenced to death *in absentia* by a court martial.

The cockpit section of '1951' as the aircraft is readied for a sortie.

An Ethiopian pilot poses with a fully armed Su-27UB.

theory) but the proportion of fighter, trainer and strike variants is not known. In 2004, it was reported that about seven MiG-23BNs and two trainers remained in service.

The most modern fighter type operated by the EtAF is the Sukhoi Su-27. Ethiopia became the first African operator of the type and the first nation to take delivery of the Su-27s rendered surplus by the reorganisation of the Russian Air Force. The Ethiopian Air Force purchased six single-seat Su-27S *Flanker-Bs* and two Su-27UB *Flanker-Cs* in late 1998 via the Promexport arms trading agency; defence experts assessed the deal as being worth US$ 120 million. Before delivery the fighters were suitably modified for export (this involved replacement of IFF transponders, among other things) and given a check-up by the Russian Air Force's ARZ No.275 aircraft repair plant in Krasnodar. The *Flankers* were delivered to Ethiopia by Russian Air Force Antonov An-22 Antey (*Cock*) transports in November 1998 and January 1999.

The current and rather elaborate FDREAF roundel (with inscriptions in Amharic) worn by some Ethiopian aircraft.

The Su-27s are operated by Sqn 5 based at Debre Zeit. One of the Su-27UBs was lost immediately after delivery, crashing at Debre Zeit AB on 6th January 1999 during a demonstration flight staged for the then President Negado Gidada and other high-ranking officials. The Russian pilot Col. Vyacheslav Myzin had attempted to execute the tailslide manoeuvre which had been the Su-27's show-stopper at various international airshows but, lacking the necessary experience, could not hold the correct attitude, causing the aircraft to flip into a spin. The crew ejected but the other Ethiopian back-seater, Flt. Lt. Abaniyeh, lost his life. Russia acknowledged the pilot's fault and supplied another Su-27 as an attrition replacement that year.

Ethiopia expressed a wish to obtain more Su-27s from Russia to make up for combat losses in the war with Eritrea (at least three EtAF *Flankers* were shot down), but these plans had to be deferred when the United Nations imposed an embargo on weapons deliveries to the belligerents in May 2000. The embargo was lifted a year later, and six ex-Russian Air Force Su-27s (four *Flanker-Bs* and two *Flanker-Cs*) were overhauled and modified to Su-27SK and Su-27UBK standard respectively by ARZ No.275 in Krasnodar and ARZ No.322 at Vozdvizhenka AB. They were delivered to Debre Zeit in 2003-04. The total delivered has been reported as 14 *Flanker-As* (three Su-27Ps and 11 Su-27SKs) and four Su-27UBs.

A batch of four Sukhoi Su-25 attack aircraft was delivered to the Ethiopian Air Force around March 2000, comprising a pair of Su-25T single-seaters and two Su-25UBK combat trainers. The two-seaters were drawn from the Russian Air Force fleet and modified in accordance with a special request from the Ethiopian Air Force. The specialised Su-25T 'tank killers', which had only seen limited service in Russia, were transferred from the Russian Air Force's 4th Combat Training & Aircrew Conversion Centre at Lipetsk-2 AB.

Pre-delivery modifications were carried out by the Russian Air Force's ARZ No.121 at Kubinka AB near Moscow. After that, the *Frogfeet* were airlifted to Ethiopia by Russian Air Force An-22 and Il'yushin IL-76MD *Candid-B* transports; this was probably one of the last such delivery flights performed by the Antey. Since acquiring the four aircraft, Ethiopia has used them in combat against Eritrean insurgent groups. According to press reports, a total of ten Su-25s were supplied to Ethiopia.

The Ethiopian Air Force operated 16 ex-Soviet Air Force Antonov An-12Bs and An-12BPs delivered in 1977. Originally they wore overall light grey colours standard for the type in Soviet Air Force service; with the excep-

Ethiopian Air Force An-12B '1505' with EtAF titles in English and Amharic

Well-worn An-12B '1505' shows obvious signs of previously being a quasi-civil Soviet Air Force aircraft with the Soviet flag, Aeroflot titles and registration (and tactical code, from earlier days when it had worn star insignia) painted out.

tion of three aircraft lost to enemy action or in accidents, the surviving *Cubs* were repainted in a three-tone green/brown camouflage scheme by 2000.

Other Soviet transports operated by the Ethiopian Air Force included the Antonov An-24 *Coke* airliner (the exact version(s) and number delivered is unknown, but it is known that one aircraft crashed on 21st February 1989), the An-26 *Curl* tactical transport (again, the number delivered is unknown, but one aircraft is known to have crashed on 14th January 1982) and the An-32 *Cline* (two aircraft were delivered in 1989). The Presidential flight of the EtAF operated a Yakovlev Yak-40 'Salon' (*Codling*) VIP jet for shorter trips and, reportedly, a Tupolev Tu-154 *Careless* (version unknown) for longer ones.

The Soviet helicopters delivered to the Ethiopian Air Force started with the Mil' Mi-2 *Hoplite* on the lighter end of the scale. More than 30 Mil' Mi-8T *Hip-C* transport/assault helicopters were delivered to the EtAF, starting in 1977, and were followed by more than ten Mi-17 *Hip-Hs*. Four of the Mi-8Ts came from a Hungarian order; they had never been delivered to the Hungarian Air Force and were eventually sold to Ethiopia by the local company Intertraverz. Several of the helicopters fell into enemy hands during the 1998-2000 war with Eritrea.

Ethiopia also purchased the Mil' Mi-6 *Hook* heavy-lift helicopter. Ten such helicopters were reportedly on strength with one EtAF squadron but no serials are known.

The Ethiopian Air Force operated more than 40 Mil' Mi-24A *Hind-A* and Mi-24D *Hind-D* attack helicopters delivered in 1978 (probably ex-Soviet Army Aviation aircraft). They were joined in 1988 by a batch of new Mi-35 *Hind-Es*. According to Russian sources, only 18 Ethiopian *Hinds* remained operational in early 1995.

On 3rd December 1998 Russia announced the sale of ten second-hand Mi-24s to Ethiopia. The helicopters, which were part of a large weapons package handled by the Promexport arms export agency, were scheduled to arrive in Addis Ababa on 10th-23rd December. Ethiopia had declined to deal with Rosvo'oruzheniye (= Russian Armament), the other then-current Russian arms exporter – possibly because the latter had already delivered Russian weapons to Eritrea. (In November 2000 Rosvo'oruzheniye and Promexport merged into the Rosoboronexport (= Russian Defence Export) agency in order to eliminate unnecessary internal competition.) That same year Belarus supplied two Mi-24V *Hind-Es* and Mi-24P *Hind-Fs* from ex-Soviet stocks, plus a further two Mi-24Ps in 2000. Eleven Mi-24s were reportedly on strength with the FDREAF in November 2000. Part of the fleet was lost to ground fire during the 1998-2000 war.

According to some sources, the Ethiopian Navy took delivery of two Mil' Mi-14 *Haze* amphibious helicopters. These were most certainly ex-Soviet Navy machines, as new-build Mi-14s were only supplied to the Soviet Union's closest allies.

After-sale support had never been a strong point with Soviet aircraft, especially in such far-off places as Africa. As a result, much of the fleet was grounded for lengthy periods by lack of engines and other key components. Not wishing to depend on the Soviet Union for heavy maintenance, the Ethiopian Air Force established a repair facility known as the Aircraft Heavy Maintenance Centre. This is now known as the Dejen Aviation Engineering Co. (DAVEC) and handles all Soviet aircraft types in service with the EtAF.

Ethiopian Air Force at war

Since the 1974 revolution, the Ethiopian Air Force has repeatedly waged wars with enemies within and outside. One of them was the Ogaden War with neighbouring Somalia. As mentioned earlier, Somalia's leader, Gen. Siad Barré, took advantage of the turmoil in Ethiopia to try and annexe the borderside Ogaden province which Somalia regarded as its own because of the large Islamic Somali population. With support from Somalia, an insurgency had developed to such an extent that by May 1977 the rebels controlled several towns. An incursion by the Somali Army began on 13th July 1977 and by the end of the month 60% of the Ogaden was occupied.

Both belligerents used Soviet weapons, resulting in a rather improbable 'dog eat dog' (or rather 'MiG eat MiG') situation. The Soviet Union, while supplying arms to both sides, persisted in negotiating a ceasefire. At the pinnacle of his army's success Siad Barré, incensed that his troops were being halted in their advance by an enemy with Soviet-supplied weapons, expelled 6,000 Soviet advisors from Somalia, demanding that they leave by 13th November 1977. As a result, Soviet aid to Somalia was discontinued and that to Ethiopia appreciably increased. A massive airlift of arms and aircraft was set in motion on 25th October with Cuban pilots and over 10,000 soldiers following. The reinforcements arrived in the nick of time just as the Somalis launched their final assault on 28th November. The battle to capture Harar (not to be confused with Harare in Zimbabwe!) was gaining momentum, and on 3rd February 1978 the Ethiopians launched a counteroffensive. A series of ferocious air strikes by Cuban-flown MiG-17F and MiG-21MF fighter-bombers supported the Ethiopian army, now reinforced by 6,000 Cuban soldiers, which attacked the Somalis who were relentlessly driven back.

In early 1978 the Ethiopian Mi-24As got their baptism of fire in the Ogaden War. Flown by Cuban pilots, the Mi-24s raided Somalian

EtAF An-12BP '1516', again with traces of the former Soviet registration.

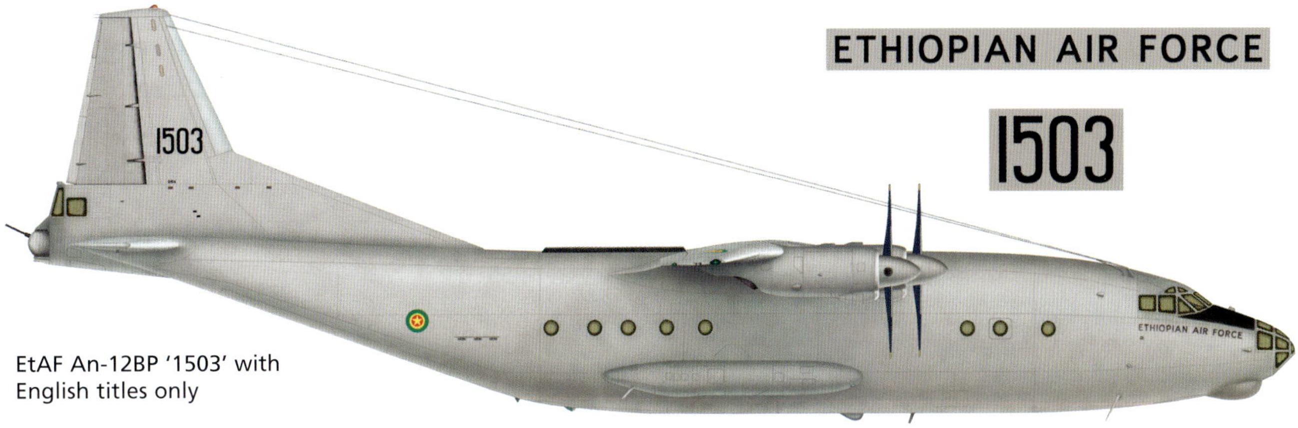

EtAF An-12BP '1503' with English titles only

positions with virtual impunity, knocking out armoured vehicles and artillery batteries. Somalian air defences were too weak to offer any real opposition to the choppers.

On 9th March 1978, the Somali leader ordered his forces to withdraw, the last unit leaving Ethiopia on 15th March. This action signalled the end of the Ogaden War.

As far as can be ascertained, four Ethiopian MiG-21MFs were shot down by Somali Air Corps fighters of the same type, accounting for half of the EtAF losses in the conflict. The Ethiopians claimed 11 'kills' (all scored by Northrop F-5As flown by Israeli mercenary pilots). It is interesting to note that the last Ethiopian aerial victory was on 3rd August 1977; presumably the EtAF had then decided to concentrate on ground attack.

On the home front, what had started as a guerrilla campaign in 1961 had developed into a full-scale civil war in the 1980s. When the Ethiopian government forces had regrouped, they had to wage war with not only the EPLF but also the Tigrean People's Liberation Front (TPLF), the Ethiopian People's Revolutionary Democratic Front (EPRDF) led by Meles Zenawi and other rebel factions that had proliferated since Mengistu seized power. The reinforced Ethiopian army attacked the rebels on 15th May 1978 and was partially effective in southern Eritrea, reopening communications to Asmara and Massawa. Several MiG-21MFs flown by Cuban pilots joined a squadron of North American F-86Ds in attacking troop concentrations of the EPLF and helped the army retain control of the Red Sea Ports. However, the EPLF was not destroyed and it regrouped to defeat the Ethiopians at Afabet the following year. Attacks by MiG-21MFs and other EtAF aircraft continued, with ever-increasing use of napalm bombs, but the Eritreans were remarkably resilient. In spite of a willingness to accept a huge number of casualties in their attacks, the Ethiopians were unable to defeat the EPLF. According to available sources, two Cuban MiG-21MF pilots

Ethiopian Air Force An-12BP '1503' at Djibouti-Ambouli AB in 1992.

An Ethiopian Air Force Mi-24A; note the Russian 'Danger' stencil

An EtAF Mi-35 with exhaust/air mixers fitted and UB-32A FFAR pods

(1st Lt. Raúl Hernandez Vidal and Col. Benigno Gonzalez Cortéz) were shot down by AA fire over Eritrea; the former pilot lost his life.

The Soviet Union maintained a presence in the conflict. More than 40 Mi-24As were thrown into battle against the EPLF guerrillas. The choppers were used in the close air support role, their main weapon being S-5 FFARs. Soviet military advisors and technicians assisted the government forces in training crews and maintaining the choppers. *'Mi-35s were delivered from the USSR in 1988,* – Maj.

Two EtAF Mi-24As with intake filters; note the differently painted Mi-24 in the background.

An Ethiopian Air Force Mi-35P with exhaust/air mixers and B-8V20 FFAR pods

S. A. Mel'nichenko, one of the advisors, recalled. – *The graduates of the local flying school flew them successfully. Besides flying the usual missions of the Mi-24, they had to seek and destroy separatist gunboats in the Red Sea. These fast craft were the scourge of shipping in the area, appearing out of nowhere, attacking ships moored in the harbours and disappearing just as quickly. After the helicopters had sunk eight of these boats the enemy gave up using them. In February 1989 UPK-23-250 gun pods were used with great success against tanks. An armoured convoy moving along a canyon was attacked by two groups of Mi-35s which took turns strafing it, destroying eight tanks.'*

There was no air-to-air combat in the 1980s, the rebels having no aircraft. On the other hand, they had access to Soviet-made 9K32 Strela (SA-7 *Grail*) man-portable air defence systems (MANPADS) to fend off the strafing MiG-21MFs and MiG-23BNs. In 1982 the MiGs made massive air raids on Naqfa, an EPLF stronghold which the Ethiopian Army had repeatedly but unsuccessfully tried to take, burning down most of the city with napalm bombs. These fighters were even used to destroy abandoned Ethiopian tanks and armoured vehicles to stop them from falling into enemy hands after an armoured convoy had blundered into a trap in a ravine at Adishrum between Naqfa and Afabet on 17th March 1988. Another favoured tactic of the rebels was commando raids. One such raid on Asmara airport on 20th-24th May 1984 was a spectacular success, the rebels claiming the

A well-weathered Ethiopian Air Force Mi-17 with members of the local militia in the foreground.

destruction of 16 MiG-21MFs and MiG-23BNs, six helicopters and ten other aircraft. A second raid took place on 14th January 1986 when an incredible 42 Ethiopian and Soviet aircraft, including many MiG-21MFs, were claimed to have been wrecked (though most military historians assess the actual figure to be much lower, as 'kill' statistics are notoriously optimistic).

As the conflict escalated the separatists managed to shoot down several Mi-24As. No Mi-35s were lost due to enemy action – at least until early 1990; however, one *Hind-E* was damaged beyond repair in a wheels-up landing. When spares supplies from the Soviet Union dried up most of the Ethiopian Air Force's helicopters became unserviceable. A few *Hinds* were stolen by Ethiopian defectors and used against the government forces.

Meanwhile, dissent was growing in the Ethiopian Armed Forces as many officers were opposed to obviously civilian objectives being targeted. In May 1989 many EtAF officers participated in an unsuccessful coup against President Mengistu, who purged the Air Force of dissident elements, grounding it for a while in order to do so. A series of successful offensives by the EPLF forced the EtAF to evacuate the base at Asmara, concentrating aircraft at the bases in Debre Zeit and Dire Dawa, which were overcrowded as it was. Morale in the EtAF was low, and in 1991 a total of 22 aircraft were flown by defectors to neighbouring countries.

The EPLF received considerable support from Muslim countries, but help for Ethiopia was limited to the Warsaw Pact countries and Israel. On 30th December 1989 Cuba began to withdraw troops from Ethiopia, and shortly afterwards the USSR stopped military supplies as well. Between 1984 and 1988 successive droughts brought famine, adding to the worries of the Ethiopian government and the distress of its people. Civilians suffered dreadful deprivation as both sides used famine-aid as a political weapon, but what finally broke the Derg regime was its failure to quell the Eritrean rebellion, a failure that encouraged more and more Ethiopians to join the EPRDF. Rebel troops finally entered Addis Ababa on 28th May 1991, Mengistu fleeing to Zimbabwe.

The new government headed by Meles Zenawi imprisoned all EtAF officers above the rank of captain but many pilots flew off into exile, leaving behind only 20 MiG-23BN and MiG-21*bis*. The Eritreans had captured eight

The local populace is happy about a photo opportunity with a Mi-17 in PDREAF markings as a backdrop.

MiG-21 fighters and one MiG-21UM trainer, all of which were returned to Ethiopia. Thus, in 1991 the EtAF was a broken reed and did not effectively recover until 1995 (as the FDREAF). The outcome of the war was the recognition by a new caretaker Ethiopian government of an independent Eritrea after a UN-monitored referendum in April 1993.

Actually, the bottom line was that one dictatorship (the Derg) was replaced by another – the EPRDF, aka the Woyanne regime. Several key areas of government, including the military, were dominated by the Tigrean ethnic minority. Large numbers of people belonging to other Ethiopian tribes were displaced from their fertile ancestral lands and confined in ghettos. This, along with widespread corruption and human rights violations, led to growing dissent – in the armed forces as well.

After the recognition of an independent Eritrea by Ethiopia, the border between the two countries was undemarcated, with several areas remaining in dispute. Such ambiguity could only lead to tension between the two countries. On 6th May 1998, Eritrean troops entered the disputed Badme region near the border, which resulted in a skirmish with the Ethiopian militia and security police. Eritrea claimed that several of its officials had been murdered near Badme and invaded with a large mechanised force, inciting Ethiopia to mobilise its army. More border incidents took place, each involving successively larger units.

In 1998 the conflict escalated to a full-scale war. On 12th May two Eritrean Army brigades supported by tanks and artillery attacked Badme and advanced eastward, seizing the surrounding flatland. On 5th June, Ethiopian MiG-23BN fighter-bombers redeployed to Bahir Dar attacked Asmara IAP while Eritrean Aermacchi MB.339FD light strike aircraft bombed Mekelle, the capital of Tigré Province, from where a squadron of MiG-21s operated, lightly damaging a MiG-21 and a Mi-8. The following day, two MiG-21MFs strafing Asmara and a nearby Eritrean Air Force base saw no aircraft on site but encountered heavy anti-aircraft fire; one MiG was shot down and the other damaged, and the second wave of the assault was called off. Two MiG-23BNs were reportedly also lost In June 1998. Mounting civilian casualties in bombed cities gave rise to a US-brokered agreement stipulating that neither side would attack 'population centres'. Fighting continued but there were no major thrusts by either side until more aircraft were obtained.

Russian help provided to the Ethiopians reportedly included the refurbishment of about a dozen MiG-21 fighters (variously reported as MiG-21MFs or MiG-21*bis*), as well as the supply of ten more MiG-23BNs. By the end of the year, eight Su-27s and several hundred R-27 air-to-air missiles had been supplied. Alarmed by the amount of Russian support and the large number of East European 'volunteer' pilots, the Eritrean President Isaias Afeworki made the dramatic announcement that any mercenaries shot down and captured in his country would be beheaded.

For its part, Eritrea purchased ten MiG-29s from the Ukraine. The stage was now set for the next round of heavy fighting which began with an Eritrean offensive near Adigrat on 5th February 1999 and an Ethiopian counteroffensive (supported by Mi-35 helicopters) near Alitena, Tsorona and Zalambessa two days later. In preparation for Operation *Sunset*, which began on 23rd February, FDREAF An-12Bs were used as makeshift bombers. On 25th February two FDREAF MiG-23BNs strafing Eritrean positions were intercepted by two ErAF MiG-29s; the Russian-flown Su-27s flying top cover destroyed both of them with R-73 missiles, but not before one MiG-23BN had been shot down by two Eritrean R-27R AAMs.

The following day, two MiG-21s and a MiG-23BN were strafing Eritrean positions when an Eritrean MiG-29UB appeared on the scene, shooting down both MiG-21s with R-73 missiles. In turn, it was bounced by two Su-27s whose missiles failed to score a direct hit, but the *Fulcrum* inexplicably disappeared and has not been seen since. If the kill could be confirmed Capt. Esther Tolossa would be the first female jet pilot to have destroyed an enemy jet fighter in aerial combat. As a result of these engagements the FDREAF gained and retained complete mastery of the air – Eritrean aircraft did not appear over the battlefield any more. A belated attempt to tip the scales took place on 18th March 1999 after the Ethiopians launched Operation *Westwind*. FDREAF MiG-21MF and MiG-23BN fighter-bombers were attacked by Eritrean MiG-29s, one of which was downed by the escorting Su-27s.

The intensive ground attack operations by the FDREAF incurred significant losses not only to ground fire but also as a result of the wear and tear imposed by desert operations. The use of poorly trained pilots also played a role; on 20th April 1999 a MiG-21 flown by such a pilot hit a power line pylon and crashed onto a suburb of Arba Minch in southern Ethiopia, killing eight civilians and injuring 14. Both sides recognised the need to take stock and rest their weary troops, so the war entered a relatively quiet period.

This time Eritrea had at its disposal some 200 of the latest Russian 9K38 Igla (SA-18 *Grouse*) MANPADS. At least three Ethiopian fixed-wing aircraft and two Mi-35 helicopters fell victim to these IR-homing/optically-guided missiles in early 1999, the losses being officially attributed to 'AA artillery'. One of the Mi-35s was shot down near Bure on 14th February 1999, killing the crew. Ten days later another Ethiopian *Hind* was reportedly downed in the Badme region. Eritrea claimed that two of the chopper's three crewmen were 'Russian mercenaries' on the grounds that they did not have the 'dog tags' worn by Eritrean military personnel. However, this cannot be proved, since the bodies were so badly burnt that it was impossible to tell even their skin colour (quite apart from the fact that many people indiscriminately refer to persons originating from any former Soviet republic as 'Russians' – especially in a negative context).

Refreshed and re-armed, the Ethiopian Army launched its final offensive, Operation Hard Wind, in May 2000. An addition to the FDREAF inventory at this stage was the delivery of eight Su-25T dedicated anti-tank aircraft; these were used for night attacks against armour and high-value targets, such as command posts. Instead of employing the usual frontal attack, the new offensive skirted the Western flank of the Eritrean defences and, having broken through and appearing to be attacking Adigrat, suddenly wheeled east towards Areza to link up with an attack on Adi Quala. Only resolute Eritrean counter-attacks saved the day but militarily the Ethiopians were in a strong position. The only reason for their failure to administer the *coup de grâce* was the fear of offending the international community by appearing to annexe the whole of Eritrea rather than being satisfied with only the disputed border territories. On 18th May, two Ethiopian MiG-21s were intercepted by an Eritrean MiG-29, whose pilot, having missed with his missiles, shot one *Fishbed* down and damaged the other with his 30-mm cannon. Immediately afterwards, however, his aircraft was destroyed by R-73 missiles fired by an EtAF Su-27. (Another account of the incident says that no aircraft were shot down at all – the MiG-29 simply took off too late to intercept the attackers.)

For some time the African Union, the European Union and Libya had been trying to

Known Ethiopian aircraft of Soviet/Russian origin

Type	Serial	C/n	F/n	Notes
An-12B (+)?	1501?	?		No reports
An-12B (+)	1502	401802		
An-12BP (+)	1503	4342009?		Grey overall. Tashkent-built? Also reported as Voronezh-built c/n 402009, but this is Russian Navy '28 Yellow'
An-12B (+)?	1504	?		
An-12BP (+)	1505	5342907		Grey overall; ex-Siviet Air Force CCCP-1****
An-12B (+)	1506	4342002?		Ex-CCCP-*****. Destroyed by enemy action Tesenni 15-1-1984
An-12B (+)?	1507	?		
An-12B (+)?	1508	?		
An-12B (+)?	1509	?		Damaged beyond repair at Addis Ababa-Haile Selassie Int'l before 5-1984, derelict
An-12B (+)?	1510	?		
An-12B (+)?	1511	?		
An-12BP (+)	1512	5343206		
An-12B (+)?	1513	?		
An-12B (+)?	1514	?		
An-12B (+)?	1515	?		
An-12B (+)?	1516	?		Grey overall
An-32	1551	1310?		
An-32	1552	1405?		Crashed before 2007
UTI-MiG-15	1414	?		Ex-South Yemen Alr Force, sand/green camouflage. Wfu Dire Dawa by 2008
UTI-MiG-15	1415	?		Ex-South Yemen Alr Force, sand/green camouflage. Wfu Dire Dawa by 2008

MiG-17F	1401	?		Ex-South Yemen AIr Force, sand/green camouflage. Wfu Dire Dawa by 2008
MiG-17F	1402	?		Ex-South Yemen AIr Force, sand/green camouflage
MiG-17F	1403	?		Ex-South Yemen AIr Force, sand/green camouflage. Wfu Dire Dawa by 2008
MiG-17F	1404	?		Ex-South Yemen AIr Force, sand/green camouflage
MiG-17F	1405	?		Ex-South Yemen AIr Force, sand/green camouflage
MiG-17F	1406	?		Ex-South Yemen AIr Force, sand/green camouflage
MiG-17F	1407	?		Ex-South Yemen AIr Force, sand/green camouflage
MiG-17F	1408	?		Ex-South Yemen AIr Force, sand/green camouflage
MiG-17F	1409	?		Ex-Soviet Air Force '02', natural metal finish. Wfu Dire Dawa by 2008
MiG-17F	1410	?		Ex-South Yemen AIr Force, sand/green camouflage
MiG-17F	1411	?		Ex-South Yemen AIr Force, sand/green camouflage
MiG-17F	1412	?		Ex-South Yemen AIr Force, sand/green camouflage
MiG-17F	1413	?		Ex-South Yemen AIr Force, sand/green camouflage
MiG-21R	1454	?	?	
MiG-21R	1455	?	?	Also reported as MiG-21MF. Captured by EPLF at Asmara
MiG-21R	1461	?	?	Also reported as MiG-21MF. Captured by EPLF at Asmara
MiG-21R	1462	?	?	
MiG-21R	1463	?	?	
MiG-21R	1464	?	?	Sand/brown camouflage. Also reported as MiG-21MF. Captured by EPLF at Asmara
MiG-21R	'1083'	?	?	Gate guard at Debre Zeit AB, fake serial (see MiG-21*bis*)
MiG-21MF	1041	?	?	Operated 1977-78, then stored
MiG-21MF	1042	?	?	Operated 1977-78, then stored
MiG-21MF	1043	?	?	Gate guard at Debre Zeit AB
MiG-21MF	1044	?	?	Operated 1977-78, then stored
MiG-21MF	1045	?	?	Operated 1977-78, then stored
MiG-21MF	1046	?	?	Operated 1977-78, then stored
MiG-21MF	1047	?	?	Operated 1977-78, then stored
MiG-21MF	1048	?	?	Operated 1977-78, then stored
MiG-21MF	1049	?	?	Operated 1977-78, then stored
MiG-21MF	1050	N750*****	?	Operated 1977-78, then stored
MiG-21*bis*	1051	N750*****	?	Also reported as MiG-21MF
MiG-21*bis*	1052	N75051511	75-33-**?	
MiG-21*bis*	1053	N750*****	?	
MiG-21*bis*	1054	N750*****	?	Also reported as MiG-21MF
MiG-21*bis*	1055	N750*****	?	
MiG-21*bis*	1056	N750*****	?	
MiG-21*bis*	1057	N750*****	?	
MiG-21*bis*	1058	N750*****	?	Also reported as MiG-21MF. Captured by EPLF at Asmara
MiG-21*bis*	1059	N750*****	?	Based Debre Zeit 2000
MiG-21*bis*	1060	N750*****	?	Also reported as MiG-21MF
MiG-21*bis*	1061	N750*****	?	
MiG-21*bis*	1062	N750*****	?	
MiG-21*bis*	1063	N750*****	?	
MiG-21*bis*	1064	N750*****	?	Also reported as MiG-21MF
MiG-21*bis*	1065	N750*****	?	Also reported as MiG-21MF. Captured by EPLF at Asmara
MiG-21*bis*	1066	N750*****	?	
MiG-21*bis*	1067	N750*****	?	
MiG-21*bis*	1068	N750*****	?	
MiG-21*bis*	1069	N750*****	?	
MiG-21*bis*	1070	N750*****	?	
MiG-21*bis*	1071	N750*****	?	
MiG-21*bis*	1072	N750*****	?	
MiG-21*bis*	1073	N750*****	?	
MiG-21*bis*	1074	N750*****	?	
MiG-21*bis*	1075	N750*****	?	
MiG-21*bis*	1076	N750*****	?	

MiG-21*bis*	1077	N750*****	?	
MiG-21*bis*	1078	N750*****	?	
MiG-21*bis*	1079	N750*****	?	
MiG-21*bis*	1080	N750*****	?	Also reported as MiG-21MF
MiG-21*bis*	1081	N750*****	?	Shot down near Tugwojale, Somalia, 7-1981, pilot killed
MiG-21*bis*	1082	N750*****	?	Also reported as MiG-21MF. Captured by EPLF at Asmara
MiG-21*bis*	1083	N750*****	?	Also reported as MiG-21MF. Captured by EPLF at Asmara
MiG-21*bis*	1084	N750*****	?	
MiG-21*bis*	1085	N750*****	?	
MiG-21*bis*	1086	N750*****	?	
MiG-21*bis*	1087	N750*****	?	Also reported as MiG-21MF. Refurbished by IAI 1997; noted 1998
MiG-21*bis*	1088	N750*****	?	Wfu by 2003
MiG-21*bis*	1089	N750*****	?	
MiG-21*bis*	1090	N750*****	?	
MiG-21*bis*	1091	N750*****	?	Also reported as MiG-21MF
MiG-21*bis*	1092	N750*****	?	Wfu by 2003
MiG-21*bis*	1093	N750*****	?	
MiG-21*bis*	1094	N750*****	?	
MiG-21*bis*	1095	N750*****	?	Wfu by 2003
MiG-21*bis*	1096	N750*****	?	
MiG-21*bis*	1097	N750*****	?	
MiG-21*bis*	1098	N750*****	?	Captured at Asmara, taken over by Eritrean AF
MiG-21*bis*	1099	N750*****	?	
MiG-21*bis*	1100	N750*****	?	
MiG-21*bis*	1101	N750*****	?	
MiG-21*bis*	1102	N750*****	?	
MiG-21*bis*	1103	N750*****	?	
MiG-21*bis*	1104	N750*****	?	
MiG-21*bis*	1105	N750*****	?	
MiG-21*bis*	1106	N750*****	?	
MiG-21*bis*	1107	N750*****	?	
MiG-21*bis*	1108	N750*****	?	
MiG-21*bis*	1109	N750*****	?	
MiG-21*bis*	1110	N750*****	?	
MiG-21*bis*	1111	N750*****	?	
MiG-21*bis*	1112	N750*****	?	
MiG-21*bis*	1113	N750*****	?	
MiG-21*bis*	1114	N750*****	?	Noted mid-1980s
MiG-21*bis*	1115	N750*****	?	
MiG-21*bis*	1116	N750*****	?	
MiG-21*bis*	1117	N750*****	?	
MiG-21*bis*	1118	N750*****	?	
MiG-21*bis*	1119	N750*****	?	
MiG-21*bis*	1120	N750*****	?	
MiG-21*bis*	1121	N750*****	?	
MiG-21*bis*	1122	N750*****	?	
MiG-21*bis*	1123	N750*****	?	
MiG-21*bis*	1124	N750*****	?	Noted mid-1980s
MiG-21*bis*	1125	N750*****	?	
MiG-21*bis*	1126	N750*****	?	
MiG-21*bis*	1127	N750*****	?	Captured by EPLF at Asmara
MiG-21*bis*	1128	N750*****	?	
MiG-21*bis*	1129	N750*****	?	Captured by EPLF at Asmara
MiG-21*bis*	1130	N750*****	?	Flown to Sudan by defector 6-6-1989
MiG-21*bis*	1131	N750*****	?	
MiG-21*bis*	1132	N750*****	?	
MiG-21*bis*	1133	N750*****	?	

MiG-21*bis*	1134	N750*****	?	
MiG-21UM	1001	?	?	
MiG-21UM	1002	?	?	Natural metal finish
MiG-21UM	1003	?	?	
MiG-21UM	1004	?	?	
MiG-21UM	1005	?	?	
MiG-21UM	1006	?	?	
MiG-21UM	1007	?	?	
MiG-21UM	1008	?	?	
MiG-21UM	1009	?	?	
MiG-21UM	1010	?	?	
MiG-21UM	1011	?	?	
MiG-21UM	1012	?	?	Captured by EPLF at Asmara
MiG-21UM?	1013	?	?	Also reported as MiG-21MF
MiG-21UM?	1014	?	?	Also reported as MiG-21MF; based Debre Zeit 1986
MiG-23MLA	1801	03903*****	?	
MiG-23MLA	1802	03903*****	?	
MiG-23MLA	1803	03903*****	?	
MiG-23MLA	1804	03903*****	?	
MiG-23MLA	1805	03903*****	?	
MiG-23MLA	1806	03903*****	?	
MiG-23MLA	1807	03903*****	?	
MiG-23MLA	1808	03903*****	?	
MiG-23MLA	1809	03903*****	?	
MiG-23MLA	1810	03903*****	?	
MiG-23MLA	1811	03903*****	?	
MiG-23MLA	1812	03903*****	?	
MiG-23MLA	1813	03903*****	?	
MiG-23MLA	1814	03903*****	?	
MiG-23MLA	1815	03903*****	?	
MiG-23MLA	1816	03903*****	?	
MiG-23MLA	1817	03903*****	?	
MiG-23MLA	1818	03903*****	?	
MiG-23UB	1201	B103****	?	
MiG-23UB	1202	B103****	?	
MiG-23UB	1203	B103****	?	Reported in error as MiG-23BN based at Dire Dawa
MiG-23UB	1204	B103****	?	
MiG-23UB	1205	B103****	?	Tan/green/brown camouflage
MiG-23UB	1206	B103****	?	Wfu 2005
MiG-23UB	1207	B103****	?	Reported in error as MiG-23BN based at Dire Dawa
MiG-23UB	1209	B103****	?	
MiG-23BN	1225	03932*****	?	
MiG-23BN	1251	03932*****	?	Used in 1998 Eritrean War
MiG-23BN	1252	03932*****	?	
MiG-23BN	1253	03932*****	?	
MiG-23BN	1254	03932*****	?	
MiG-23BN	1255	03932*****	?	
MiG-23BN	1256	03932*****	?	
MiG-23BN	1257	03932*****	?	No SPS-141 jammer; extra IRCM flare dispensers
MiG-23BN	1258	03932*****	?	
MiG-23BN	1259	03932*****	?	
MiG-23BN	1260	03932*****	?	
MiG-23BN	1261	03932*****	?	
MiG-23BN	1262	03932*****	?	
MiG-23BN	1263	03932*****	?	
MiG-23BN	1264	03932*****	?	Destroyed at Asmara

MiG-23BN	1265	03932*****	?	Wfu; sand/green/dark brown camouflage
MiG-23BN	1266	03932*****	?	No SPS-141 jammer
MiG-23BN	1267	03932*****	?	Destroyed at Debre Zeit
MiG-23BN	1268	03932*****	?	
MiG-23BN	1269	03932*****	?	
MiG-23BN	1270	03932*****	?	No SPS-141 jammer; based at Debre Zeit, used in 1998 Eritrean War
MiG-23BN	1271	03932*****	?	
MiG-23BN	1272	03932*****	?	
MiG-23BN	1273	03932*****	?	
MiG-23BN	1274	03932*****	?	Based at Asmara 1983; taken over by Eritrean Air Force
MiG-23BN	1275	03932*****	?	
MiG-23BN	1276	03932*****	?	Based at Debre Zeit 1983; tan/green/brown camouflage
MiG-23BN	1277	03932*****	?	
MiG-23BN	1278	03932*****	?	
MiG-23BN	1279	03932*****	?	
MiG-23BN	1280	03932*****	?	
MiG-23BN	1281	03932*****	?	
MiG-23BN	1282	03932*****	?	
MiG-23BN	1283	03932*****	?	
MiG-23BN	1284	03932*****	?	
MiG-23BN	1285	03932*****	?	Sand/green/dark brown camouflage. Extra IRCM flare dispensers; used in 1998 Eritrean War
MiG-23BN	1286	03932*****	?	Shot down 14-11-1989
MiG-23BN	1287	03932*****	?	
MiG-23BN	1288	03932*****	?	Seen 1983
MiG-23BN	1289	03932*****	?	Based at Debre Zeit
MiG-23BN	1290	03932*****	?	Extra IRCM flare dispensers; used in 1998 Eritrean War
MiG-23BN	1291	03932*****	?	
MiG-23BN	1292	03932*****	?	
MiG-23BN	1293	03932*****	?	
MiG-23BN	1294	03932*****	?	
MiG-23BN	1295	03932*****	?	
MiG-23BN	1296	03932*****	?	
MiG-23BN	1297	03932*****	?	
MiG-23BN	1298	03932*****	?	
MiG-23BN	1299	03932*****	?	
MiG-23BN	1300	03932*****	?	Possibly captured by EPLF and became Eritrean Air Force ER-01
MiG-23BN	1301	03932*****	?	
MiG-23BN	1302	03932*****	?	
MiG-23BN	12...	0393207556	3750	Extra IRCM flare dispensers on rear fuselage
Mi-2	776	?		
Mi-2	777	?		
Mi-8T	850	368**?	?	
Mi-8T	851	368**?	?	
Mi-8T	852	368**?	?	
Mi-8T	855	36809	?	
Mi-8T	857	368**?	?	
Mi-8T	858	368**?	?	
Mi-8T	859	368**?	?	
Mi-8T	861	368**?	?	
Mi-8T	862	36821	?	
Mi-8T	863	36822	?	
Mi-8T	864	36823	?	
Mi-8T	865	36824	?	

Mi-8T	866	368**?	?	
Mi-8T	1601	?	?	
Mi-8T	1602	?	?	
Mi-8T	1603	?	?	
Mi-8T	1620	?	?	Captured by Eritreans, to Eritrean AF; serial also reported as Mi-24A c/n 2201305!
Mi-8T	?	10451	?	Diverted from Hungarian Air Force order; d/d 11-1997
Mi-8T	?	10452	?	Diverted from Hungarian Air Force order; d/d 11-1997
Mi-8T	?	10453	?	Diverted from Hungarian Air Force order; d/d 11-1997
Mi-8T	?	10454	?	Diverted from Hungarian Air Force order; d/d 11-1997
Mi-17	869	?	?	
Mi-17	2001	324M10	?	
Mi-17	2004	324M**?	?	
Mi-17	2006	324M**?	?	Captured by Eritreans, to Eritrean AF
Mi-17	2008	324M**?	?	Captured by Eritreans, to Eritrean AF
Mi-17	2009	324M**?	?	
Mi-17	2011	324M**?	?	Green camouflage, new insignia
Mi-17	2012	324M**?	?	
Mi-17	2013	324M**?	?	
Mi-17	2017	324M**?	?	
Mi-24A	1616	?	?	
Mi-24A (late)	1621	?		Sand/tan camouflage
Mi-24A	1624	?	?	
Mi-24A?	1628	?	?	
Mi-35	2101	?	?	
Mi-35	2105	?	?	
Mi-35	2106	?	?	
Mi-35	2108	?	?	Destroyed 18-3-1999
Mi-35P	2123	?	?	Assigned to UNAMID (Sudan), coded UNO793
Mi-35P	2125	?	?	Assigned to UNAMID (Sudan), coded UNO794
Mi-35P	2128	?	?	Assigned to UNAMID (Sudan), coded UNO795
Mi-35P	2129	?	?	Assigned to UNAMID (Sudan), coded UNO796
Mi-35P	2130	?	?	Assigned to UNAMID (Sudan), coded UNO797
Su-27SK	1801?	369110*****		
Su-27SK	1804?	369110*****		
Su-27SK	1951	369110*****		
Su-27SK	1952	369110*****		
Su-27SK	1953	369110*****		
Su-27SK	1954	369110*****		
Su-27SK	1955	369110*****		
Su-27SK	1956	369110*****		
Su-27UBK	1901	963104*****	?	
Su-27UBK	1907	963104*****	?	
Su-27UBK	1909	963104*****	?	
Su-27UBK	1944	963104*****	?	
Su-27SK	?	36911027715		Russian Air Force/582nd Fighter Regiment '05 Blue'; possibly overhauled by ARZ No.275 and sold abroad (to Ethiopia?)
Su-27SK	?	36911028104		Russian Air Force/404th Fighter Regiment '01 Blue'; possibly overhauled by ARZ No.275 and sold abroad (to Ethiopia?)
Su-27SK	?	36911028106		Russian Air Force/404th Fighter Regiment '02 Blue'; possibly overhauled by ARZ No.275 and sold abroad (to Ethiopia?)
Su-27UBK	?	96310412032	0703	Russian Air Force/60th Fighter Regiment '12 Red'; overhauled by ARZ No.322 and sold abroad (to Ethiopia?)
Su-27UBK	?	96310412043	0707	Russian Air Force/404th Fighter Regiment '72 Blue'; possibly overhauled by ARZ No.275 and sold abroad (to Ethiopia?)
Tu-154… 'Salon'	1501?	?		Clash with An-12 serial block; no reports, existence unconfirmed
Yak-40 'Salon'	1601	?		Clash with Mi-8 serial block

The sole Gambian Air Force Su-25K parked at Banjul, with personnel seeking shelter from the sun beneath its wing.

Another view of the same aircraft, showing the code '81 Blue'. Note the very weathered paintwork.

Ghana Air Force Mi-17V-5 G692 with the cargo ramp open. Note the optional wire cutters above and below the flight deck.

Ghana Air Force Mi-17V-5 G692

arrange a ceasefire. Exhausted by the war, both sides agreed to this on 18th June 2000. A 25-km (15-mile) Temporary Security Zone was established within Eritrea, enforced by United Nations peacekeeping forces; on 12th December 2000 a Peace Agreement was signed by the two governments. Neither side, especially Ethiopia, is happy with the agreement, so peace in the region remains a very fragile affair.

Interestingly, in February 2010/July 2011 five Ethiopian Mi-24Ps were seconded to the African Union/United Nations Hybrid Operation in Darfur, Sudan (UNAMID). The helicopters operated from Nyala, Sudan, and wore appropriate 'UNO' codes in addition to their regular serials.

Gambia

Gambia (again, sometimes referred to as *the* Gambia) gained independence from the United Kingdom on 18th February 1965. Peace and quiet reigned until 1981 when a coup d'état attempt was foiled. Another coup followed in 1994, and the internal situation has been unstable since then, compounded by tense relations with erstwhile ally Senegal with which Gambia had formed a confederation (Senegambia) in 1981-89.

The Gambian Air Force operates only one combat jet, a Sukhoi Su-25K *Frogfoot-A* attack aircraft serialled '81 Blue' (c/n 25508110630). The aircraft was delivered in 2002; it is based at Banjul (Yundum airport) and is one of the very last Tbilisi-built *Frogfeet*. The plans to buy additional Su-25s never materialised.

Additionally, the Gambian government flight includes a pair of Il'yushin IL-62M 'Salon' *Classic* long-haul airliners. The first of these, registered C5-GNM (c/n 3036142), was purchased around 2003 from the Central African Republic where it had flown as TL-ACL. The other aircraft, C5-RTG (c/n 1356234), is newer and was bought before November 2005 from the Uzbekistan government flight with which it had served since new as UK-86569.

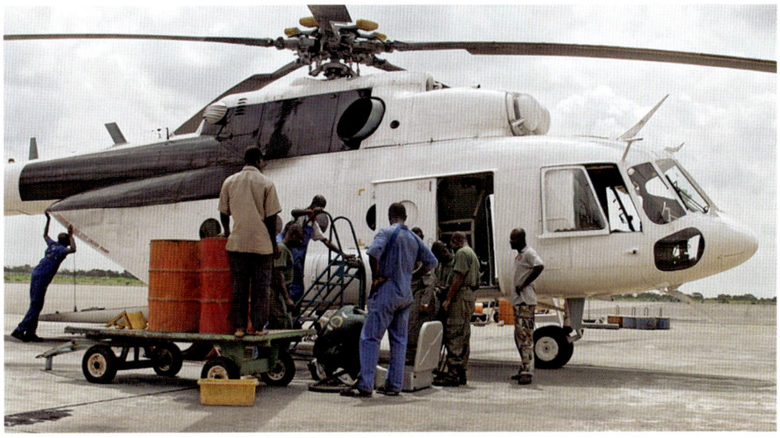

Another Ghana Air Force Mi-17V-5 in all-white colours is refuelled from a makeshift 'tanker truck'.

Ghana

Despite being on good terms with the Soviet Union, Ghana did not order Soviet combat aircraft. The only aircraft of military value supplied to this African nation by the USSR was the Antonov An-12 *Cub* transport. On 4th October 1961 Ghana Airways took delivery of an Irkutsk-built An-12A. Among other things the freighter was used for flights to the UK to pick up spares for Ghana Airways' fleet of British-built airliners. However, it obviously had military uses as well. The career of the Ghana Airways An-12 proved to be brief. In 1962 the aircraft was withdrawn from use at Accra – to quote the *Royal Air Force Flying Review* maga-

The Ghana Air Force roundel.

Known Ghanaian military and government aircraft of Soviet/Russian origin

Type	Serial/registration	C/n	F/n	Notes
An-12A (+)	9G-AAZ	024009	?	Ghana Airways livery
Mi-2P	G660	536823090		Ex-Libyan Arab Air Force (6823?). Ground instructional airframe at Accra
Mi-2P	G661	536824090		Ex-Libyan Arab Air Force (6824?). Preserved Armed Forces Museum, Kumasi
Mi-4S	9G-ABG?	15145		Also reported as 9G-ATP (which crashed in 1970!) with this c/n! Became, see next line
	9G-OTP (2)			Retired 1970
Mi-4S	9G-OTP (1)	08112		Returned to the Soviet Union, fate unknown
Mi-17V-5	G690	228M01	?	
Mi-17V-5	G691	228M02	?	Crashed 12-1-2007
Mi-17V-5	G692	228M03	?	Sand/green/light green camouflage
Mi-17V-5	G693	228M04	?	

Avia-14 Salon 3X-PRG equipped with tip tanks and featuring a non-standard window layout was the aircraft of the President of the Republic of Guinea.

The very similar Guinean Air Force roundel.

zine, *'owing to its inadequate operating economy'*. In 1963 the machine was made airworthy and returned to the Soviet Union; its subsequent fate is unknown. According to *Royal Air Force Flying Review*, in 1961 the Ghanaian Air Force received three more An-12s donated by the Soviet government. No serials are known.

Soviet helicopters in Ghana were rather more numerous. The Ghanaian Air Force operated a pair of Mil' Mi-2P *Hoplite* helicopters received from Libya. In December 2004 the Air Force took delivery of four Mil' Mi-17V-5 *Hip* medium utility helicopters, one of which has since been lost in a crash. Additionally, in the 1960s the Ghanaian government flight operated a pair of Mil' Mi-4S *Hound* VIP helicopters, the information on which is somewhat contradictory.

Guinea (Guinea-Conakry)

Guinea was a French colony until 1958 and the new Republic of Guinea was quick to create a modest air arm with assistance from the USSR. The Guinean Air Force (FAG – *Force Aérienne de Guinée*; also referred to as *Armée de l'Air Guinée*, or AAG) received eight Mikoyan/Gurevich MiG-17AS *Fresco-A* fighter-bombers around 1960 in return for letting Soviet Navy Tupolev Tu-95RTs *Bear-D* maritime reconnaissance/over-the-horizon targeting aircraft use Guinean bases as forward operating locations. Six MiG-17Fs were in service at least until late 1991. At least two Mikoyan/Gurevich UTI-MiG-15 *Midget* trainers were supplied at the same time for proficiency training. The fighters were based at Conakry-Gbessia airport.

Pilots from neighbouring Guinea-Bissau reportedly took training in these MiG-17s after initial conversion training in UTI-MiG-15s. The plan was to borrow the fighters for use against Portuguese forces in Guinea-Bissau's struggle for independence. Eventually, however, the 'Carnation Revolution' of 25th April 1974 in Portugal, one of whose goals was to grant independence to Portuguese colonies, rendered this unnecessary.

Guinean Air Force Mi-17 FAG-1009 lands at what appears to be a stadium.

Here, the same helicopter is shown resting between missions. Note the unusual position of the roundel on the tail rotor pylon.

In 1986, again as a *quid pro quo* for the Soviet use of the maintenance facilities at Conakry-Gbessia, the Guinea Air Force took delivery of eight Mikoyan MiG-21*bis Fishbed-L* fighters (alternatively reported as MiG-21MF *Fishbed-Js* or even older MiG-21PFM *Fishbed-Fs*) and one MiG-21UM *Mongol-B* trainer. The unit operating them had no number or name. Five were reported to be earmarked for restoration to airworthy condition by Russian engineers in 2001.

On 22nd April 2004 a MiG-21*bis* crashed at Kassa village near Conakry after suffering an engine failure, killing the pilot, Flt. Sgt. Name Keita. On 16th April 2007 a MiG-21UM flown by a Ukrainian contract pilot and Guinean pilot Capt. Touré Alja caught fire in mid-air and crashed onto the TV station in the middle of Conakry, narrowly missing the US Embassy; fortunately there were no fatalities. Later, a replacement aircraft was procured from the Ukraine.

As regards transports, the Guinean Air Force had at least two Antonov An-2 *Colt* utility biplanes. Four Antonov An-12BP medium transports were delivered to the Guinean Air Force. The *Cubs* were nominally owned by flag carrier Air Guinée and painted in the airline's livery to facilitate flights abroad.

In addition to five ordinary Avia-14 airliners (Il'yushin IL-14 *Crate* built under licence in Czechoslovakia) operated by Air Guinée on a lease basis, two Super Avia-14s (an improved version with circular cabin windows and

Guinea Air Force Mi-25 FAG-1003 armed with two B-8V20 FFAR pods makes a low fly-past.

Another view of the same machine. Note that, as was often the case in Africa, the Raduga LLLTV/ FLIR system, ATGM guidance antenna 'egg' and wingtip-mounted missile rails have been removed as unnecessary.

wingtip fuel tanks for longer range) were delivered new as VIP aircraft to the Presidential flight. Appropriately, they wore custom registrations – 3X-PDG and 3X-PRG (standing for *Président de [la République de] Guinée*). Eventually they, and the rest of the Avia-14s, were returned (in the case of the VIP examples, sold back) to Czechoslovakia.

Air Guinée introduced the Il'yushin IL-18 *Coot* medium-haul airliner in 1960 with several

Known Guinean military or paramilitary aircraft of Soviet origin

Type	Serial/registration	C/n	F/n	Notes
An-12BP (+)	3X-GBA	02348008		Air Guinée livery
An-12BP (+)	3X-GBB	02348009		Air Guinée livery
An-12BP (+)	3X-GBC	7345001		Ex-Aeroflot Soviet Airlines CCCP-11030? Air Guinée livery
An-12BP (+)	3X-GBD	02348301		Air Guinée livery
Super Avia-14S	3X-PDG	014105	201	Ex-OK-OZC. Became, see next line
	3X-GAE			Sold, to Czechoslovak AF '4105'
Super Avia-14S	3X-PRG	015101	203	Ex-OK-OZB. Became, see next line
	3X-GPA			Sold, to Czechoslovak AF '5101'
IL-18D	3X-GAT	188010604		Ex-Chosonminhang (North Korean Airlines) '834', D/D 8-8-1968; named 'Conakry'. WFU Conakry ?-1986
IL-18D	3X-GAX	187009803		Ex-Yugoslav Government Flight IL-18D 'Salon' YU-AIB, bought 3-1978. Crashed Conakry 3-9-1978
IL-18D	3X-GOC	?		Leased from Aeroflot?, dates unknown
IL-18D	3X-GOD	186009601		Leased from Aeroflot 1968-69, ex/to CCCP-75437
IL-18D	3X-GOE	?		Leased from Aeroflot?, dates unknown
IL-18D 'Salon'	3X-GOF	187010103		Leased from Aeroflot (government flight) 3-1968 to 8-1968, ex/to CCCP-75453
IL-18V	3X-KKN	181003703		Out-of-sequence registration (KKN for City of Kankan. D/D ?-1961; became, see next line
	3X-GAB			Crashed Casablanca 9-7-1967
IL-18V	3X-LBE	181003705		Out-of-sequence registration (LBE for City of Labé). D/D ?-1961; became, see next line
	3X-GAA			Returned and WFU Moscow-Sheremet'yevo by 10-77; refurbished and delivered to Aeroflot Moscow Territorial CAD/Domodedovo UAD/212th Flight) as CCCP-74299 by 4-1979
IL-18V	3X-NZE	181003704		Out-of-sequence registration (NZE for City of N'zerekore). D/D ?-61; became, see next line
	3X-GAC			Returned, to Aeroflot as CCCP-75428
IL-18V	75800	182004205?		Leased from Aeroflot 1963-1964, ex/to CCCP-75800
IL-18V	75803	182004304		Leased from Aeroflot 1963, ex/to CCCP-75803
IL-18V	75841	182005302		Leased from Aeroflot 1965, ex/to CCCP-75841
UTI-MiG-15	121	?		
MiG-17AS	513 Red	?		Dark green camouflage
MiG-21*bis*	302 Red			Green/dark green/brown camouflage
MiG-21*bis*	662 Red			Sand/green camouflage
MiG-21*bis*	361 Red?			
MiG-21UM	120 Red			Green/dark green/brown camouflage. Crashed 2007
MiG-21UM	250 Red?			Sand/green camouflage; possibly serialled 256/258/259 Red
Mi-6	3X-GPD	?		
Mi-8S	3X-GDF	?	?	Government flight; reported in error as Mi-17
Mi-8S	3X-GDI	?	?	Air Guinée livery
Mi-8S	3X-GDT	?	?	Government flight
Mi-8S	3X-GVB	?	?	Government flight; delivered not later than 1975; reported in error as Mi-17
Mi-17-1V	FAG-1009	?	?	Green/brown camouflage
Mi-24V	FAG-1003	?	?	ATGM guidance equipment removed

Guinea-Bissau Air Force MiG-21MF '75 Red' in a hangar at Bissau-Bissalanca. The aircraft had been disused for years when the picture was taken.

aircraft leased from Aeroflot Soviet Airlines pending delivery of its own *Coots*. Though mostly used on scheduled international flights and Hadj tours from Conakry, occasionally the airliners carried President Sésé Sekou Touré on his trips abroad. The transport squadron of the Guinean Air Force reportedly also used the Air Guinée IL-18s from time to time; 75803 and 75841 were reported as Air Force operated in 1982. According to the 28th November/4th December 2000 issue of *Flight International*, two IL-18s were still operational in Guinea then!

A number of Mil' Mi-8MTV-1 *Hip-H* utility helicopters was delivered to the Guinean Air Force. Additionally, the government flight operated four civil-registered Mi-8S *Hip-C* VIP helicopters, one of which wore Air Guinée colours. A single Mil' Mi-6 *Hook* heavy-lift helicopter was placed on the Guinean register and apparently nominally operated by Air Guinée; yet, such a machine would obviously have military applications as well. The Mil' Mi-24V *Hind-E* attack helicopter was supplied to Guinea in post-Soviet times, a single example having been seen in August 1999.

Little is known about the operations of Guinean Air Force aircraft. It is known that in April 1971 Guinean MiG-17s overflew Freetown, the capital of neighbouring Sierra Leone, during unrest in the city; in September that year they took part in the Guinean military intervention in Sierra Leone, and two of the *Frescos* crashed, killing the pilots. In late 2000 the tables were turned – Guinea came under attack from the Rally of Democratic Forces of Guinea (RFDG – *Rassemblement des Forces Democratiques de Guinée*) and Revolutionary United Front (RUF) militias based in Sierra Leone and Liberia. In January 2001 two Guinean MiG-21s supported the Guinean Army's counterattack on the RFDG guerrillas that had seized the town of Guekedou in southern Guinea.

Guinea-Bissau

After achieving independence on 24th September 1973, the former colony of Portuguese Guinea became the Republic of

A Guinea-Bissau Air Force Mi-8TV.

Another view of the same machine, showing the lack of a serial.

Known Guinea-Bissau Air Force aircraft of Soviet origin			
Type	**Serial/registration**	**C/n**	**Notes**
An-2	09	?	
IL-18V	RA-75431	180002003	Leased from M. M. Gromov Flight Research Institute, 1992
IL-18V	RA-75851	182005501	Leased from M. M. Gromov Flight Research Institute, 1992
UTI-MiG-15	34 Red	?	Natural metal finish. Wfu Bissau-Bissalanca by 1991
UTI-MiG-15	?	022528	Czechoslovak-built (Aero CS-102); ex-East German AF '114 Black'
MiG-17F	42 Red	?	1975
MiG-17F	43 Red	?	Overall turquoise finish. Wfu Bissau-Bissalanca by 1991
MiG-17F	91 Red	?	Overall turquoise finish. Wfu Bissau-Bissalanca by 1991
Lim-5	?	1C 0608	Ex-East German Air Force '444 Red'
Lim-5	?	1C 0727	Ex-East German Air Force '600 Red'
Lim-5	?	1C 0717	Ex-East German Air Force '700 Red'
MiG-21MF	31 Red?	?	Wfu Bissau-Bissalanca by 1991
MiG-21MF	75 Red	?	Green/dark green/brown camouflage. Wfu Bissau-Bissalanca by 1991

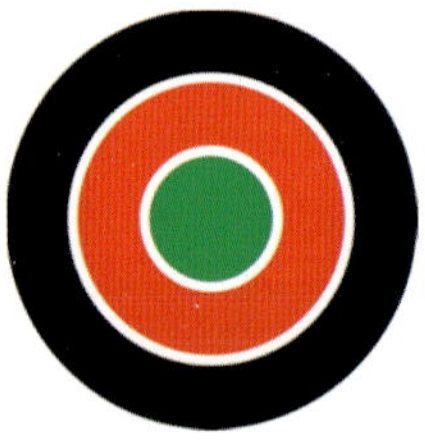

Guinea-Bissau. Shortly afterwards the Guinea-Bissau Air Force (*Força Aérea da Guinea-Bissau*) was established with Soviet assistance. Its small fleet included a squadron of Mikoyan/Gurevich MiG-17F *Fresco-C* fighters based at Bissalanca. The first three of these were delivered in 1973; two years later the fleet had grown to at least five. Three PZL Lim-5s (the Polish licence-built version of the MiG-17F) were purchased from East Germany in 1980; the rest may have been loaned from the Guinea Air Force (see above). According to *Flight International,* however, only five MiG-17Fs were ever delivered. At least two of these fighters were lost in crashes; the remaining three were still in service in late 1991.

For lead-in and proficiency training, two Mikoyan/Gurevich UTI-MiG-15 *Midget* trainers were delivered in 1973 and augmented by a third (ex-East German Air Force) example in 1980. The type was retired in 1986.

Reports suggest that the MiG-17Fs were replaced by six Mikoyan MiG-21MF *Fishbed-J* fighters (alternatively reported as MiG-21*bis* *Fishbed-Ls*) and at least one MiG-21UM *Mongol-B* trainer delivered from the USSR in the second half of the1980s.

The transport element of the Guinea-Bissau Air Force included an unknown number of Antonov An-2 *Colt* utility biplanes. Two Il'yushin

5Y-EDM, one of the four Mi-17s operated by the Kenyan Police. This one is a VIP-configured Mi-172.

Known Kenya Air Force and Kenyan Police aircraft of Russian origin				
Type	**Serial/registration**	**C/n**	**F/n**	**Notes**
Mi-172	5Y-EDM	404M01	?	Police. D/D 1999
Mi-17	5Y-UKW	404M02	?	Police. D/D 1999. Crashed Kapsabet, Kenya, 11-5-2009
Mi-17	5Y-STA	404M03	?	Police
Mi-17	5Y-SFA	404M04	?	Police
Mi-171	KAF-1101	594896*****	?	
Mi-171	KAF-1102	594896*****	?	
Mi-171	KAF-1103	594896*****	?	Crashed near Juja, Kenya, 4-4-2011

IL-18V *Coot* airliners were wet-leased from the Mikhail M. Gromov Flight Research Institute (LII) in Zhukovskiy, Russia, in the autumn of 1992. The lessor was some government organisation, judging by the fact that both aircraft wore 'Republica da Guinée Bissau' titles. Both were returned on 16th December 1992.

The Mil' Mi-8 was the only Soviet helicopter delivered to Guinea-Bissau. One Guinea-Bissau Air Force *Hip* lacking a serial was seen in-country in 1985, and another reportedly crashed on 25th February 1988.

Lesotho

The Lesotho Police operates a pair of Mil' Mi-2 *Hoplite* light helicopters which were donated by the Libyan Arab Air Force around 1987. Judging by the c/n commencing 56…, at least one of the Mi-2s was built in an armed assault version (Mi-2PPK or something of the sort).

Additionally, an airline called Anton Air International operated four Antonov An-12 *Cub* freighters, which may have had some military applications.

The Lesotho Air Force roundel.

Known Lesotho paramilitary aircraft of Soviet origin			
Version	**Registration**	**C/n**	**Notes**
An-12B (+)	7P-ANA	2340709	Demilitarised. Ex-RA-11038, leased from Aviakompaniya Pilot ?-1997
An-12BP (+)	7P-ANB	4342404	Ex-RA-11760, leased from Aviakompaniya Pilot ?-1997. To Trans Air Congo as TN-AGH
An-12BP	7P-ANC	9346608	Ex-RA-11658, leased from Aviakompaniya Pilot ?-1997
An-12…	TN-AGF	?	Operated in 1998
Mi-2	PMU-7	567033031	Police Mobile Unit; ex-Libyan Arab Air Force ('7033'?)
Mi-2	PMU-8	?	Police Mobile Unit

Kenya

The only Soviet type militarily operated by Kenya is the Mil' Mi-17 *Hip-H* medium helicopter. In 1999 the Kenya Police Air Wing took delivery of two Kazan' Helicopters-built Mi-8s bought new from Russia – a Mi-172 VIP helicopter and a regular Mi-17. Both were retrofitted with loud-hailers under the tailboom and an optoelectronic observation system under the nose. In 2001 they were augmented by two more similarly equipped Mi-17s. The choppers operated from Nairobi-Wilson airport and wore a distinctive blue/white colour scheme – initially anonymous, later with 'Kenya Police' titles.

In 2010 the Kenya Air Force, too, took delivery of three new Mil' Mi-171s built by the Ulan-Ude Aircraft Production Association.

Liberia

Liberia does not have a proper air force, but the Liberian armed forces include an Air Reconnaissance Unit (ARU) based at Monrovia's Roberts International airport (formerly James Spriggs Payne airport). Among other things, it operated two Mil' Mi-17 *Hip-H* helicopters, one of which (serialled ATU-003, with a wrong prefix) was withdrawn from use by October 2003.

The ARU reportedly also had three Mil' Mi-2 *Hoplite* light helicopters serialled LAF-05, 001 and 002. However, the latter two were police helicopters and thus would more logically be assigned to the co-located Justice Air Wing (and have serials prefixed JAW-).

Libya

Libya was an Italian colony until 1943 when Allied troops ousted the Italians and Germans. It was then jointly administered by Great Britain and France (each covering a separate area) on behalf of the United Nations until 24th December 1951 when it became the independent United Kingdom of Libya. On 1st September 1969 King Idris I was deposed in a military coup d'état led by Col. Muammar Abu al-Qaddafi and the Libyan Arab Republic was created, Qaddafi becoming 'Brother, Leader and Guide of the Revolution' – in short, head of state, and all Libyan assets were nationalised. Accordingly the former Royal Libyan Air Force was quickly transformed into the Libyan Arab Air Force (LAAF, or *al Quwwat al-Jawwiya al-Arabiya al-Libiya*; often rendered as the Libyan Arab Republic Air Force or LARAF). Hence Great Britain and the USA had to hastily evacuate the military bases they had established in northern Libya in the 1950s. In 1977, after a reform of the organs of government, the country's official name was changed to the Libyan Socialist People's Arab Jamahiria (the latter word is Arabic for 'power of the masses').

Of course, the ultra-revolutionary colonel immediately found favour with the Soviet government, which saw the September revolution as an opportunity to gain another foothold in Africa. The Soviet Union and Libya concluded an economic co-operation agreement in 1972 and a trade agreement in 1974, followed up in 1985 by a long-term economic co-operation programme. This inevitably meant the supply of weapons to the Libyan Armed Forces in general and the LARAF in particular. In due course Libya became one of the largest recipients of Soviet military hardware in general, and in Africa in particular.

Deliveries of combat aircraft to the LAAF began in the 1970s. These were largely state-of-the-art types, but a few obsolescent Mikoyan/Gurevich UTI-MiG-15 *Midget* trainers were also supplied. They were used in order to reduce wear and tear on the more advanced and expensive combat trainers. By 1987 the UTI-MiG-15s had been withdrawn.

The Libyan Arab Air Force roundel and fin flash used in 1977-2011.

MiG-21UM '113' was among the Libyan Arab Air Force aircraft seen at the LAVEX-2007 airshow at Tripoli-Mitiga (Umm Aitiqah). Note the position of the roundel on the nose.

Here the same aircraft taxies at Tripoli-Mitiga, carrying a drop tank.

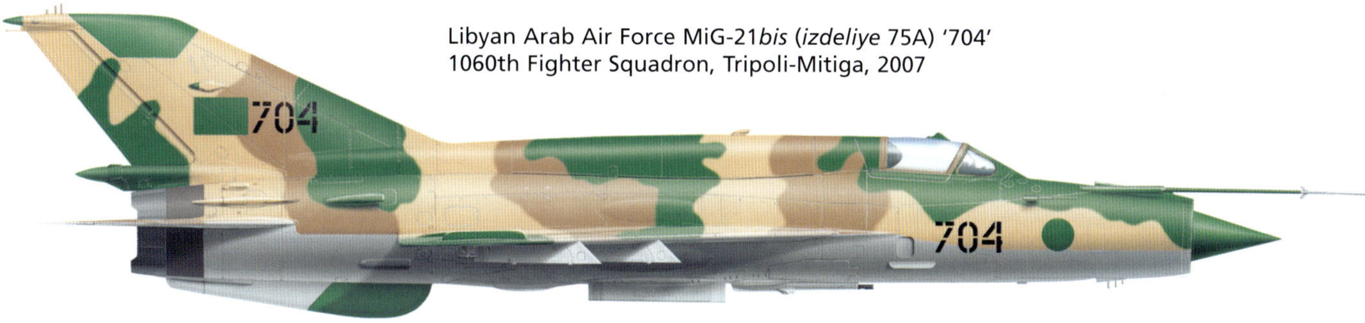

Libyan Arab Air Force MiG-21*bis* (*izdeliye* 75A) '704'
1060th Fighter Squadron, Tripoli-Mitiga, 2007

Around 1975 the LAAF took delivery of 25 Mikoyan MiG-21UM (*izdeliye* 69) *Mongol-B* trainers, followed by 50 MiG-21MF (*izdeliye* 96F) *Fishbed-J* fighters; these figures are disputed as being too high by some sources. The Libyan top command did not rate the MiG-21 too highly because of its short range and low warload, preferring the Dassault Mirage IIID. Hence the MiG-21MF and MiG-21UM were not used operationally; instead, their mission was to train the so-called 'Force 14' – the cadre of the Air Force of the future Palestinian State which, according to the plans of the Arab strategists, would be formed once Israel had been occupied. (Sure. Dream on.) Both types were subsequently used by the LAAF but 30 of the MiG-21MFs were transferred to Syria in 1982 to make up for the heavy losses in the battles with Israel in the Beka'a Valley.

Starting about 1980, Libya received 94 (?) MiG-21*bis* (*izdeliye* 75A) *Fishbed-L* fighters; these equipped three squadrons of the 1060th Fighter Regiment based at Gamal Abdel Nasser AB at El Adem near Tobruk, leaving an

Freshly overhauled MiG-21*bis* '704' was also among the LAAF aircraft seen at LAVEX-2007.

MiG-21*bis* '800' was captured by the anti-Qaddafi rebels at Benghazi and included into the Free Libyan Air Force.

Libyan Arab Air Force MiG-23UB '8425' 1023rd Interceptor Squadron, Tripoli-Mitiga, 2009

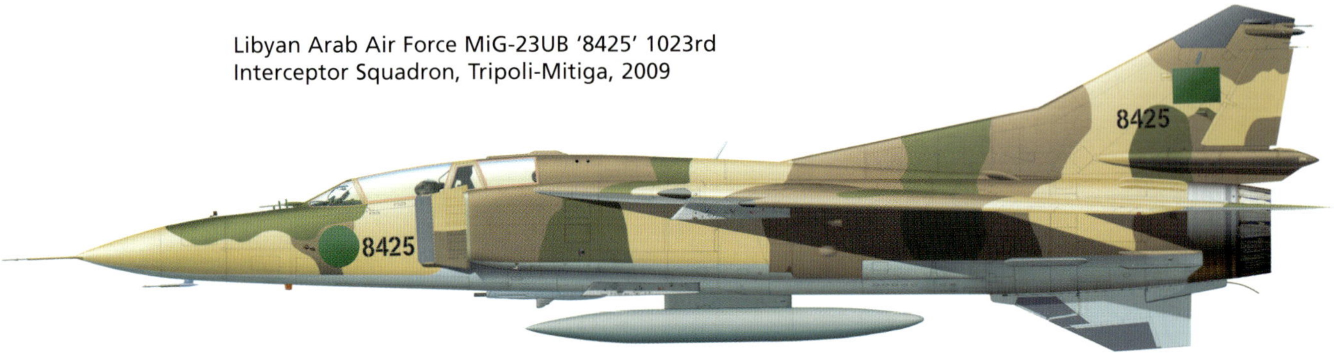

adequate number in storage for spares. An Israeli intelligence report of 3rd April 2006 stated that Libya had 33 operational *Fishbed-Ls* plus 25 in storage at that time. As late as 2007 the MiG-21*bis* and MiG-21UM were still active with at least one LAAF squadron, No.1021 at El Adem (Gamal Abdel Nasser AB); some reports suggest one more unit at Okba Ben Nafi.

The Libyan Arab Air Force was one of the first customers for the Mikoyan MiG-23MS *Flogger-E* 'swing-wing' fighter. The Libyan pilots took their training for the type in the Soviet Union (at the Krasnodar Military Pilot

LAAF MiG-23UB '8423' and MiG-23MLAE-2 '6117' make a spectacular flypast with the wings set 45° at the LAVEX-2009 air-show. The trainer was lost in a fatal crash on 8th October 2009 during the show.

Another Llibyan MiG-23MLAE-2, '7102', at rest.

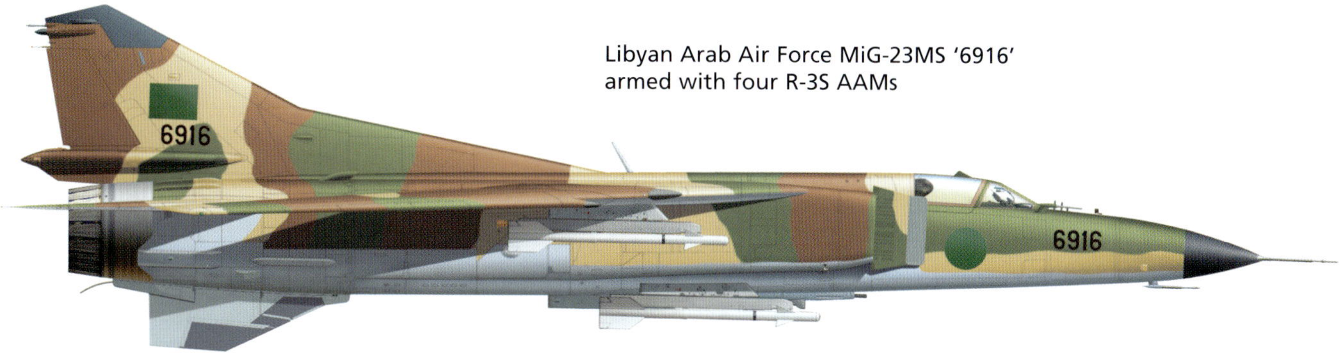

Libyan Arab Air Force MiG-23MS '6916' armed with four R-3S AAMs

Libyan Arab Air Force MiG-23MF '0206' armed with two R-23 AAMs

Libyan Arab Air Force MiG-23UB '8272' delivered in 1982

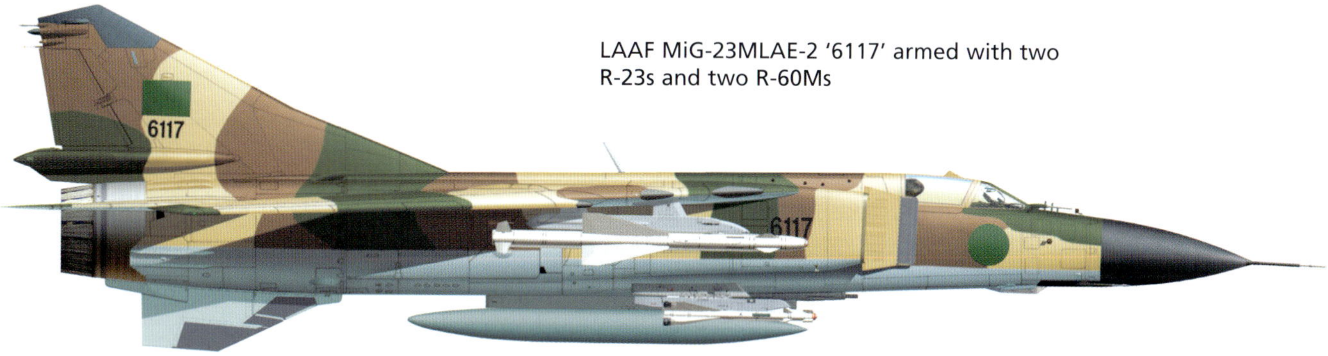

LAAF MiG-23MLAE-2 '6117' armed with two R-23s and two R-60Ms

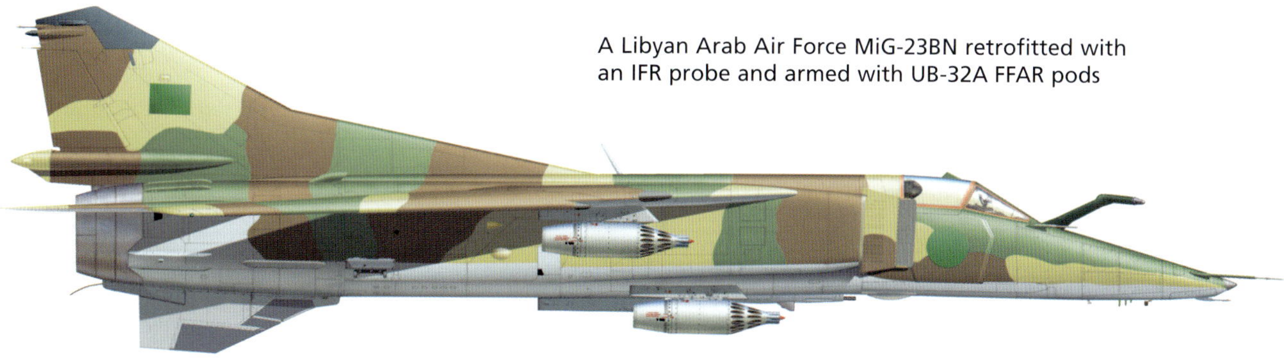

A Libyan Arab Air Force MiG-23BN retrofitted with an IFR probe and armed with UB-32A FFAR pods

MiG-23MLAE-2 '6472' after being taken over by the Free Libyan Air Force

A Libyan soldier examines the tail section of a FLAF MiG-23MLAE-2 ('6130') shot down over Benghazi on 19th March 2011.

The Free Libya Air Force roundel and fin flash

College, which specialised in coaching foreign trainees). Deliveries began in 1974, and by the end of 1975 eighty of these fighters plus twenty MiG-23UB *Flogger-C* trainers had been received; other reports suggest the total number was only 54). In similar manner to the MiG-21, some twenty MiG-23MSs were donated to Syria in 1982 as combat attrition replacements.

In the 1980s, it was reported that the LARAF had received as many as 200 MiG-23MF *Flogger-B* interceptors, making it the largest operator of the type outside the Soviet Union. However, this is obviously an error; while the figure is probably correct, it encompasses *all MiG-23 variants delivered* – or at least the fighter and trainer versions.

The most advanced fighter variant supplied to Libya was the MiG-23MLAE *Flogger-G* (not the MiG-23MLD *Flogger-K* as often reported, since the Libyan examples have the short dorsal fin but lack the MiG-23MLD's extra dog-tooth on the wing glove fairings), 48 of these have been delivered. For years the general belief was that the Libyan *Flogger-Gs* were equipped with N003E Ametist (Amethyst) fire control radars; not until 2009 did a close look at them reveal that they, in fact, have the more advanced N008E radar and thus are actually MiG-23MLAE-2s (*izdeliye* 23-22B). It is not entirely clear, though, whether they had this radar since new or had been upgraded. Fact is, several LAAF MiG-23MLAEs have been upgraded and refurbished by the Russian Air Force's Aircraft Repair Plant No.121 at Kubinka AB near Moscow in the 2000s (they have been described as 'upgraded to MiG-23MLD standard, with the exception of the automatic wing sweep adjustment based on flight speed – the pilot still has to select the wing sweep

Its paintwork faded and its serial obliterated, this 'war weary' LAAF MiG-23BN retrofitted with an IFR probe from a Mirage F.1 was non-operational by the outbreak of the Libyan Civil War and did not participate in the fighting.

An excellent shot of an immaculate Libyan MiG-25PD ('6716') intercepted over the Mediterranean. The aircraft carries two R-40TD and four R-60M AAMs.

This view shows that '6716' had been retrofitted with IRCM flare launchers in over-wing strakes akin to those of the MiG-23MLD.

manually'). Speaking of which, Libyan MiG-23s have also been refurbished in the Ukraine by Odesaviaremsevis in Odessa and the Chugooyev Aircraft Repair Plant.

Conversion and proficiency training was performed on MiG-23UB *Flogger-Cs*; 40 such aircraft were delivered. On 7th October 2009 one of the MiG-23UBs crashed inexplicably during a demonstration flight at the LAVEX-2009 airshow at Tripoli-Mitiga (also rendered as Umm Aitiqah, aka Okba ben Nafi AB) about 5 km (3.1 miles) east of the capital. As the aircraft made an aileron roll – possibly at too low speed, the nose dropped and the jet entered a vertical dive, coming down in the Suk al-Jowhoa residential area just outside the base. The pilots, Col. Yahya and Col. Najeeb Ashour Hamdey, did not eject and were killed. Technical problems were ruled out; 1023 Sqn CO Col. Buknass had flown the same aircraft a day earlier and said it was fully serviceable.

The MiG-23s wore a sand/green/brown camouflage scheme. To confuse the casual observer, some of the MiG-23MSs initially wore the same red/white/black roundels and fin flashes as their Egyptian sister ships (Egypt and Libya formed a federation in 1972-77), the only difference being in the serials. In 1977, when the Jamahiria was introduced, the 'pan-Arabic' insignia gave place to the instantly recognisable solid green roundels and fin flashes.

By 2004, seven LAAF interceptor squadrons were equipped with MiG-23 variants but many aircraft were not operational. The MiG-23MS served with the Air Defence Command's 1023rd Interceptor Squadron at Tripoli-Terrabalus, which later moved to Tripoli-Mitiga, and the 1050th Interceptor Squadron at Al Bumbah-North AB. There are also reports that they were operated by the 1060th Sqn at Benghazi (Benina AB), 1090th Sqn at Misurata

Libyan Arab Air Force MiG-25PD '6716' armed with R-40TD and R-60 AAMs; note the flare launchers

LAAF MiG-25RBK '499' adorned with nose art reflecting its reconnaissance role

and an unknown squadron at El Beida (Labraq AB). The MiG-23MFs were assigned to the 1023rd Sqn at Tripoli-Mitiga and the 1090th Sqn at Misurata (Tripoli Air Defence Sector), two squadrons (including the 1040th Interceptor Squadron) at Benghazi/Benina AB (Benghazi Air Defence Sector) and three squadrons (including the 1060th Interceptor Squadron) at El Adem/Gamal Abdel Nasser AB (Tobruk Air Defence Sector). The MiG-23MLAE saw service with the 1023rd Sqn, 1040th Sqn (which had moved to Benghazi in the meantime), 1050th Sqn, 1060th Sqn and 1090th Sqn. The MiG-23UB trainers were distributed

Libyan Arab Air Force MiG-25PD '7811'
Tripoli-Mitiga, 2009

Libyan Arab Air Force MiG-25PU '207'
Tripoli-Mitiga, 2009

between the interceptor and fighter-bomber squadrons.

In the late 1970s, when the Mikoyan MiG-25 was cleared for export, the LAAF received a sizeable number of MiG-25PD *Foxbat-D* interceptors, MiG-25RBK *Foxbat-B* reconnaissance/strike aircraft and MiG-25PU *Foxbat-C* trainers. The aircraft were assigned to the Air Defence Command, serving with the 1025th Sqn at El Jufrah-Khun AB, 1055th Sqn at Sirte (Ghurdabiya AB) and an unidentified squadron at Sabqa. Interestingly, at least two of the MiG-25PDs were retrofitted with tandem IRCM flare launchers in tall fairings supplanting the standard inboard boundary layer fences (in line with the inboard missile pylons).

The LAAF had a strong requirement for strike aircraft as well. Among other things, it placed an order for 50 Mikoyan MiG-23BN *Flogger-H* fighter-bombers; the first of these were shipped to Libya in the summer of 1974 and the last of the 48 eventually delivered arrived in 1976. The MiG-23BNs served with the 1124th Fighter-Bomber Squadron at Tripoli-Terrabalus (later at Tripoli-Mitiga) and two unidentified fighter-bomber squadrons at Labraq and al-Bumbah. Following the example of the Iraqis, the Libyans have retrofitted at least one MiG-23BN with a fixed in-flight refuelling probe of the type fitted as standard to the Dassault Mirage F.1, allowing it to work with the Algerian Air Force's IL-78E tanker. The work was performed by the German company Intec Technical Trade & Logistic GmbH, which also modified several LAAF Lockheed C-130H Hercules transports to tanker/transport configuration. It is not known how many MiG-23BNs were eventually modified.

Part of the MiG-23 fleet was destroyed on the ground by a Chadian attack on the Libyan

Opposite page: Looking rather weathered and worn, LAAF MiG-25PD '7708' is seen in storage at Tripoli-Mitiga in company with several Aero L-39ZA Albatros trainers.

Opposite page, below: MiG-25RBK '499' is shown at the same location – and in similar condition. The nose art was on the port side only.

Surrounded by Albatrosses and Let L-410UVP-T transports, LAAF MiG-25PU '207' is shown at Mitiga. Two more *Foxbats* – MiG-25PDs '994' and '7811' – are visible beyond.

Front view of a Libyan Arab Air Force Su-22 carrying four PTB-1150 drop tanks.

base Maaten-as-Sarah (aka El Khufra AB) during the Libyan involvement in the Chadian civil war (see below) and by a US air raid on Tripoli in April 1986. It is not known how many MiG-23BNs, if any, were still airworthy in 2004; none were flyable by 2009 anyway.

Mikoyan was not the only supplier of strike aircraft to Libya. The LAAF reportedly took delivery of more than 100 Sukhoi Su-20 *Fitter-C*, Su-22 *Fitter-F* and Su-22M *Fitter-H* fighter-bombers and Su-22UM3 *Fitter-G* trainers (the exact proportion is unknown). 80 were reportedly still on the inventory in 2004; other sources state a more modest figure of about 50 aircraft in 2000. The fighter-bombers were operated by the 1032nd Sqn at Tripoli-Mitiga

Libyan Arab Air Force Su-22 '2514' with R-3S missiles; in the 1970s the LAAF serials were in Arabic characters

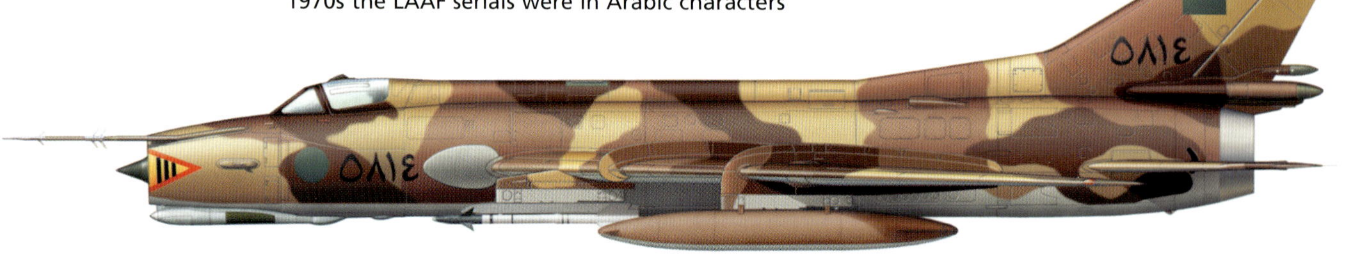

(Okba ben Nafi AB), the 1022nd Sqn at Sirte (Ghurdabiya AB) and a further squadron at Ghadames. However, by 1998 most of these aircraft were probably unserviceable because spares supplies were cut off in 1992 due to UN sanctions. The lifting of the sanctions in 1999 made it possible to refurbish some of the *Fitters*; in 2009 the Su-22M3s were still very much alive, participating in the LAVEX-2009 airshow at Tripoli-Mitiga in October.

LARAF Su-22s wore a tan/green 'Middle Eastern' camouflage and originally had serials in Arabic characters. These were later replaced by three-digit serials in European numerals; unlike most LAAF types, the serials are in high-visibility white.

Another important strike asset was the Sukhoi Su-24MK *Fencer-D* tactical bomber, for which Libya became the second customer. In 1989, six Su-24MKs were dismantled and delivered by Soviet Air Force Antonov An-22 transports to Tripoli-Mitiga, where they were reassembled by Soviet personnel. A further nine were subsequently delivered, bringing the *Fencer* fleet to 15. Further sales were halted when Libya failed to complete payment, citing a drastic fall in oil income as the reason.

The bombers saw service with the 1124th Sqn at Sirte (Ghurdabiya AB) and wore the *Fencer's* standard export camouflage scheme of two shades of brown on the upper surfaces and pale blue undersides. In early 2004 twelve Su-24MKs reportedly remained in service.

In 1974 Libya requested the delivery of the Tu-22 supersonic heavy bomber. Negotiations on the deal continued until 1976, and training of Libyan personnel at Zyabrovka AB in Belorussia began that same year. Despite reports of as many as 24 aircraft being delivered, in 1977 the LAAF received no more than nine – eight Tu-22B *Blinder-A* bombers and a single Tu-22U *Blinder-C* trainer. The aircraft, which were sourced from several Soviet Air Force regiments, were six to eight years old,

This Libyan Su-22M was stranded at ARZ No.558 in Baranovichi, its refurbishment being delayed due to UN sanctions. The serial '74 Yellow' is temporary, not the actual LARAF serial.

Another Libyan 'expatriate' at Baranovichi – Su-22M3 '109 Yellow' with PTB-1150 drop tanks.

LAAF Su-22UM3 '109 Yellow' on overhaul at Baranovichi

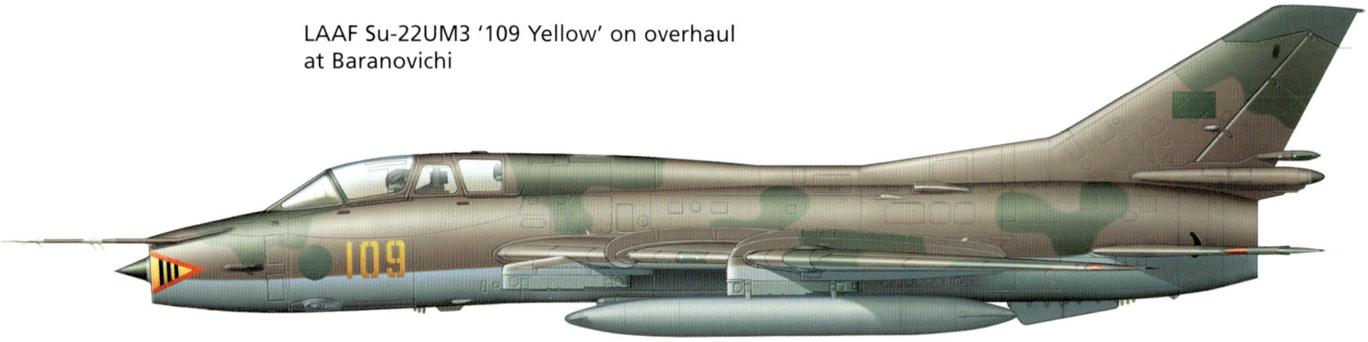

with 100-200 hours' total time since new; they were refurbished and converted to Tu-22B standard (this means stripping them of nuclear capability) by the Kazan' aircraft factory which had built them.

The Libyan Tu-22s wore tan/green camouflage and originally sported the 'pan-Arabic' roundels and fin flashes which were replaced by the green ones after 1977. Unlike other LARAF aircraft, the *Blinders* apparently wore

Libyan Arab Air Force Su-22M '307 White' at Tripoli-Mitiga during the LAVEX-2009 airshow. The Su-22Ms carry a squadron badge below the fin flash.

The pilot of '307 White' prepares to start the engine for a demo flight at the show.

Libyan Arab Air Force Su-22M '318 White', 1032 Sqn, Tripoli-Mitiga AB, 2009

no serials; at least, none have been reliably identified.

Originally the Tu-22s were based at Tripoli-Mitiga; however, the airport is located right on the coast of the Mediterranean, and soon the bombers started suffering from corrosion in the humid and salty air. Therefore the Libyans commissioned the Soviet Union to build a new base for the *Blinders* at El Jufrah oasis in Fezzan Province, about 1,000 km (621 miles)

Two views of Su-22M '318 White', which was also present at LAVEX-2009 and showed an obviously fresher finish. White serials were rarely seen on Libyan combat aircraft.

Above: A pair of Su-22Ms performing at the LAVEX-2009 airshow. Note the sun-bleached paintwork on '311 White'.

Below: LAAF Su-24MKs also took part in the demonstration flights at LAVEX-2009. Here, '37 Black' displays its colour scheme in a banked turn. Note the badge of No.1124 Sqn on the fin.

Bottom: '36 Black', another 1124 Sqn Su-24MK, with the wings at 45°. Note the difference in camouflage patterns between individual aircraft.

south of Tripoli; the dry desert climate would eliminate the corrosion problem. The Soviet specialists did an excellent job, designing and constructing an excellent base that outclassed even the ones at home – because the Libyan government gave all the money that was needed. El Jufrah-Khun AB was equipped with excellent maintenance facilities and excellent accommodation for the personnel, which Soviet airbases could not boast. In particular, the four-man chalets for the tech staff were equipped with air conditioners – a must, given that in the summer season the daytime temperatures could reach 70°C (158°F).

Soviet technical specialists were stationed full time at El Jufrah to help maintain the bombers and oversee the repairs if necessary – and the necessity often arose, as undershoots and overruns on landing were frequent. Due to the daytime heat the work took place from 5 am to 11 am and from 4 pm to whatever-it-takes-to-get-the-job-done.

Having studied in the Soviet Union, the Libyan pilots spoke fairly good Russian and had an excellent command of Russian bad language. They were not particularly avid flyers; the general attitude seemed to be 'okay, I'll fly if I must'. Accordingly the Tu-22Bs logged only 50-60 hours each in Libyan service, being

Libyan Arab Air Force Su-24MK '35 Black', 1024th Sqn, with PTB-3000 drop tanks on the wing glove pylons

largely a 'big stick' meant for intimidating the neighbours.

One of the Tu-22Bs was wrecked in an off-field forced landing soon after delivery, damaging the landing gear; another was reportedly flown to Iraq. Two more *Blinders* were shot down during the civil war in Chad. The survivors were grounded in the mid-1990s because Russian technical support was cut off in 1992. In the early 2000s, however, one of the Tu-22Bs was briefly returned to 'airworthy' status for the sake of a single flypast over Tripoli on 2nd March (the day of the Libyan Jamahiriya). The inverted commas have been used intentionally; after the aircraft had sat unrefuelled and unattended for eight years the rubber of the fuel tanks had deteriorated, and when the bomber landed it was leaking like a sieve. It is said that the first thing the pilot did after the mission was to kneel and pray ardently, so glad he was to stay alive.

In 2003 the LAAF invited Russian specialists to determine if the six surviving *Blinders* could be restored to active status. The aircraft were in surprisingly good condition; the problem was that spares for the Tu-22 were hard to come by. The factory team did manage to find the necessary spares, but eventually the plan to ship the Tu-22s to Russia for refurbishment via Kaliningrad did not materialise.

In 2004 the Libyan Arab Air Force leased at least one Antonov An-2 *Colt* utility biplane from Russia for search and rescue duties. The aircraft, which is painted white with a yellow tail and yellow cheatline, sports no markings other than the serial.

The LAAF's principal tactical transport is the Antonov An-26 *Curl* turboprop twin. Starting in 1982, at least 29 late-production examples were delivered, serving with the 1226th Sqn at Tripoli-Mitiga. Originally the An-26s wore a green/brown camouflage scheme with the serial writ large on the nose and fin (at the latter location it was often supplemented by the legend 'LAAF'). Later, some of the *Curls* were ostensibly placed on the civil register and operated by Libyan Arab Air Cargo, a commercial division of the LAAF. Initially these aircraft wore a plain all-white livery with crudely applied titles, but subsequently a smart green/white livery was introduced, exploiting the Libyan flag colour of the Qaddafi era. Interestingly, there are considerable variations in colour schemes between individual aircraft – some have a cheatline while others do not. Equally interestingly, the quasi-civil An-26s still carry their serials in small type on the nose. At least three Libyan An-26s have been sold or donated to the air forces of Chad (not later than 2008), Somalia (1990), Sudan (after 1984), Niger (June 1997) and Namibia (two aircraft in 2002). Easy come, easy go?

A quartet of Antonov An-32 *Cline* twin-turboprop tactical transports was acquired in the Ukraine; two of these are second-hand aircraft bought from the Antonov Design Bureau in December 2005, the other two were delivered new in 2006. All four are outfitted as An-32P Firekiller waterbombers with large water tanks low on the centre fuselage sides for fire-fighting, but may be used for transport duties as well. The aircraft are civil-registered and likewise listed in the Libyan Arab Air Cargo fleet but wear a slightly different livery with green and red trim and a distinctive tail logo.

Another Antonov type operated by the LAAF's Transport Command is the An-72 *Coaler-A* twin-turbofan STOL transport. Five such aircraft serve with the 1274th Sqn at Tripoli-Mitiga. For a while, most of them wore the colours of either Libyan Arab Air Cargo or a multi-national airline named Sin-Sad Airlines set up in 2000 as an affiliate of the CEN-SAD organisation (Community of Sahel-Saharan States). Part of the fleet was registered in Liberia; however, a year later there was a change of management at the Liberian Civil Aviation Authority and the registrations were cancelled in May 2001, the An-72s moving to the Equatorial Guinean register.

A closer look at a Libyan Tu-22B wearing pre-1977 red/white/black 'pan-Arabic' roundels. No serial is visible.

Additionally, the demonstrator of the An-74TK-200 *Coaler-B* commercial convertible version was leased from the Khar'kov State Aircraft Production Enterprise, retaining its Ukrainian registration. Unfortunately this aircraft crashed fatally on 23rd April 2006 while flying a humanitarian airlift mission from Tripoli to N'djamena, Chad. The LAAF placed an order for two new-build An-74TK-200S ambulance aircraft, but the outbreak of the Libyan Civil War in February 2011 meant that the jets could not be delivered, and eventually the manufacturer found a new customer for them. Finally, a single An-74TK-300D – a VIP-configured example of the An-74TK-300 version in which the *Coaler*'s distinctive overwing engine nacelles are replaced by conventional underslung ones, no longer making use of the

Below: This Tu-22B wears the new solid green LAAF insignia.

Bottom: Here, two Libyan Tu-22Bs are intercepted by an Italian Air Force Lockheed F-104S.

Coanda effect to boost short-field performance – was delivered in April 2009.

In 1980-85 the LAAF took delivery of 25 factory-fresh Il'yushin IL-76 heavy transports. Five of these were IL-76M *Candid-Bs* built with a tail gunner's station. Initially they wore the full beige/white/black livery of the flag carrier Libyan Arab Airlines; in late 1980, however, two IL-76Ms flying supplies to North Yemen reportedly carried full military markings. The other 20 were commercial IL-76T/IL-76TD *Candid-As* operated by Jamahiria Air Transport (the commercial division of the LAAF) in an all-white livery with only a Libyan flag on the tail and black titles (in Arabic to port and in English to starboard). Later, however, the white IL-76T/TDs also sported Libyan Arab Airlines titles and logos!

By 1991 Jamahiria Air Transport was renamed Libyan Arab Air Cargo, and it is this airline which nominally operates the IL-76s as of this writing. Most of the aircraft belong to the 1276th Sqn at Tripoli-Mitiga; a few are operated by a detachment of the 1276th Sqn at El Jufrah-Khun AB. Some *Candid-As* were seen with 'Libyan Air Cargo' (without the 'Arab') or 'Libyan Arab Air Transport' titles. Later, most of them gained the green/white livery. In the 2000s, three second-hand IL-76TDs built in 1990-91 were acquired as attrition replacements; one of them briefly wore a military serial prefixed LAF in spite of the Libyan Arab Air Cargo livery before adopting a more fitting civil identity.

The LAAF also operates a single Il'yushin IL-78E tanker in Libyan Arab Air Cargo colours. On reflection, this livery is not too surprising, since the *Midas-A* can be reconfigured as a transport by removing the extra fuel tanks and UPAZ-1A hose drum units. This aircraft inherited the registration of an IL-76M destroyed by the USAF in 1986 (5A-DLL).

The most capable transport aircraft operated by the LAAF is the Antonov An-124-100 Ruslan (*Condor*) heavy-lifter. Two such aircraft were bought second-hand in early 2002, making Libya one of only two non-CIS operators of the type. Both aircraft wear the green/white livery of Libyan Air Cargo (modified by adding a pale blue belly and a tail logo) and perform commercial cargo flights, generating revenue, when not called upon to haul military materiel. Among other things, in 2005 one of the two An-124s was jointly operated with the Belgian airline Skylink. Since 2008 both aircraft are operated for FLATAM, a Franco-Libyan joint venture that was the worldwide representative of Libyan Air Cargo, specialising in the carriage of outsize and heavy loads, notably to destinations in Africa and the Middle East. When the Libyan Civil War broke out, both aircraft were apparently stranded outside Libya, thus escaping destruction by NATO air strikes or opposition guerrillas.

VIP transportation duties were performed by three Il'yushin IL-62M 'Salon' *Classic* long-haul airliners acquired second-hand from Russia and assigned to the 1276th Sqn. The aircraft were operated by Sin-Sad Airlines or Trans African Airlines, two of them being registered in Equatorial Guinea and the Central African Republic for a while before being transferred to the Libyan register. One of the IL-62Ms has since been transferred to Libyan Air Cargo, gaining the carrier's full green/white livery. On 29th March 2006 a Libyan IL-62M overran runway 14L at Moscow-Domodedovo airport after landing long and broke into three sections. There were no casualties but the aircraft was a total loss.

As regards rotary-wing types in LAAF service, the smallest dimensionally was probably the largest numerically. No fewer than 94 Mil' Mi-2 *Hoplite* utility helicopters were delivered to Libya in 1979-81. Of these, 54 were in the Mi-2T utility version, 42 were passenger-configured Mi-2Ps, and eight represented an armed assault version (Mi-2PPK or something of the sort). The unit operating them remains unknown (although it's rather surprising that such a large quantity should remain unnoticed!). At least three of the helicopters have been transferred to other nations (two to Ghana and one to Lesotho).

In the early 1980s the LAAF purchased about 40 Mil' Mi-8T *Hip-C* helicopters in unarmed utility configuration. They are in service with the 1328th Sqn at Tripoli-Mitiga and helicopter detachments at Benghazi (Benina AB), Sirte (Ghurdabiya AB) and El Adem (Gamal Abdel Nasser AB). Because of the frequent dust storms the Mi-8Ts have been retrofitted with Mi-17 style air intake dust filters installed via adapters. Two of the Mi-8Ts were transferred to the Namibian Defence Force Air Wing in 2002. Somewhat surprisingly, Libya did not order the more capable 'second-generation' versions of the Mi-8 in large numbers. Only two *Hip-Hs* – a Kazan'-built Mi-8MTV-1 and an Ulan-Ude-built Mi-171 – have been identified to date; both are operated by the Libyan Air Ambulance.

Libya was one of the first foreign customers for the Mil' Mi-24 attack helicopter,

A Libyan Arab Air Force Tu-22B in pre-1977 markings

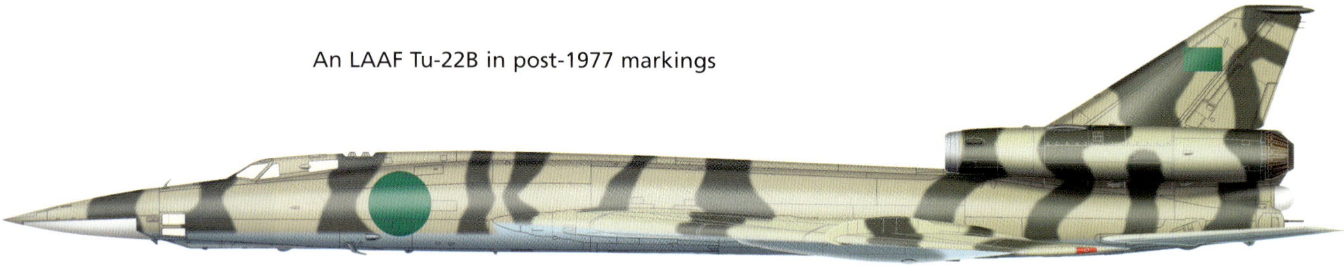

An LAAF Tu-22B in post-1977 markings

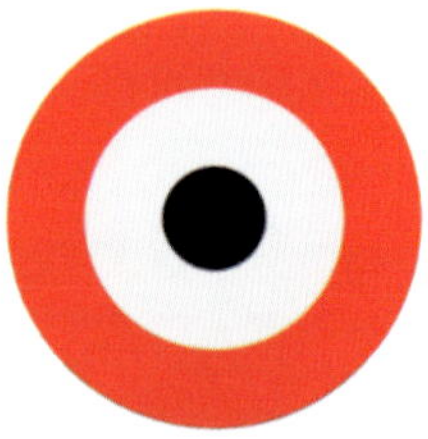

The pre-1977 Libyan Arab Air Force roundel and fin flash

starting with the Mi-24A *Hind-A* in the late 1970s, which was followed by the Mi-25 *Hind-D* and the Mi-35 *Hind-E* in the 1980s. According to Russian sources, the LAAF had 65 Mi-24As and Mi-25s on strength in early 1995, some of which were based at Ras-Lanouf, Bombah and Misurata (the latter two bases hosted flying schools with Polish instructors). By 1998 most of these helicopters were probably unserviceable because spares supplies were cut off in 1992 due to UN sanctions. However, once the sanctions had been lifted on 5th April 1999, Russia and the Ukraine began rebuilding military co-operation with Libya. This included 'reanimation of dead crocodiles' – a number of Libyan Mi-35s were

This photo illustrates the camouflage colours worn by the Libyan Tu-22s.

A Libyan Tu-22B is intercepted by F-4B-22-MC BuNo 152247/ 'NM-103', a VF-51 aircraft from USS *Franklin D. Roosevelt*.

refurbished in the Ukraine by the Aviakon repair facility in Konotop.

Having a long coastline and seaports on the Mediterranean, Libya needed anti-submarine warfare assets. Hence the Libyan Navy bought a batch of 30 Mil' Mi-14PL *Haze-A* amphibious ASW helicopters, becoming the largest foreign customer for this highly specialised derivative of the Mi-8. The helicopters are operated by the 1314th Sqn at Tripoli-Mitiga and an unidentified squadron at Misurata. Two of them have been converted to Mi-14PZh civil fire-fighting configuration by the Sevastopol' Aircraft Repair Plant in the Ukraine, wearing a red/blue colour scheme and the same logo as the An-32Ps. A long hose with a wire mesh filter at the end can be lowered from the rear of the fuselage pod when the helicopter is at the hover, allowing the Mi-14PZh to refill its water tank and make another drop. After the Libyan Civil War, however, one of the two waterbombers has been further modified for aerial ambulance work.

It may be mentioned here that in February 2010 Libya ordered a Russian arms and services package worth US$ 1.3 billion. The package included 12 Sukhoi Su-35S 'Generation 4++' single-seat multi-role fighters (the latest development of the Su-27 *Flanker* family), four Sukhoi Su-30MK2 *Flanker-F* two-seat multi-role fighters and six Yakovlev Yak-130 *Mitten* advanced trainers. However, the fall of the Qaddafi regime in October 2011 means the deal is off – the new government (the National Transitional Council) aided and abetted by the western world is surely not going to honour any commitments made by the enemy, especially the defeated enemy.

Originally Libyan An-26s wore camouflage colours with proper LAAF insignia and titles, illustrated here by 8201.

The LAAF in action

It seems that Muammar Qaddafi had been on a warlike footing ever since he seized power in the September revolution of 1969. Since then Libya has waged war against its neighbours on a regular basis, and of course the LAAF was heavily involved in the military campaigns.

In July 1977, outraged by Egyptian President Anwar Sadat's visit to Israel in order to discuss terms for a détente, thousands of Libyan protesters instituted a 'March on Cairo' and Colonel Muammar Qaddafi ordered an attack on Egyptian border posts. In response, an Egyptian armoured convoy headed for the frontier and on 21st July 1977 Egyptian Air Force (EAF) Su-20 fighter-bombers escorted by MiG-21MFs attacked Libyan radar stations and army bases. Shortly afterwards that day, similar formations raided the LAAF bases at El Adem (Gamal Abdel Nasser AB) and Al-Kurta, destroying seven Libyan fighters before they could take off. Libya retaliated by an offensive on nearby Egyptian army positions. The following day, 22nd June, a second attack was made on El Adem but the Libyans were expecting this, and two Dassault Mirage 5DE

In due course the An-26s were transferred to Libyan Arab Air Cargo, gaining civil identities but retaining their LAAF serials, as illustrated here by the 'flying dog' (5A-DOG/8303, named 'Albayda').

An-26 5A-DOE in standard Libyan Air Cargo livery

fighters shot down one of the escorting MiG-21MFs. On the same day and on 23rd June LAAF jets attacked Egyptian targets but were ambushed by EAF MiG-21MFs; three (some sources suggest four) Libyan Mirages and one MiG-23MS were shot down. Shortly afterwards a cease-fire was agreed. A popular uprising against the government in Egypt did not take place, much to Qaddafi's chagrin.

As is often the case, there were several breakdowns in the ceasefire, the most significant taking place in 1979 when two MiG-23MSs had a dogfight with two Egyptian MiG-21MFs. However, the older MiG-21MFs had the advantage of being armed with the latest US AIM-9P Sidewinder air-to-air missiles and Maj. Sal Mohammad shot down one MiG-23MS; the other one escaped.

In 1979 the Libyan Tu-22B bombers had their baptism of fire when a two-year war broke out between Tanzania and Uganda. Realising that he was losing the war, the Ugandan dictator Idi Amin Dada turned to Libya for help. Qaddafi responded immediately by sending a pair of *Blinders* to Nakasongola AB in Uganda in early March 1979. From there they made a bomb raid on the Tanzanian city of Mwanza on the night of 1st April 1979, targeting the fuel dump at the local airbase. However, the Libyans' aim was bad and the bombs missed the base altogether, landing in a nearby game reserve and killing a large number of antelope. Similarly, another strike against Tanzania was so badly aimed that the bombs landed near Nyarubanga in neighbouring Burundi!

Meanwhile, there was civil war in neighbouring Chad, where the predominantly Muslim north decided to secede from the Christian south. Backed by Libya, the Chadian National Liberation Front (FROLINAT – *Front de Libération Nationale du Tchad*) raised rebellion and advanced towards the Chadian capital of N'djamena. Taking advantage of the situation, in 1975 Libya annexed the long-coveted Aouzou Strip – a piece of disputed territory about 100 km (62 miles) wide in northern Chad. The government forces lacked an effective air defence and Libyan Tu-22Bs could hit any target within Chad with impunity. That

These two An-26s dumped at Tripoli-Mitiga will never fly again – at least the one on the left, which wears Libyan Arab Airlines colours.

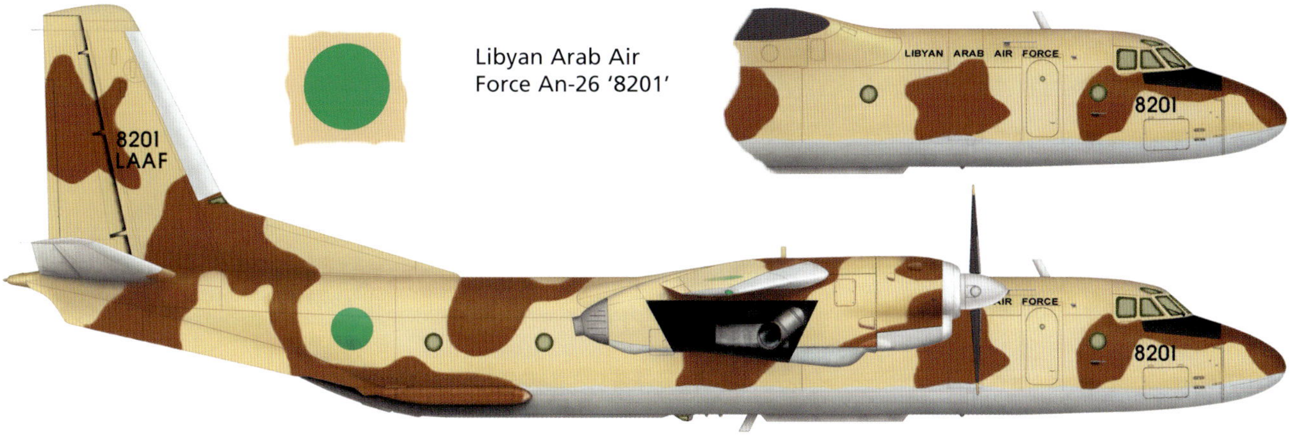

Libyan Arab Air Force An-26 '8201'

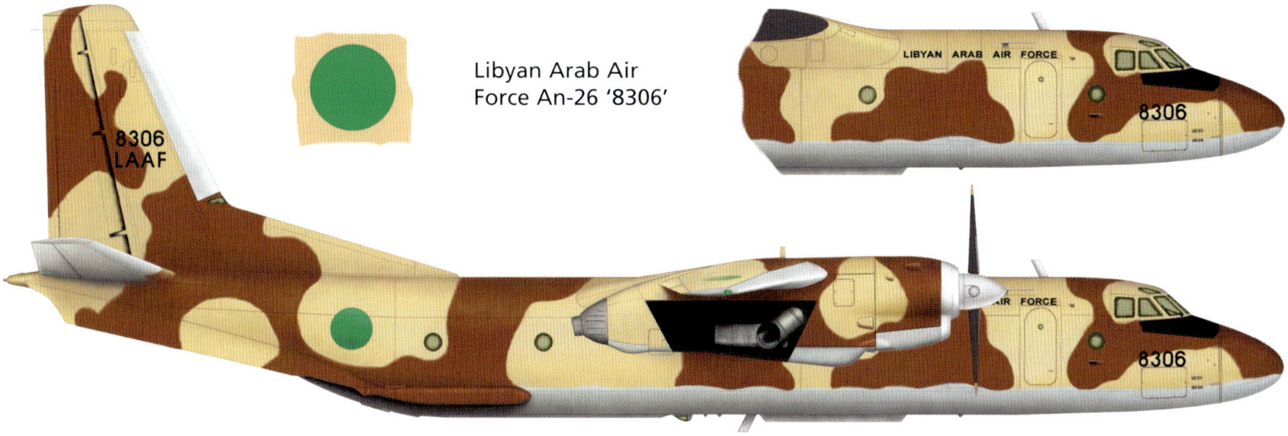

Libyan Arab Air Force An-26 '8306'

same year a coup d'état took place in Chad; the Supreme Military Council led by Gen. Félix Malloum succeeded in ensuring non-support of the FROLINAT from Sudan and Egypt – but not from Libya.

In July 1977 the FROLINAT rebels headed by Goukouni Oueddei launched a new successful offensive. By then the civil war had engulfed the entire country, which was divided along the 16th parallel. Left with no allies except France, Malloum was compelled to sign a tenuous ceasefire with FROLINAT on 27th March 1978. He became the new president of Chad in August 1978. Six moths earlier the Northern Armed Forces (FAN – *Forces Armées du Nord*) faction led by Hissène Habré, had

The exigencies of the UN sanctions (and the need to overhaul aircraft in house) occasionally led to hybrids like this. An-26 '8203' seen at Aircraft Maintenance Centre 003 at Tripoli-Mitiga has had the tail surfaces and outer wings scavenged from sister ship '8308', which was apparently reduced to spares.

Despite being registered in Equatorial Guinea, An-72 3C QTA was operated by the LAAF in Libyan Arab Air Cargo colours

split off from FROLINAT; now Habré saw his chance to gain real power. He made a deal with Malloum and switched sides, becoming Prime Minister while the FAN was integrated into the Chadian Army.

Soon, however, Habré raised a mutiny, and the FAN found itself at war with the government forces again. Neither side could win a decisive victory. Unable to stabilise the situation, Malloum resigned in March 1979, ceding the power to the Transitional National Council headed by Oueddei, with Habré as Minister of Defence. Yet, again peace did not last long; street fighting in N'djamena resumed between the pro-Western Habré and the pro-Libyan factions. The LAAF Tu-22Bs were used again at this stage; for example, on 9th October 1980 they bombed targets near N'djamena. Also in October 1980, the Libyan Mi-24As and Mi-25s drew first blood in the battle of N'djamena. Libyan forces seized the city in December, and Habré fled to Sudan; Oueddei had thus won a temporary victory.

However, Habré was not put off all that easily. In January 1981, after Oueddei had made a state visit to Tripoli, Muammar Qaddafi announced his plan to merge Libya and Chad into the 'Islamic Republic of Sahel'. This was greeted with extreme hostility by everyone, and eventually Qaddafi dropped the idea. As soon as the Libyan forces pulled out, Habré launched a counter-offensive and recaptured N'djamena in 1982; now it was Oueddei's turn to flee – to Libya, of course.

The *Blinders* saw action again in July 1983 when Oueddei made another attempt to oust Habré, who received full official military backing from France. Between July and September Libyan Tu-22Bs, MiG-23BNs, Su-22s and Mi-24As/Mi-25s flew several sorties against Chadian government troops near Fadah, Faya-Largeau (Habré's home town) and Umm Chalouba during the northerners' offensive. The *Hinds* also raided Habré's bases in Sudan when the Kalaït enclave was occupied by the Libyans.

The LAAF did not use the Tu-22B's potential to the full – it was common practice to carry only ten FAB-250 bombs amounting to a bomb load of just 2,500 kg (5,510 lb). On the

An-72 LAAF-723 wears the livery of Libyan Arab Air Cargo (albeit minus titles).

Stripped down to bare metal, this LAAF An-72 is in the midst of an overhaul at Tripoli-Mitiga. An-26 5A-DOF/ 8302 'Ben Weleed' in full Libyan Air Cargo colours is parked alongside.

other hand, the Libyans occasionally used the Tu-22's 23-mm tail gun for strafing enemy troops on the move, or would simply 'boom' them by making a supersonic dash at low altitude.

In 1984 the French expeditionary force was withdrawn in keeping with a Franco-Libyan accord – which Libya did not intend to observe, as it turned out. After a short period of relative peace and quiet, war flared up again in early 1986. France came into the picture again, launching Operation *Epervier* (Sparrowhawk) in order to keep the Libyans at bay; this included the deployment of air defence assets to Chad. This allowed the Chadian government forces to advance, recapturing territories in the north (including Faya-Largeau). When luck turned against Oueddi in late 1986/early 1987 the Libyan Mi-24As/Mi-25s were used to defend Bardaï, Zouar and Falah; one helicopter was shot down near the latter town on 3rd January 1987.

On 17th February 1986 a solitary LAAF Tu-22B took off from Sebha AB in Libya; at 0700 hrs local time it attacked N'djamena International airport with impunity, cratering the runway with three 500-kg (1,102-lb) high-explosive bombs and rendering it unusable for 36 hours. As an excuse for being caught off guard, the French air defence forces protecting N'djamena claimed that the bomber had either used a civil airway, pretending to be an airliner, or followed a wadi (dry river-bed) at ultra-low level, climbing to 3,000-5,000 m (9,840-16,400 ft) immediately before the attack.

Tu-22Bs and MiG-23BNs supported the Libyan army offensive near the Tibesti Plateau in the Aouzou Strip between October 1986 and March 1987. However, in March Habré's troops launched a counter-offensive, seizing the LAAF's forward base at Ouadi-Doum in northern Chad. The assorted aircraft captured there included two *Blinders* and three Mi-25s in reasonable condition. Habré turned the hel-

An-72 LAAF-724 has borrowed the starboard engine's front cowling from a sister ship in Sin-Sad colours.

Operated by the Libyan government, 5A-CAA is one of the few examples of the An-74TK-300 variant featuring underslung engines instead of overwing nacelles.

icopters over to France; in turn, the French transferred one *Hind* to Great Britain and another to the USA. Infuriated by this setback, Qaddafi ordered an air raid on Ouadi-Doum to destroy the captured aircraft, but the effect was minimal.

8th August 1987 brought more frustration when Chadian forces overran Aouzou, an LARAF base in the Aouzou Strip, destroying one Mi-24 on the ground. Libyan forces made several counterattacks, including bombing raids by Tu-22Bs. One of the bombers was shot down by a shoulder-launched SAM; some sources claim it was a Soviet-made 2K12 Koob (Cube; SA-6 *Gainful*) SAM battery captured from the Libyans.

On 5th September 1987 a force of about 2,000 Chadian commandos made a surprise raid 110 km (68 miles) into Libyan territory, seizing Maaten-as-Sarah AB from where the LAAF was staging attacks against Chad. They took the Libyans by surprise, killing 1,713 of the 2,500-strong garrison. Two MiG-23BNs and a Mi-25 attempting to take off were shot down and 22 other aircraft, including two MiG-21s, were captured or destroyed along with other military equipment. Knowing that Libya would not let that pass, the French expeditionary force (Task Force *Epervier*) was placed on heightened alert. Sure enough, Qaddafi ordered a retaliatory air raid on N'djamena by a pair of Tu-22Bs on 7th September. This time, however, the Frenchmen were ready for action. A Centaure air defence radar detected the bomber, which had followed a designated airway and then abruptly left it over Niger, heading south-east towards N'djamena. A pair of French Air Force Dassault Mirage F.1 fighters flying combat air patrol (CAP) was too far away to intercept the suspicious aircraft, which was flying exceptionally fast and not responding to IFF interrogation and was thus coded hostile. Once the target had been positively identified as a Tu-22, the French Army's 405th Missile Regiment at Camp Dubut near N'Djamena armed with Raytheon MIM-23 Hawk surface-to-air missiles was ordered to engage the target. A single missile was fired and scored a direct hit, blowing the bomber apart. This was the first time since the Second World War that the French had shot down an enemy aircraft. A ceasefire was agreed on 11th September 1987.

Col. Qaddafi had 'ideological differences' with another African dictator – the Sudanese General Gaafar Muhammad an-Nimeiry who had supported Habré during the war. Quite simply, Nimeiry had sided with the wrong guy and should be punished. Hence in mid-August 1984 a Libyan Tu-22B bombed Omdurman, Nimeiry's home town, killing five civilians and injuring dozens more. However, in 1985 Nimeiry was unseated in a coup d'état and the new government of Field Marshal Abdel Rahman Suwar al-Dahab was decidedly pro-Libyan. In fact, Western sources state that in March 1986 the Sudanese government borrowed several *Blinders* from the LAAF to make raids against Nimeiry's supporters who had holed up in southern Sudan. Other LAAF aircraft loaned to Sudan to help the Khartoum government in its struggle against the Sudanese People's Liberation Army (SPLA) insurgency include MiG-23MSs and MiG-25RBKs, which were forward deployed to Khartoum and repainted in Sudanese insignia.

In 1980 a defecting Libyan MiG-23MS pilot attempted to fly from Tripoli-Mitiga to Calabria, southern Italy, but crashed fatally in so doing. A little later another MiG-23MS pilot defected successfully to Egypt; his aircraft was later despatched to the USA for evaluation.

Libya also had several brushes with the USA stopping short of all-out war. In 1973, without any international consultation (and against the views of the Soviet Union), Muammar Qaddafi unilaterally imposed a new limit on Libyan territorial waters, which then included almost the whole of the Gulf of Sidra and part of the US Navy Sixth Fleet's exercise zone, and subsequently declared a 'line of death', the crossing of which would invite a military response. Each year one of the American aircraft carriers was engaged in air defence exercises in this area, the pilots of the carrier wing firing at target drones; on three occasions over the past seven years these exercises had been held in the Gulf of Sidra. The USA refused to accept this Libyan move and began Freedom of Navigation operations, and confrontations followed.

On 16th September 1980 fifteen Libyan MiG-23MSs challenged a US Air Force/55th Strategic Reconnaissance Wing Boeing RC-135U ELINT aircraft serialled 64-14847 over the Gulf of Sidra and reportedly damaged it before US Navy fighters came to the rescue, running them off. Full details of the incident are not known.

In July 1981 the US administration again announced its intention of holding Sixth Fleet exercises in the gulf, *'so as to return this unlawfully appropriated part of the sea to freedom'*. On 18th August 1981, the opening day of the two-day exercise, LAAF MiG-23MS fighters, MiG-25RBK reconnaissance/strike aircraft and Su-22M fighter-bombers, together with French *Armée de l'Air* Dassault Mirage F.1s, periodically entered the area designated for the air exercises – apparently probing the Americans' reaction; 74 such flights were logged. Almost every one of the Libyan aircraft involved was intercepted at the edge of the exercise area by US Navy Grumman F-14A Tomcat fighters launching from the aircraft carrier USS *Chester W. Nimitz* which prevented the Libyans from approaching the American ships under threat of attack.

On 19th August, taking advantage of the early sunrise at that time of year, four pairs of LAAF aircraft were returning to base after patrolling the exercise area. At 0700 hrs local time a pair of F-14As of the VF-41 'Black Aces' squadron scrambled from USS *Chester W. Nimitz* to intercept a pair of Su-22Ms which had taken off from Ghurdabiya AB near Sirte, the nearest Libyan air base to the gulf, two minutes earlier. The lead Tomcat wearing the tactical code 'AJ 102' (BuNo 160383) was flown by pilot Cdr. Henry Kleeman and weapons systems operator Lt. Dave Venlet; the second aircraft coded 'AJ 107' (BuNo 160385) was flown by pilot Lt. Larry Muczynski and WSO Lt. (JG) James Anderson. After a long flat take-off the Su-22Ms headed north, climbing to 7,000 m (22,965 ft) and cruising at 850 km/h (528 mph). The Libyan shore radar stations picked up the American aircraft and tracked their manoeuvres.

At a distance of about 15 km (9.3 miles), the pilots of both Tomcats spotted the Su-22Ms almost at the same moment. The Americans flew in a loose echelon starboard formation with a separation of 3 km (1.86 miles) and the wingman flying 50 m (164 ft) above his flight leader. The Libyans used a Soviet-style formation, with the wingman 1,500 m (0.93 miles) to the side of his leader and 1 km (0.62 miles) behind him. 14 seconds later the commander of the American pair began a standard manoeuvre to the left, climbing towards the Libyans. This was a typi-

IL-62M 5A-DNY wears full Libyan Air Cargo livery and is named 'Ghadamis'.

All-white Libyan Arab Airlines IL-76TD 5A-DNS visiting Prague-Ruzyne airport in the 1980s.

cal fighter manoeuvre with a 180° turn by the leader that would bring him up behind the aircraft he was intercepting. The Libyans guessed the Americans' intention, repeating their manoeuvre almost concurrently, and a classic dogfight ensued. In the midst of a port turn the F-14 crews suddenly saw the lead Su-22M launch an R-13M missile. However, the missile had been fired on the edge of its envelope and missed, self-destructing harmlessly. This was the Libyans' fatal mistake, and the Americans retaliated immediately. The *Fitters* were then behind the lead F-14A which took evasive action, pulling 6 Gs in a sharp U-turn. In spite of his failure to score a 'kill', the stubborn Libyan did not break away, counting on an opportunity to repeat the attack – this time against the American wingman.

At this point Henry Kleemann got behind the Libyan wingman, who was trying to keep formation with his leader. He had only to keep the silhouette of the Su-22M in his sight and press the launch button. But the Su-22 pilot was no rookie; realising the danger, he opened up his throttle and made an anti-missile manoeuvre into the sun, thus depriving Kleemann of an opportunity to fire his heat-seeking AIM-9 Sidewinder missiles. Yet the experienced American pilot took his time, waiting until the target was no longer against the disc of the sun and the seeker head of his missile could distinguish the target's heat signature from the heat of the sun. Only then did he launch his Sidewinders from a distance of 1,350 m (4,430 ft). The Libyan was entering a steep dive when the missile hit his rear fuselage. The Su-22M burst into flames, exploding a second later.

The leader of the Libyan pair did not break and run; he wanted revenge. Right in front of him Muczynski's F-14 was making a left turn; in order to nail him the Libyan had to 'get

A more recent image of an IL-76T in full Libyan Air Cargo livery.

In contrast, Libyan Air Cargo did not bother to repaint this recently acquired IL-76TD, 5A-DNX; even the titles are applied sloppily.

inside' his turn, if only for an instant. Coming closer to the target, the Libyan doggedly sought an opportunity to fire. Realising the danger of this moment, Muczynski tightened his turn still more and was soon in a position to attack. Launched from a distance of 1,100 m (3,600 ft), the Sidewinder ripped away the starboard side of the Su-22M's jet-pipe and the rear fuselage side as far as the wing, the aircraft disintegrating in mid-air.

According to one of the Libyan pilots who took part in the skirmish, the Libyans had no intention of attacking anybody and found themselves unexpectedly attacked by the Americans while on a routine training flight. Indeed, the idea of attacking F-14A interceptors with fighter-bombers intended for quite a different role seems ridiculous and very doubtful. If Qaddafi had indeed decided to 'teach the Americans a lesson', he would have chosen more suitable aircraft which were purpose-built for air combat – the MiG-21*bis*, MiG-23MS, MiG-25PD or even the Dassault Mirage F.1ED. It seems far more likely that the Americans got fed up with the nosy Libyans snooping around the carrier task force (CTF) and took the opportunity to 'punish' them.

Another incident that took place on 4th January 1989 was an instant replay of the first one. The aircraft carrier USS *John F. Kennedy* had deployed to the Gulf of Sidra; on that day the carrier task force was operating some 130 km (80 miles) north of Libya, with two pairs of F-14As escorting a group of Grumman A-6 Intruder attack aircraft on exercise south of Crete. The southernmost CAP station was taken by two Tomcats from VF-32; the leader was BuNo 159610 (coded 'AC 207') flown by pilot Cdr. Joseph Bernard Connelly and WSO Cdr. Leo F. Enwright, his wingman being BuNo 159437 (coded 'AC 202') flown by pilot Lt. Hermon C. Cook III and WSO

Starting life as a true IL-76M, 5A-DZZ became an IL-76M 'Falsie' with no gunner's station. Note the Jamahiria Air Transport titles.

Bombs away! These stills from footage taken by the thermal imager of a USAF F-111F attacking Tripoli-Mitiga on the night of 15th April 1986 show the last moments of four Libyan Arab Airlines IL-76s. The system's seeker swivels through 180° as the F-111 passes over the target, hence the different aspects.

Lt. Cdr. Steven Patrick Collins. Due to the high tensions caused by the CTF's presence, the pilots had been warned to expect hostile action by the Libyans. Sure enough, at 1150 hrs local time a Grumman E-2C Hawkeye from the VAW-126 Airborne Early Warning Squadron alerted the F-14 crews that two pairs of Libyan MiG-23s (to be precise, MiG-23MSs – *Auth.*) had taken off from Bumbah AB and were heading directly towards the carrier. The F-14s acquired the first pair of MiG-23s on radar when they were 133 km (82 miles) away at 3,000 m (9,840 ft). According to the American pilots' reports, they turned away from the head-on approach to show the Libyans they were not attempting to engage; yet the MiG-23s changed course with a firm intention of intercepting them. The scenario was repeated four more times, the MiGs getting steadily closer; the F-14 crews had been authorised to fire first if threatened, so they did not need to wait for the Libyans to attack.

Shortly after 1201 hrs the lead F-14A fired the first AIM-7M Sparrow missile at the MiG-23s, which were 26 km (16 miles) away, but the missile failed to track because of a wrong switch-setting. A second Sparrow was launched at 19 km (11.8 miles) but it also went ballistic. As the MiG-23s continued to approach, at 11 km (6.8 miles) range the Tomcats split and the *Floggers* went for the wingman, while the lead F-14 circled to get on their tails. At 9 km (5.5 miles) range the wingman fired a third Sparrow and downed one of the Libyan fighters. Meanwhile, the other F-14 gained the rear quadrant on the other MiG, destroying it with an AIM-9 Sidewinder at 2.8 km (1.74 miles) range. Both Libyan pilots ejected successfully and parachuted into the sea, but the Libyan Air Force was unable to recover them.

Libya had long had a reputation for sponsoring international terrorism and had supported extremist groups in the Palestinian territories and Syria, being strongly anti-Western and anti-Israel. Muammar Qaddafi indicated that he would continue to support the outlawed Red Army Faction in Germany, the Red Brigades in Italy, and the Irish Republican Army as long as the European governments supported anti-Qaddafi Libyans. A bombing attack on the 'La Belle' nightclub in West Berlin by Libyan secret agents on 5th April 1986 was the last straw for the US administration, which had long nurtured plans to eliminate Qaddafi. After several days of diplomatic talks with European and Arab partners, on 14th April President Ronald Reagan ordered a strike on Libya known as Operation *El Dorado Canyon*. On the night of 15th April eighteen General Dynamics F-111F strike aircraft of the USAF's 48th Tactical Fighter Wing on temporary deployment to RAF Lakenheath in England set course for Libya, supported by four EF-111A Raven ECM aircraft of the 20th TFW on TDY to RAF Upper Heyford. At the same time fifteen Grumman A-6, LTV A-7 Corsair II and McDonnell Douglas F/A-18 Hornet attack aircraft and EA-6B Prowler ECM aircraft launched from the aircraft carriers USS *Saratoga*, USS *America* and USS *Coral Sea* on station in the Gulf of Sidra. At 0200 hrs local time they struck five targets, including Qaddafi's Bab al-Azizia residence in Tripoli, Tripoli-Mitiga and Benghazi-Benina airbases, as well as radar and anti-aircraft sites.

The Libyan air defences were caught by surprise. There were no reports of Libyan aircraft attacking the US bombers, and the Americans lost only a single F-111 which was shot down by a Libyan SAM over the Gulf of Sidra. As a result of the raid, four Jamahiria Air Transport IL-76TDs were destroyed at Tripoli-Mitiga, as were many MiG-23MSs and MiG-23BNs and several helicopters at

Two views of Libyan Air Cargo An-124-100 5A-DKN 'Sabrata' undergoing heavy maintenance at the Aviant aircraft factory at Kiev-Svyatoshino.

The other Libyan An-124-100, 5A-DKL 'Susa', parked at Tripoli-Mitiga.

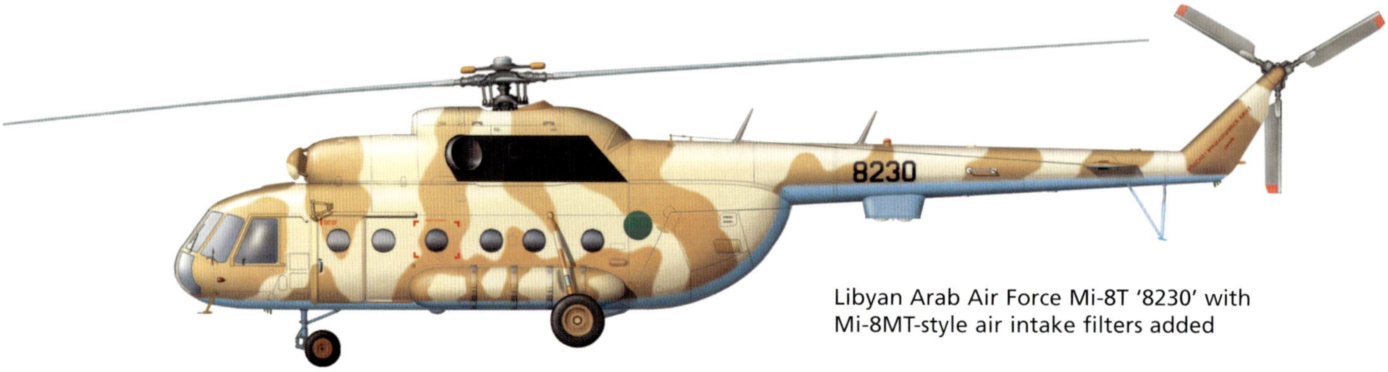

Libyan Arab Air Force Mi-8T '8230' with Mi-8MT-style air intake filters added

Benghazi, and many others were damaged. Apparently the Americans used cluster bombs filled with shrapnel against the airbases, as the IL-76s at Tripoli were so full of holes that they looked like a sieve. However, the American plan to kill Qaddafi failed, as the Libyan leader had been forewarned by a telephone call, evacuating the Bab al-Azizia residence in the nick of time before the bombs dropped. Moreover, some bombs landed off-target, causing collateral damage to diplomatic and civilian sites in Tripoli and Benghazi. The death toll amounted to some 300 Libyans (including the personnel at the bombed airbases) and two USAF pilots from the downed F-111.

Operation *El Dorado Canyon* provoked worldwide outrage, being regarded as a wanton aggression in violation of international law and the UN Charter. However, despite denouncing the attack and taking political steps, the Soviet Union made it clear that it would not provide additional help to Libya beyond resupplying basic weaponry. It made no attempt to militarily intimidate the United States, despite the ongoing American operations in the Gulf of Sidra and its previous knowledge that the United States might launch an attack. Fact is, Qaddafi had stepped out of line far too often, verbally attacking the policy agendas and ideology of the Soviet Union, and his participation in various international interventions often conflicted with Soviet goals in various areas. In a situation where the Soviet Union was apparently attempting to lead a subtle diplomatic effort that could improve its global status, close association with a loose cannon like Qaddafi became a liability.

On 15th April 1992 the United Nations imposed sanctions on Libya, including a ban on international flights. These were intended to pressure the Libyan government into extraditing the terrorists who had blown up Pan American Airlines Boeing 747-121A N739PA *Clipper Maid of the Seas* over Lockerbie, Scotland, on 21st December 1988 with the tragic death of all 259 occupants plus 11 on the ground. Thus Libyan aircraft were theoretically either locked inside the country, many of them grounded by lack of spares, or stranded

Strikingly painted Mi-8MTV-1 '2860' is operated by the flying ambulance service.

outside; for example, Libyan Arab Air Cargo IL-76TD 5A-DNA sat for several years at Moscow-Bykovo, undeliverable after refurbishment at Aircraft Repair Plant No.402, while sister ship 5A-DNG was stored at Khartoum, Sudan. In reality, however, on at least one occasion a Jamahiria Air Transport IL-76TD (5A-DNS) was seen visiting Paris-Le Bourget in March 1998 in contravention of the sanctions. The latter were lifted on 5th April 1999 after the two suspects involved in the bombing had been handed over for trial by a Scottish court.

Being temporarily unable to send its aircraft to Russia or the Ukraine for refurbishment, the LAAF established an overhaul facility at Tripoli-Mitiga known as Maintenance Centre 003, reducing the Air Force's reliance on overseas contractors. It handles such types as the An-26 and An-72.

Revolution, V 2.0

While not officially holding any post in the Libyan government since 1977, Muammar Qaddafi was in fact ruling his country in a hard-line manner. Fearing a military coup (as the Armed Forces had perpetually been a nest of vipers conspiring against him), in 1991 he implemented a military 'reform', abolishing the regular Armed Forces and forming several brigades (so-called *qateeb*) instead; each brigade was commanded by one of his sons. Widespread corruption and nepotism in Qaddafi's inner circle, as well as declining living standards and large-scale human rights violations, led to mounting protests in Libya. On 15th February 2011, protesters in Benghazi clashed with riot police, hurling rocks and Molotov cocktails; the police responded with rubber bullets, tear gas and water cannons. In

Top: These sand-coloured LAAF Mi-8Ts make an interesting comparison with the camouflaged example below.

Above: Mi-8T '8334' (again with intake filters added) sporting a sand/ tan desert camouflage is seen here at LAVEX-2009.

A Libyan Arab Air Force Mi-8T runs up the engines. The tailboom is black with soot, making the serial illegible. Note the Mi-8MT-style intake filters.

two other cities, Bayda and Zintan, police allegedly used deadly force against rioters demanding the removal of Qaddafi and torching government buildings. After that, the protests escalated into a rebellion that spread across the country, the insurgents taking up arms; a full-blown civil war had begun. This time Qaddafi was unable to nip the insurgency in the bud, like he had done with the 2006 riot in Benghazi. On 27th February the anti-Qaddafi forces created the National Transitional Council (NTC); this was not so much an interim government, but rather an organ co-ordinating the actions of the rebel forces in different towns and the opposition's political 'face' and mouthpiece for the outside world.

The rebels were a motley crew composed primarily of civilians, such as teachers, students, lawyers and oil workers, plus a contingent of former Libyan Army personnel who, disenchanted with Qaddafi, had defected to join the rebels. The anti-Qaddafi forces also included extremist groups (some of them affiliated to the al-Qaeda terrorist network), such as the Libyan Islamic Fighting Group and the Obaida Ibn Jarrah Brigade. The rebels denied al-Qaeda membership, although indications of this came not only from Qaddafi's administration but also from US intelligence reports. The rebel forces were commanded by General Abdul Fatah Younis.

In early March Qaddafi's troops began an offensive, retaking several coastal cities captured by the rebels and mounting an attack on the rebels' main stronghold, Benghazi. The LAAF was actively involved, fighter-bomber pilots being tasked with destroying the ammunition dumps in the rebel strongholds of Benghazi and Ajdabiya. Alarmed by reports of casualties among civilians, the United Nations Security Council passed several resolutions, one of which established a 'no-fly zone' over Libya in order to prevent attacks on civilians. On 19th March a multi-state coalition began a military intervention in Libya to implement UN Security Council Resolution 1973 taken in response to events during the Libyan civil war. In particular, French Air Force, Italian Air Force, US Air Force and British Royal Air Force flew strike sorties in Libya and enforced the 'no-fly zone'; US Navy and Royal Navy ships launched cruise missiles and enforced a naval blockade to prevent the Qaddafi regime from receiving weapons via Mediterranean seaports.

The Tripoli government announced a ceasefire but failed to uphold it, though it subsequently accused rebels of violating the ceasefire when they continued to fight as well. Throughout the conflict, rebels rejected gov-

With the An-148-100 airliner prototypes in the background, three LAAF Mi-35s (858, 860 and 861) freshly overhauled by Aviakon await redelivery at Kiev-Gostomel' in 2010.

Libyan Arab Air Force Mi-35 '861'

ernment offers of a ceasefire and efforts by the African Union to end the fighting because they would not settle for anything less than the removal of Qaddafi.

Soon after the outbreak of the war the rebels had their own air arm known as the Free Libyan Air Force (FLAF); it was equipped with aircraft captured from the LAAF or surrendered by defecting pilots, including Col. Ali Atiyya, who defected and joined the revolution on 13 March 2011. These included several (four?) MiG-21*bis* fighters and at least two MiG-21UM trainers flown to rebel-held Benghazi on 17th March, at least nine MiG-23MLAE fighters (four captured at Tobruk, four at Misratah and one at al-Abrak, although some of them were apparently not airworthy) and a MiG-23UB trainer captured at Benghazi. On 16th March three Su-22Ms were delivered to Benghazi by defecting LAAF pilots. One An-26 transport was seized by the rebels at Kufrah on 17th February and a further two were captured at Tripoli, though they were probably unflyable; one IL-76TD was seized by the United Arab Emirates authorities in Dubai on 11th August and subsequently transferred to the FLAF, and another was captured in Tripoli. Helicopters included several Mi-2s captured at Misratah, several Mi-8Ts and at least two Mi-17s, three Mi-14s, and at least two Mi-35s – all seized at Benghazi (two more Mi-35s were destroyed at the same location on 23rd February). On 6th March the crew of a Su-24MK bomber defecting to the rebels attempted to land on a highway near Ras-Lanouf; unfortunately for them, the rebels misunderstood their intentions and greeted the bomber with a hail of heavy machine-gun fire, shooting it down.

The rebels put the windfall (pardon the expression) of aircraft to good use, employing them against the loyalists. The Free Libyan Air

Here, Mi-35 '861' is seen after redelivery with drop tanks. The nose machine-gun is not yet fitted.

A fine shot of Libyan Navy Mi-14PL LC1406 taxying for take-off.

The same helicopter in flight, showing the ASW version's towed MAD 'bird' at the rear of the fuselage pod.

LC1417 was one of the two Libyan Mi-14s converted into Mi-14PZh waterbombers and wearing this striking livery. Note the fire-fighting service logo and titles.

Here, LC1406 is pictured at rest.

Force had its baptism of fire on 14th March, when two MiG-21s, four MiG-23s and two Mi-35s flew sorties from Tobruk, attacking pro-Qaddafi troops between Brega and Ajdabiya. The fighters carried four 250-kg (551-lb) bombs each; the helicopters were armed with four rocket pods giving a total of 128 57-mm unguided rockets and carried a full load of 12.7-mm ammunition for the nose-mounted machine-gun, plus two gunners manning 7.62-mm machine-guns in the troop cabin.

On 15th March four MiG-21s, a MiG-23 and a Mi-35 attacked a loyalist armoured convoy advancing near Brega and Ajdabiya; the intelligence had been provided by rebel ground forces. The following day a pair of Mi-35s attacked a troop convoy including low-loaders with tanks.

On 17th March the MiG-21s and MiG-23s attacked and sank two of the four loyalist ships that were attempting a landing assault on the eastern coast at Zouwaïtina north of Adjabiya. The helicopters, however, did not fly – a loyalist saboteur had put them out of action by shooting at them with a machine-gun and stealing the cyclic pitch lever (!) on one of the choppers. That same day the Free Libyan Air Force suffered its first loss when MiG-21*bis* crashed after take-off from Benghazi airport due to a malfunction. (Some sources say a MiG-21 was shot down at Quaifiya, a suburb of Benghazi, the pilot ejecting safely; this might be the same aircraft.)

A FLAF MiG-23UB was shot down on 19th March 2011 during the Second Battle of Benghazi (the loyalists' attempt to recapture the city); again, this was a 'friendly fire' inci-

LC1417 dumps its load of water through the former torpedo bay doors. Note the stowed hose allowing the helicopter to refill the water tank at the hover.

dent. The pilots, Col. Mohammed Mbarak al-Okaili and Col. Mahdi Rashid Mohammad as-Samin, ejected too late and were killed. However, the loyalists' attempts proved futile – the rebels captured most of their tanks, dispersing the infantry; that same evening NATO aircraft delivered a strike against Qaddafi's artillery that had been shelling the city.

On 21st March a LAAF fighter pilot, Mohammad Mohtar, was ordered to bomb rebel positions near Ajdabiya, Instead, he made a kamikaze attack on the Bab al-Azizia government residence, causing extensive damage. This led to rumours that Khamis Qaddafi, one of the leader's sons, suffered heavy burns and died in hospital shortly afterwards. Later, the rumours turned out to be false – Qaddafi Jr. had survived the attack unharmed.

It should be noted that the NATO's support of the rebels occasionally backfired. On 2nd April someone supplied the NATO strike aircraft pilots with incorrect information, and instead of Qaddafi's troops it was a rebel armoured convoy that got hit. The tanks had been the rebels' most valued weaponry and had just been restored to working condition.

On 5th April the government forces began a new offensive, forcing the rebels back all the way to Ajdabiya. On 9th April a FLAF Mi-35 serialled '854 Black' and piloted by Capt. Hussein Mohammad al-Warfalli (the helicopter squadron commander) attacked loyalist positions in Ajdabiya. However, the government forces had received timely warning of the assault and the Mi-35 was shot down by heavy machine-gun fire, the pilot losing his life. Other FLAF pilots killed in action included Col. Fahri Sallaby, Col. Mohammed al-Abiyi, helicopter pilots Ibrahim Badr Mohammed Dahil, Mohammed Fathi Sallamy and others – 12 in all. On the other hand, the rebels claimed to have shot down several LAAF Mi-35s.

Also on 9th April a FLAF MiG-23 fighter was intercepted by NATO aircraft and escorted back to its base. Later, FLAF MiG-21s were used to escort transport aircraft bringing supplies to rebel-held cities besieged by government troops in case LAAF fighters should attempt an attack.

Encouraged by the NATO intervention, on 20th August rebel forces began a coastal offensive, moving westward and regaining the ground they had lost weeks before. Two days later they were at the gates of Tripoli, and by 24th August they had largely taken control of the entire capital, mopping up the remaining Qaddafi loyalists. However, when the rebels stormed the government compound at Bab al-Azizia, it turned out that Qaddafi and the members of his family had fled from the residence to his home town of Sirte. On 16th September 2011 the National Transitional Council was recognised by the United Nations as the legal representative of Libya, replacing the Qaddafi government.

After the fall of Tripoli, the outcome of the civil war was obvious. On 29th August the rebels captured the city of Ghadames near the borders of Tunisia and Algeria; several of Qaddafi's family members fled to Algeria but the deposed leader stayed, deciding to fight to the last. In September the rebels besieged the stronghold of Bani Walid, where Qaddafi's elder son Saïf al-Islam was thought to be hiding. In October the rebels began a final onslaught against Sirte, which was fiercely defended by the remnants of the loyalist forces. The city finally fell on 20th October 2011; Qaddafi attempted to flee, but his vehicular convoy was attacked and largely destroyed by a French Air Force strike called in by radio, whereupon the rebels captured the hated ex-leader and executed him there and then. Finally, on 23rd October 2011 the National Transitional Council declared 'the liberation of Libya' and the official end of the war. That day three FLAF Mi-14s conducted a flypast at the declaration of national liberation in Benghazi conducted by NTC Chairman Mustafa Abdul Jalil.

As one might imagine, hostilities did not cease immediately; not only were there pockets of resistance by former Qaddafi loyalists but the local militia and tribes started bickering and fighting for power. The most serious clashes erupted in late January 2012. Thus, on 23rd January Bani Walid was captured by local fighters, who subsequently established a non-NTC town council due to the NTC's refusal to co-operate with tribal leaders. More clashes occurred in Sabha, where FLAF MiG-21*bis* and MiG-21UM jets were deployed to engage the insurgents.

On 20th June 2012 the FLAF Chief of Staff, Saqr Geroushi, announced plans for the rebuilding of the Libyan Air Force. The plans included proposals for the purchase of French, British and American aircraft – Dassault Rafale and Dassault Mirage F.1 fighters, Eurofighter Typhoons, Lockheed C-130 Hercules transports and Boeing Vertol CH-47 Chinook helicopters. Thus, no more sales of Russian aircraft to Libya are to be expected henceforth.

Known Libyan Arab Air Force aircraft of Soviet/Russian origin				
Type	**Serial/ registration**	**C/n**	**F/n**	**Notes**
An-2	211 Black	1G 23908		Ex-RA-02281, leased from Polyot/Flight, D/D 2004
An-26	8201	11805		Became, see next line
	5A-DOX			Donated to Sudan Air Force as 7755/ST-ALU
An-26	8202	11801?		
An-26	8203	11809		Became, see next line
	5A-DOW/8203			Libyan Arab Air Cargo, green cheatline and green tail, named 'Misratah'
An-26	8204	?		Donated to Somalian Air Force as 6O-204 in 1990
An-26	8205	12301		Reportedly became, see next line
	5A-DNV?			Clash with IL-76TD (see below)! Became, see next line
	5A-DOP			Libyan Arab Air Cargo
An-26	8206	12303		
An-26	8207	12306		Became, see next line
	5A-DOA/8207			Libyan Arab Air Cargo, named 'Wadi Jaref'
An-26	8208	27312307		Received wings and tail surfaces of An-26 '8308' (see below) during overhaul at plant No.003
	5A-DOB			Libyan Arab Air Cargo
An-26	8209	12308		Became, see next line
	5A-DOC/8209			Libyan Arab Air Cargo, grey/white fuselage (no cheatline), green tail, named 'Waddan'
An-26	8210	12309		
An-26	8211	12403?		
An-26	8212	12406		Became, see next line
	5A-DOD/8212			Libyan Arab Air Cargo, green cheatline and green tail, named 'Waw Alnamoos'
An-26	8213	12407		
An-26	8214	12504		Became, see next line
	5A-DOZ/8214			Libyan Arab Air Cargo. Damaged in belly landing Atoqah 31-10-2007
An-26	8301	13003?		C/n not confirmed for this serial; see next line
	5A-DOE			Libyan Arab Air Cargo. C/n confirmed. Operated by Libyan Arab Air Cargo
An-26	8302	130-07		Became, see next line
	5A-DOF/8302			Libyan Arab Air Cargo, all-white c/s; later green cheatline and green tail, named 'Ben Weleed
An-26	8303	130-08		Became, see next line
	5A-DOG/8303			Libyan Arab Air Cargo, green cheatline and white tail, named 'Albayda'
An-26	8304	130-09		
	5A-DON/8304			Libyan Arab Air Cargo, green cheatline and white tail
An-26	8305	13102		
An-26	8306	13103		Registration 5A-DOI assigned but possibly reservation only as aircraft is wfu Tripoli-Mitiga
An-26	8307	13104		Became, see next line
	5A-DOJ			Libyan Arab Air Cargo
An-26	8308	13105?		Dismantled, wings and tail surfaces used to rebuild An-26 '8208'
An-26	8309	13106		Became, see next line
	5A-DOK			Sin-Sad Airlines. Crashed al El Jufrah-Khun AB 2008
An-26	8310	13107		
An-26	8311	13108		Became, see next line
	5A-DOL			Libyan Arab Air Cargo
An-26	8312	13109?		C/n not confirmed for this serial; see next line
	5A-DOV			Libyan Arab Air Cargo, c/n confirmed
An-26	8313	13110		Sold to Chad Air Force as TT-LAP
An-26	8314	13201		C/n not confirmed for this serial; see next line
	5A-DOU			Libyan Arab Air Cargo, c/n confirmed
An-26	8315	13202		Previously reported as c/n 13205. Became, see next line
	5A-DOH			Libyan Arab Air Cargo. Became, see next line
	5A-DOQ			Libyan Arab Air Cargo
An-26	8316	13205?		
An-32P	5A-DRC	0703		Ex-Antonov Design Bureau UR-48093, based on An-32 *sans suffixe*

An-32P	5A-DRD	1306		Ex-Antonov Design Bureau UR-48004, based on An-32 *sans suffixe*
An-32P	5A-DRE	3507		Based on An-32B
An-32P	5A-DRF	3602		Based on An-32B
An-72	LAAF-072	365.720.20.337	0301	Ex-Soviet Air Force CCCP-72900. All-white, no titles. Became, see next line
	5A-DSH			Operated by Allibia Airlines. All-white c/s
An-72	EL-ALL	365.720.80.789	0809	Ex-Enimex ES-NOL. Operated by Sin-Sad Airlines. Became, see next line
	3C-QQS			Nominally West Africa Air; Libyan Arab Air Cargo c/s, no titles. Became, see next line
	5A-DNL (2)			See IL-76TD below! Libyan Arab Air Cargo, seen 2006. Became, see next line
	LAAF-721			Libyan Arab Air Cargo c/s, no titles; seen 2007. Repainted as, see next line
	LAF-721			Grey c/s, seen 2011
An-72	TL-ACV	365.720.80.777	0802	Ex-3D-RTV; illegal TL- registration worn in 1998-2002. All-white c/s, no titles. Became, see next line
	9XR-RV			All-white c/s, no titles. Became, see next line
	LAAF-722			All-white c/s, no titles
An-72	TL-ADF	365.720.30.430	0402	Ex-Russian Air Force RA-72905, bought 2003; Libyan Arab Air Cargo, all-white c/s, no titles. Became, see next line
	LAAF-723			Libyan Arab Air Cargo c/s, no titles; seen 2005
An-72	3C-QTA	365.720.60.653	0610	Ex-Russian Air Force RA-72929, bought 2003 and initially operated as such! Libyan Arab Air Cargo c/s, no titles. Repainted as, see next line
	LAAF-724			Libyan Arab Air Cargo c/s, no titles; seen 2005
An-74TK-200	UR-74038	365.470.97.933	1605	Leased by LAAF from the Khar'kov State Aircraft Production Enterprise. Crashed near Kousséri, Cameroun, 23-4-2006
An-74TK-300D	5A-CAA	365.470.121.1080	2203	Ex-UR-PAV, D/D? 23-4-2009; operated by Libyan Civil Aviation Authority, green/white c/s, no titles
[An-74TK-200S]		365.720.98.973	1901	Ordered but undeliverable; to Turkmenistan AF
[An-74TK-200S]		365.470.98.976	1903	Ordered but undeliverable; to Turkmenistan AF
An-124-100	5A-DKL	19530502761	03-01	Ex-UR-82066, named 'Susa'
An-124-100	5A-DKN	19530502792	03-02	Ex-RA-82003, named 'Sabrata'. Stored Kiev-Gostomel'
IL-62M 'Salon'	3C-QQR	4648414	4801	Ex-EL-ALM, ex-RA-86711. Became, see next line
	5A-DKT			Stored Moscow-Domodedovo
IL-62M 'Salon'	3C-QQZ	3052657	5205	Ex-EL-ALZ, ex-RA-86553. Became, see next line
	TL-ABW			Became, see next line
	5A-DNY			Full Libyan Air Cargo c/s, named 'Ghadamis'
IL-62M 'Salon'	5A-DKR	4053514	5301	Ex-RA-86554. Crashed Moscow-Domodedovo 29-3-2006
IL-76M	5A-DKK	0003423675	1709	Damaged beyond repair in hard landing at Sheba in 1985
IL-76TD	5A-DKS	1033418584	9006	Ex-Aviaenergo RA-76843; Sin-Sad Airlines
IL-76M	5A-DLL (1)	093421612	1603	Destroyed Tripoli-Mitiga 15-4-1986 by USAF strike; see IL-78E below!
IL-76M	5A-DMM	0003423679	1710	Reported destroyed Tripoli 15-4-1986 by USAF strike
IL-76TD	5A-DNA	0023439140	2905	
IL-76TD	5A-DNB	0023437086	2802	
IL-76TD	5A-DNC	0023437084	2801	
IL-76TD	5A-DND	0033445299	3305	
IL-76T	5A-DNE	0013432952	2408	
IL-76T	5A-DNF	0033445302	3306	Destroyed Tripoli-Mitiga 15-4-1986 by USAF strike
IL-76T	5A-DNG	0013432961	2501	
IL-76TD	5A-DNH	0033446356	3409	
IL-76T	5A-DNI	0013430878	2210	
IL-76T	5A-DNJ	0013430869	2208	
IL-76T	5A-DNK	0013430882	2301	
IL-76T	5A-DNL (1)	0033447357	3410	Destroyed Tripoli-Mitiga 15-4-1986 by USAF strike; see An-72 above
IL-76TD	5A-DNO	0043451509	3808	
IL-76TD	5A-DNP	0043451516	3809	
IL-76TD	5A-DNQ	0043454641	4201	DBR on landing Bamako, Mali, 17-10-2007, scrapped on site
IL-76TD	5A-DNS	0023439145	2907	All-white c/s, Libyan Airlines titles/logo
IL-76TD	5A-DNT	0023439141	2906	
IL-76TD	5A-DNU	0043454651	4203	

IL-76TD	5A-DNV	0043454645	4202	
IL-76TD	5A-DNW	0043450479?	3710?	Not in fleet list 1999 – returned to the Soviet Union? (reported as c/n 0043456700, f/n 4305, to CCCP-76493 but this is doubtful!)
IL-76TD	5A-DNX	1033414480	8710	Ex-UN 76024. Libyan Air Cargo, grey/white c/s
IL-76M	5A-DRR	083415469	1208	D/D 15-3-79
IL-76TD	5A-DRS	1033414474	8709	Ex-Atlant Soyuz RA-76367
IL-76M	5A-DZZ	093416501	1306	Delivered as 'true' IL-76M with tail gunner's station; later converted to IL-76M 'Falsie' with IL-76T-style commercial tailcone; Libyan AIr Cargo green/white livery but Jamahiria Air Transport titles!
IL-76TD	LAF 110	1003403063	7706	Ex-Samara Airlines RA-76798. Basic Libyan Arab Air Cargo c/s. Became, see next line
	5A-DRT			Libyan Arab Air Cargo
IL-78E	5A-DLL (2)	0093493799	7010	Libyan Arab Air Cargo c/s
MiG-21MF	601	?	?	Force 14
MiG-21MF	602	?	?	Force 14
MiG-21MF	603	?	?	Force 14
MiG-21MF	604	?	?	Force 14
MiG-21MF	605	?	?	Force 14
MiG-21MF	606	?	?	Force 14
MiG-21MF	607	?	?	Force 14
MiG-21MF	608	?	?	Force 14
MiG-21MF	609	?	?	Force 14
MiG-21MF	610	?	?	Force 14
MiG-21MF	611	?	?	Force 14
MiG-21MF	612	?	?	Force 14
MiG-21MF	613	?	?	Force 14
MiG-21MF	614	?	?	Force 14
MiG-21MF	615	?	?	Force 14
MiG-21MF	616	?	?	Force 14
MiG-21MF	617	?	?	Force 14
MiG-21MF	618	?	?	Force 14
MiG-21MF	619	?	?	Force 14
MiG-21MF	620	?	?	Force 14
MiG-21MF	621	?	?	Force 14
MiG-21MF	622	?	?	Force 14
MiG-21MF	623	?	?	Force 14
MiG-21MF	624	?	?	Force 14
MiG-21MF	625	?	?	Force 14
MiG-21MF	626	?	?	Force 14
MiG-21MF	627	?	?	Force 14
MiG-21MF	628	?	?	Force 14
MiG-21MF	629	?	?	Force 14
MiG-21MF	630	?	?	Force 14
MiG-21MF	631	?	?	Force 14
MiG-21MF	632	?	?	Force 14
MiG-21MF	633	?	?	Force 14
MiG-21MF	634	?	?	Force 14
MiG-21MF	635	?	?	Force 14
MiG-21MF	636	?	?	Force 14
MiG-21MF	637	?	?	Force 14
MiG-21MF	638	?	?	Force 14
MiG-21MF	639	?	?	Force 14
MiG-21MF	640	?	?	Force 14
MiG-21MF	641	?	?	Force 14
MiG-21MF	642	?	?	Force 14
MiG-21MF	643	?	?	Force 14
MiG-21MF	644	?	?	Force 14
MiG-21MF	645	?	?	Force 14

MiG-21MF	646	?	?	Force 14
MiG-21MF	647	?	?	Force 14
MiG-21MF	648	?	?	Force 14
MiG-21MF	649	?	?	Force 14
MiG-21MF	650	?	?	Force 14
MiG-21*bis*	701	N750*****	?	1060 FR
MiG-21*bis*	702	N750*****	?	1060 FR
MiG-21*bis*	703	N750*****	?	1060 FR
MiG-21*bis*	704 Black	N750*****	?	Sand/green/brown camouflage; seen at LAVEX-2007 airshow, Tripoli-Mitiga
MiG-21*bis*	705	N750*****	?	1060 FR
MiG-21*bis*	706	N750*****	?	1060 FR
MiG-21*bis*	707	N750*****	?	1060 FR
MiG-21*bis*	708	N750*****	?	1060 FR
MiG-21*bis*	709	N750*****	?	1060 FR
MiG-21*bis*	710	N750*****	?	1060 FR
MiG-21*bis*	711	N750*****	?	1060 FR
MiG-21*bis*	712	N750*****	?	1060 FR
MiG-21*bis*	713	N750*****	?	1060 FR
MiG-21*bis*	714	N750*****	?	1060 FR
MiG-21*bis*	715	N750*****	?	1060 FR
MiG-21*bis*	716	N750*****	?	1060 FR
MiG-21*bis*	717	N750*****	?	1060 FR
MiG-21*bis*	718	N750*****	?	1060 FR
MiG-21*bis*	719	N750*****	?	1060 FR
MiG-21*bis*	720	N750*****	?	1060 FR
MiG-21*bis*	721	N750*****	?	1060 FR
MiG-21*bis*	722	N750*****	?	1060 FR
MiG-21*bis*	723	N750*****	?	1060 FR
MiG-21*bis*	724	N750*****	?	1060 FR
MiG-21*bis*	725	N750*****	?	1060 FR
MiG-21*bis*	726	N750*****	?	1060 FR
MiG-21*bis*	727	N750*****	?	1060 FR
MiG-21*bis*	728	N750*****	?	1060 FR
MiG-21*bis*	729	N750*****	?	1060 FR
MiG-21*bis*	730	N750*****	?	1060 FR
MiG-21*bis*	731	N750*****	?	1060 FR
MiG-21*bis*	732	N750*****	?	1060 FR
MiG-21*bis*	733	N750*****	?	1060 FR
MiG-21*bis*	734	N750*****	?	1060 FR
MiG-21*bis*	735	N750*****	?	1060 FR
MiG-21*bis*	736	N750*****	?	1060 FR
MiG-21*bis*	737	N750*****	?	1060 FR
MiG-21*bis*	738	N750*****	?	1060 FR
MiG-21*bis*	739	N750*****	?	1060 FR
MiG-21*bis*	740	N750*****	?	1060 FR
MiG-21*bis*	741	N750*****	?	1060 FR
MiG-21*bis*	742	N750*****	?	1060 FR
MiG-21*bis*	743	N750*****	?	1060 FR
MiG-21*bis*	744	N750*****	?	1060 FR
MiG-21*bis*	745	N750*****	?	1060 FR
MiG-21*bis*	746	N750*****	?	1060 FR
MiG-21*bis*	747	N750*****	?	1060 FR
MiG-21*bis*	748	N750*****	?	1060 FR
MiG-21*bis*	749	N750*****	?	1060 FR
MiG-21*bis*	750	N750*****	?	1060 FR
MiG-21*bis*	751	N750*****	?	1060 FR
MiG-21*bis*	752	N750*****	?	1060 FR
MiG-21*bis*	753	N750*****	?	1060 FR

MiG-21*bis*	754	N750*****	?	1060 FR
MiG-21*bis*	755	N750*****	?	1060 FR
MiG-21*bis*	756	N750*****	?	1060 FR
MiG-21*bis*	757	N750*****	?	1060 FR
MiG-21*bis*	758	N750*****	?	1060 FR
MiG-21*bis*	759	N750*****	?	1060 FR
MiG-21*bis*	760	N750*****	?	1060 FR
MiG-21*bis*	761	N750*****	?	1060 FR
MiG-21*bis*	762	N750*****	?	1060 FR
MiG-21*bis*	763	N750*****	?	1060 FR
MiG-21*bis*	764	N750*****	?	1060 FR
MiG-21*bis*	765	N750*****	?	1060 FR
MiG-21*bis*	766	N750*****	?	1060 FR
MiG-21*bis*	767	N750*****	?	1060 FR
MiG-21*bis*	768	N750*****	?	1060 FR
MiG-21*bis*	769	N750*****	?	1060 FR
MiG-21*bis*	770	N750*****	?	1060 FR
MiG-21*bis*	771	N750*****	?	1060 FR
MiG-21*bis*	772	N750*****	?	1060 FR
MiG-21*bis*	773	N750*****	?	1060 FR
MiG-21*bis*	774	N750*****	?	1060 FR
MiG-21*bis*	775	N750*****	?	1060 FR
MiG-21*bis*	776	N750*****	?	1060 FR
MiG-21*bis*	777	N750*****	?	1060 FR
MiG-21*bis*	778	N750*****	?	1060 FR
MiG-21*bis*	779	N750*****	?	1060 FR
MiG-21*bis*	780	N750*****	?	1060 FR
MiG-21*bis*	781	N750*****	?	1060 FR
MiG-21*bis*	782	N750*****	?	1060 FR
MiG-21*bis*	783	N750*****	?	1060 FR
MiG-21*bis*	784	N750*****	?	1060 FR
MiG-21*bis*	785	N750*****	?	1060 FR
MiG-21*bis*	786	N750*****	?	1060 FR
MiG-21*bis*	787	N750*****	?	1060 FR
MiG-21*bis*	788	N750*****	?	1060 FR
MiG-21*bis*	789	N750*****	?	1060 FR
MiG-21*bis*	790	N750*****	?	1060 FR
MiG-21*bis*	791	N750*****	?	1060 FR
MiG-21*bis*	792	N750*****	?	1060 FR
MiG-21*bis*	793	N750*****	?	1060 FR
MiG-21*bis*	794	N750*****	?	1060 FR
MiG-21*bis*	800 Black	N750*****	?	To Free Libya Air Force, seen Tammahint nr Sabha
MiG-21UM	069 Black	5169*****	?	1021 Sqn. Noted at LAVEX-2007 airshow, Tripoli-Mitiga
MiG-21UM	101	5169*****	?	Force 14
MiG-21UM	102	5169*****	?	Force 14
MiG-21UM	103	5169*****	?	Force 14
MiG-21UM	104	5169*****	?	Force 14
MiG-21UM	105	5169*****	?	Force 14
MiG-21UM	106	5169*****	?	Force 14
MiG-21UM	107	5169*****	?	Force 14
MiG-21UM	108	5169*****	?	Force 14
MiG-21UM	109	5169*****	?	Force 14
MiG-21UM	110	5169*****	?	Force 14
MiG-21UM	111	5169*****	?	Force 14
MiG-21UM	112	5169*****	?	Force 14
MiG-21UM	113 Black	5169*****	?	1021 Sqn. Sand/green/brown camouflage; seen at LAVEX-2007 airshow, Tripoli-Mitiga
MiG-21UM	114	5169*****	?	Force 14
MiG-21UM	115	5169*****	?	Force 14

MiG-21UM	116	5169*****	?	Force 14
MiG-21UM	117	5169*****	?	Force 14
MiG-21UM	118	5169*****	?	Force 14
MiG-21UM	119	5169*****	?	Force 14
MiG-21UM	120	5169*****	?	Force 14
MiG-21UM	121	5169*****	?	Force 14
MiG-21UM	122	5169*****	?	Force 14
MiG-21UM	123	5169*****	?	Force 14
MiG-21UM	124	5169*****	?	Force 14
MiG-21UM	125	5169*****	?	Force 14
MiG-23MS	712 Black	124004712		D/D 1974; Egyptian-style insignia
MiG-23MS	713 Black	124004713		D/D 1974
MiG-23MS	2145 Black	0391202145	?	D/D 1975
MiG-23MS	2162 Black	0391202162	?	D/D 1975; overhauled by Chugooyev Aircraft Repair Plant, the Ukraine
MiG-23MS	4014 Black	124004014		D/D 1974, 1023 Sqn
MiG-23MS	4711 Black	124004711		D/D 1974
MiG-23MS	4714 Black	124004714		D/D 1974
MiG-23MS	4052 Black (1)?	0391204052	?	D/D 1975
MiG-23MS	4135 Black	0391204135	?	D/D 1975 Shot down 2011
MiG-23MS	5927 Black	0391205927	?	D/D 1975
MiG-23MS	06907 Black	0391206907	?	Taken over by Free Libya Air Force
MiG-23MS	6915 Black	0391206915	?	1050 Sqn; reported in error as a MiG-23MLD
MiG-23MS	6916 Black	0391206916	?	1050 Sqn; reported in error as a MiG-23MLD. Reportedly to Sudanese Air Force
MiG-23MS	06918 Black	0391206918	?	Reportedly to Sudanese Air Force in 1988. Shot down in Sudan
MiG-23MS	6937 Black	0391206937	?	
MiG-23MS	8301 Black	0391208301	?	To Democratic Republic of Congo Air Force in 1988
MiG-23MS	09055 Black	0391209055	?	Reportedly to Sudanese Air Force in 1988
MiG-23MS	9082 Black	0391209082	?	
MiG-23MS	9818 Black	0391209818	?	Shot down 11-12-1988
MiG-23MS	9871 Black	0391209871	?	
MiG-23MS	9882 Black	0391209882	?	
MiG-23MF	0125 Black	0390220125	?	
MiG-23MF	0200 Black	0390220200	?	Taken over by Free Libya Air Force
MiG-23MF	0206 Black	0390220206	?	
MiG-23MF	0212 Black	0390220212?	?	
MiG-23MF	4052 Black (2)?	0390224052	1420*	F/n in the 14201 to 14204 range. Overhauled by Chugooyev Aircraft Repair Plant, the Ukraine
MiG-23MLAE-2	130 Black	2960326130	?	1023 Sqn; reserialled as, see next line
	6130 Black			Overhauled by Aircraft Repair Plant No.121, Russia, 2005. To Free Libya Air Force; shot down 19-3-2011
MiG-23MLAE-2	132 Black	2960326132	?	1023 Sqn; reserialled as, see next line
	6132 Black			
MiG-23MLAE-2	455 Black	2960326455	?	1023 Sqn
MiG-23MLAE-2	5397 Black	2960325397	18030	1023 Sqn
MiG-23MLAE-2	5451 Black	2960325451	?	
MiG-23MLAE-2	6117 Black	2960326117	?	1023 Sqn
MiG-23MLAE-2	6118 Black	2960326118	?	1023 Sqn
MiG-23MLAE-2	6148 Black	2960326148	?	1023 Sqn
MiG-23MLAE-2	6452 Black	2960326452	?	1023 Sqn. Overhauled by Aircraft Repair Plant No.121, Russia, 2005
MiG-23MLAE-2	6461 Black	2960326461	?	1023 Sqn. Overhauled by Aircraft Repair Plant No.121, Russia, 2005; operational 10-2011
MiG-23MLAE-2	6472 Black	2960326472	?	1023 Sqn; taken over by Free Libya Air Force
MiG-23MLAE-2	6474 Black	2960326474	?	1023 Sqn. Became, see next line
	26474 Black			
MiG-23MLAE-2	7202 Black	2960327202	18419	1023 Sqn. Overhauled by Aircraft Repair Plant No.121, Russia, 2005
MiG-23BN	8985 Black	0393208985	?	Retrofitted with IFR probe
MiG-23BN	9103 Black	0393209103	594*	
MiG-23UB	007 Black	B1038007	?	
MiG-23UB	1285 Black	B1038512	?	D/D 1985; 'reversed' serial (should have been 8512!). Overhauled by L'vov Aircraft Repair Plant, the Ukraine
MiG-23UB	7912 Black	B1037912	?	D/D 1979. Donated to Zimbabwe Air Force as 7012

MiG-23UB	7920 Black	B1037920	?	Donated to Democratic Republic of Congo Air Force as FG-200
MiG-23UB	8012 Black	B1038012	?	
MiG-23UB	8021 Black	B1038021	?	Stored L'vov Aircraft Repair Plant 7-2001
MiG-23UB	8133 Black	B1038133	?	Stored L'vov Aircraft Repair Plant 7-2001
MiG-23UB	8206 Black	B1038206	?	
MiG-23UB	8212 Black	B1038212	?	Stored L'vov Aircraft Repair Plant 7-2001
MiG-23UB	8214 Black	B1038214	?	Stored L'vov Aircraft Repair Plant 7-2001
MiG-23UB	8217 Black	B1038217	?	
MiG-23UB	8272 Black	B1038272	?	1023 Sqn; also reported in error as 8772
MiG-23UB	8423 Black	B1038423	?	1023 Sqn. Crashed Tripoli-Mitiga AB 8-10-2009, crew killed
MiG-23UB	8425 Black	B1038425	?	
MiG-23UB	8702 Black	B1038702	?	
MiG-23UB	8772 Black	B1038772	?	
MiG-25...	004	?	?	
MiG-25...	127	?	?	
MiG-25PU	207 Black	N220**207	?	Stored Tripoli-Mitiga
MiG-25RBK	407 Black	N02020407	?	Crashed between 1983 and 1986
MiG-25RBK	485 Black	N02020485	?	
MiG-25RBK	499 Black	N02020499	?	Stored Tripoli-Mitiga
MiG-25RBK	504 Black	N02020504	?	
MiG-25RBK	525 Black	N02020525	?	Crashed 31-8-1983
MiG-25PD	702 Black	N840**702	?	
MiG-25PD	731 Black	N840**731	?	
MiG-25PD	903 Black	N840**903	?	Stored Tripoli-Mitiga
MiG-25PU	994 Black	N220**994	?	
MiG-25...	1035	?	?	
MiG-25...	2145 Black	?	?	
MiG-25...	2708?	?	?	Reported as such but possible misquote of 7708!
MiG-25PD	6001 Black	N840*6001	?	
MiG-25PD	6008 Black	N840*6008	?	
MiG-25PD	6012 Black	N840*6012	?	
MiG-25PD	6015 Black	N840*6015	?	
MiG-25PD	6716 Black	N840*6716	?	Retrofitted with overwing IRCM flare launchers
MiG-25PD	7003 Black	N840*7003	?	
MiG-25PD	7029 Black	N840*7029	?	Retrofitted with overwing IRCM flare launchers. Stored Tripoli-Mitiga
MiG-25PD	7706 Black	N840*7706	?	Stored Tripoli-Mitiga
MiG-25PD	7708 Black	N840*7708	?	Stored Tripoli-Mitiga
MiG-25PD	7811 Black	N840*7811	?	
Mi-2P	5940	535940019		
Mi-2P	5941	535941019		
Mi-2P	5942	535942019		
Mi-2P	5943	535943019		
Mi-2P	5944	535944019		
Mi-2P	5945	535945019		
Mi-2P	5946	535946019		
Mi-2P	5947	535947019		
Mi-2P	5948	535948019		
Mi-2P	5949	535949019		
Mi-2P	6121	536121059		
Mi-2P	6122	536122059		
Mi-2P	6123	536123059		
Mi-2P	6124	536124069		
Mi 2P	6125	536125069		
Mi-2P	6126	536126069		
Mi-2P	6127	536127069		
Mi-2P	6128	536128069		

Mi-2P	6129	536129069		
Mi-2P	6130	536130069		
Mi-2P	6745	536745080		
Mi-2P	6746	536746080		
Mi-2P	6747	536747080		
Mi-2P	6748	536748080		
Mi-2P	6749	536749080		
Mi-2P	6750	536750080		
Mi-2P	6801	536801080		
Mi-2P	6802	536802080		
Mi-2P	6803	536803080		
Mi-2P	6804	536804080		
Mi-2P	6805	536805080		
Mi-2P	6806	536806080		
Mi-2P	6820	536820090		
Mi-2P	6821	536821090		
Mi-2P	6822	536823090		To Ghana Air Force as G660
Mi-2P	6823	536824090		To Ghana Air Force as G661
Mi-2P	6824	536824090		
Mi-2P	6825	536825090		
Mi-2P	6826	536826090		
Mi-2P	6827	536827090		
Mi-2P	6828	536828090		
Mi-2P	6829	536829090		
Mi-2T	7003	517003021		
Mi-2T	7004	517004021		
Mi-2T	7005	517005021		
Mi-2T	7006	517006021		
Mi-2T	7007	517007021		
Mi-2T	7008	517008021		
Mi-2T	7009	517009021		
Mi-2T	7010	517010021		
Mi-2T	7011	517011021		
Mi-2T	7012	517012021		
Mi-2T	7013	517013021		
Mi-2T	7014	517014021		
Mi-2T	7015	517015021		
Mi-2T	7016	517016021		
Mi-2T	7017	517017031		
Mi-2T	7018	517018031		
Mi-2T	7019	517019031		
Mi-2T	7020	517020031		
Mi-2T	7021	517021031		
Mi-2T	7022	517022031		
Mi-2T	7023	517023031		
Mi-2T	7024	517024031		
Mi-2T	7025	517025031		
Mi-2T	7026	517026031		
Mi-2T	7027	517027031		
Mi-2...	7028	567028031		
Mi-2...	7029	567029031		
Mi-2...	7030	567030031		
Mi-2...	7031	567031031		
Mi-2...	7032	567032031		
Mi-2...	7033	567033031		To Lesotho Police as PMU-7
Mi-2...	7034	567034031		
Mi-2...	7035	567035031		

Mi-2T	7036	517036041		
Mi-2T	7037	517037041		
Mi-2T	7038	517038041		
Mi-2T	7039	517039041		
Mi-2T	7040	517040041		
Mi-2T	7041	517041041		
Mi-2T	7042	517042041		
Mi-2T	7043	517043041		
Mi-2T	7044	517044041		
Mi-2T	7115	517115041		
Mi-2T	7116	517116041		
Mi-2T	7117	517117041		
Mi-2T	7118	517118041		
Mi-2T	7119	517119041		
Mi-2T	7120	517120051		Crashed ?-5-1993
Mi-2T	7121	517121051		
Mi-2T	7122	517122051		
Mi-2T	7123	517123051		
Mi-2T	7124	517124051		
Mi-8T	8101 Black	333**	?	
Mi-8T	8102 Black	333**	?	
Mi-8T	8104 Black	333**	?	
Mi-8T	8105 Black	333**	?	Sand/tan camouflage, Libyan Arab Air Force titles
Mi-8T	8108 Black	333**	?	
Mi-8T	8109 Black	333**	?	Red/white c/s
Mi-8T	8111 Black	333**	?	Red/white c/s
Mi-8T	8113 Black	333**	?	Sand camouflage
Mi-8T	8114 Black	333**	?	
Mi-8T	8216 Black	333**	?	
Mi-8T	8217 Black	333**	?	
Mi-8T	8218 Black	333**	?	
Mi-8T	8220 Black	33320?	?	
Mi-8T	8222 Black	33322?	?	
Mi-8T	8223 Black	33323?	?	Sand camouflage
Mi-8T	8224 Black	33324?	?	Green/brown camouflage; medevac (red crescent added to roundel)
Mi-8T	8229 Black	33329	?	Sand/tan camouflage
Mi-8T	8230 Black	33330	?	Sand camouflage
Mi-8T	8331 Black	33330	?	
Mi-8T	8332 Black	33330	?	
Mi-8T	8334 Black	333**	?	Sand/tan camouflage
Mi-8T	8335 Black	333**	?	Taken over by Free Libyan Air Force. Crashed on take-off at Murzuq 11-4-2012
Mi-8T	8340 Black	333**	?	Sand/tan camouflage
Mi-8MTV-1	2860 Black	95463	?	Libyan Air Ambulance, red/white c/s; ex-CCCP-25431
Mi-171	2861 Black	59489607570	04**	Libyan Air Ambulance, red/white c/s
Mi-14PL	LC1401	01301	?	
Mi-14PL	LC1402	01302	?	
Mi-14PL	LC1403	01303	?	
Mi-14PL	LC1404	01304	?	
Mi-14PL	LC1405	01305	?	
Mi-14PL	LC1406	01306	?	To Free Libya AF, captured at Benghazi (Benina AB) 23-2-2011
Mi-14PL	LC1407	01307	?	
Mi-14PL	LC1408	01308	?	
Mi-14PL	LC1409	01309	?	
Mi-14PL	LC1410	01310	?	
Mi-14PL	LC1411	01311	?	

Mi-14PL	LC1412	01312	?	
Mi-14PL	LC1413	01313	?	
Mi-14PL	LC1414	01314	?	
Mi-14PL	LC1415	01315	?	
Mi-14PL	LC1416	01316	?	Converted to, see next line
Mi-14PZh				Waterbomber
Mi-14PL	LC1417	01317	?	Converted to, see next line
Mi-14PZh				Waterbomber. Taken over by Free Libya AF, used for ambulance duties
Mi-14PL	LC1418	01318	?	
Mi-14PL	LC1419	01319	?	
Mi-14PL	LC1420	01320	?	
Mi-14PL	LC1421	01321	?	
Mi-14PL	LC1422	01322	?	
Mi-14PL	LC1423	01323	?	
Mi-14PL	LC1424	01324	?	
Mi-14PL	LC1425	01325	?	
Mi-14PL	LC1426	01326	?	
Mi-14PL	LC1427	01327	?	
Mi-14PL	LC1428	01328	?	
Mi-14PL	LC1429	01329	?	
Mi-14PL	LC1430	01330	?	
Mi-24A	1066 Black	?		
Mi-24A (late)	2027 Black	?		
Mi-24A	2102 Black	3202102?		
Mi-24A	2105 Black	3202105?		
Mi-24A	2108 Black	3202108?		
Mi-25?	103	***103?		
Mi-25	302 Black	013302	?	Captured by Chadian forces at Ouadi-Doum 3-87; to France, later to the USA, became *8-0616
Mi-35	853 Black	280853	?	Destroyed on the ground at Benghazi (Benina AB) by Free Libya rebels 23-2-2011
Mi-35	854 Black	280854	?	To Free Libya AF, captured at Benghazi (Benina AB) 23-2-2011. Shot down by government troops near Ajdabiya 9-4-2011
Mi-35	858 Black	280858	?	
Mi-35	860 Black	280860	?	
Mi-35	861 Black	280861	?	
Mi-35	863 Black	280863	?	
Su-22	5814	?		Serial in Arab characters
Su-22M	1506	?		Serial in Arab characters
Su-22M	74 Yellow	49407		Overhauled by ARZ No.558, Baranovichi
Su-22M3	304 White	45304?		1022 Sqn. Reported in error as Su-22UM-3K
Su-22M3	307 White	45312		
Su-22M3	311 White	47402		
Su-22M3	318 White	39702?		1032 Sqn
Su-22M3	?	47607		
Su-22UM3	109 Yellow	17532385705		Overhauled by ARZ No.558, Baranovichi
Su-22UM3	?	17532381905		Shot down 23-2-2011
Su-22UM3	?	17532383001		
Su-24MK	35 Black	416045*******	?	1124 Sqn
Su-24MK	36 Black	416045*******	?	1124 Sqn
Su-24MK	37 Black	416045*******	?	1124 Sqn
Su-24MK	38 Black	416045*******	?	1124 Sqn. Shot down by Free Libya rebels 5-3-2011
Su-24MK	39 Black?	416045*******	?	Existence not proven but likely
Su-24MK	40 Black	416045*******	?	1124 Sqn

Malagasy Air Force MiG-17F '242 Red'; note the squadron badge ahead of the serial.

Madagascar (Malagasy Republic)

Until 1958 Madagascar was a French colony. Shortly after the country achieved independence, an air transport wing was formed with French assistance; presently it developed into the Malagasy Air Force (*Armée de l'Air Malgache*, or *Tafika Anabakabaka* in the native language. A dispute with France in 1975 led to the withdrawal of French aid and the Soviet Union and its allies promptly stepped in, offering the usual 'humanitarian aid' of the time – including weapons.

Eight Mikoyan/Gurevich MiG-17F *Fresco-C* fighters were delivered (reportedly from North Korea), along with an unknown number of Mikoyan/Gurevich UTI-MiG-15 *Midget* trainers, and these aircraft together with some MiG-17s formed the *Escadrille de Chasse* (Fighter Squadron, or EdeCh). Only half of them were reportedly operational by late 1991 (some sources say none remained in service by 1991). The aircraft were based at Ivato; however, some of them were apparently transferred to Arivonimamo AB near Antananarivo.

In 1978 North Korea supplied eight MiG-21PFM (*izdeliye* 94A) *Fishbed-F* fighters and a single MiG-21U (*izdeliye* 66) Mongol-A trainer, and these aircraft joined the MiG-17s

'138 Black', a Malagasy Air Force MiG-21*bis*, sits at Tananarive in unserviceable condition. The type was probably involved in counter-insurgency operations in Madagascar.

Malagasy Air Force MiG-21UM '291 Black' and MiG-21*bis* '211 Black' hangared at Ivato AB.

Malagasy Air Force MiG-21*bis Fishbed-L* '211 Black'
Escadrille de Chasse, Tananarive-Ivato, 2005

with the EdeCh. There are unconfirmed reports of deliveries of MiG-21 variants prior to the substantiated arrival of about 12 MiG-21*bis* (*izdeliye* 75B) *Fishbed-L* and at least two MiG-21UM (*izdeliye* 69A) *Mongol-B* trainers from the Soviet Union. All surviving MiG-21s were placed in storage around 2000 and scrapped in 2007.

The transport element of the Malagasy Air Force consisted chiefly of Antonov An-26 *Curl* turboprops introduced around 1980. Eight examples are known to have been delivered; all of them were quasi-civil, with registrations in the 5R-M** block. For government VIP transportation, two Yakovlev Yak-40 *Codling* trijets (a VIP-configured Yak-40 'Salon' and a convertible Yak-40K) were delivered in 1978-79.

The rotary-wing element included at least one Mil' Mi-2U *Hoplite* training helicopter and at least six Mil' Mi-8 *Hips*, three of which were government VIP aircraft.

There are stories in circulation that the Malagasy Air Force was used to attack insurgents. If they are correct, then it is highly likely that MiG-21MF fighter-bombers were used against them but no records can be found.

This rather drab-looking Mi-8PS (5R-MOG) was used by the government of Madagascar.

Known Malagasy Air Force aircraft of Soviet origin				
Type	**Serial/registration**	**C/n**	**F/n**	**Notes**
An-26	5R-MUG	?		
An-26	5R-MUK	?		Wfu Ivato by 8-1996
An-26	5R-MUL	?		Wfu Ivato by 8-1996
An-26	5R-MUM	?		Wfu Ivato by 4-1995
An-26	5R-MUN	?		Wfu Ivato by 4-1995
An-26	5R-MUO	?		Wfu Ivato by 4-1995
An-26	8305	8305		Ex-Kuban' Airlines RA-26672, bought 1998. Reportedly to, see next line
	5R-MUP			Wfu Ivato by 5-2006
An-26B	5R-MVP	10208		Ex-RA-26021
MiG-17F	113 Red	?		WFU Arivonimamo
MiG-17F	242 Red	?		Last seen 1987

MiG-17F	933 Red	?		WFU Arivonimamo
MiG-21*bis*	94	N750*****	?	Noted Ivato 2005
MiG-21*bis*	108 Black	N750*****	?	Silver overall. Derelict Ivato
MiG-21*bis*	130 Black	N750*****	?	
MiG-21*bis*	138 Black	N750*****	?	Silver overall, large Tafika Anabakabaka titles. Derelict Ivato
MiG-21*bis*	164 Black	N750*****	?	Noted Ivato 2004
MiG-21*bis*	169 Black	N750*****	?	Noted Ivato 2004
MiG-21*bis*	173 Black	N750*****	?	Noted Ivato 2004
MiG-21*bis*	204 Black	N750*****	?	Noted Ivato 2005
MiG-21*bis*	211 Black	N750*****	?	Noted Ivato 2005
MiG-21*bis*	233 Black	N750*****	?	Noted Ivato 2005
MiG-21*bis*	255 Black	N750*****	?	Grey overall, large Tafika Anabakabaka titles
MiG-21*bis*	266 Black	N750*****	?	Noted Ivato 2005
MiG-21*bis*	291 Black	N750*****	?	Grey overall, large Tafika Anabakabaka titles
MiG-21UM	339 Black	?	?	Grey overall, large Tafika Anabakabaka titles
MiG-21UM	355 Black	?	?	
Mi-2U	5R-MVG	548450123		Wfu Ivato
Mi-8T	5R-MOD	33501	?	Derelict Ivato by 4-2008
Mi-8	5R-MOE	33502	?	
Mi-8	5R-MOF	33503	?	
Mi-8PS	[5R-M]OG	33504	?	Grey/black c/s. DBR 4-11-2002, wreck dumped Ivato
Mi-8PS	5R-MVF	105107	?	Ex-Aviabaltika LY-HAX
Mi-8PS	5R-MVX	105108	?	Ex-Aviabaltika LY-HAY. Possibly crashed near Antananarivo 20-10-2002
Yak-40 'Salon 1st Class'	5R-MUA	9840859		D/D 23-12-1978. Sold to Interpipe Group, the Ukraine, by 2002 as UR-88309
Yak-40K/'Salon'	5R-MUB	9940760		Sold to Interpipe Group by5-1998 as UR-88310

Mali

When the former French colony of Mali proclaimed independence on 22nd September 1960, all equipment deliveries from the Western world were promptly cut off and the government of the newly-established republic turned to the Soviet Union for assistance – which was readily extended in the hope of securing another political beachhead in Africa. The Mali Air Force (*Force Aérienne de la Republique du Mali* – FARMA) was formed in 1961. Its aircraft had alphanumeric serials with the same TZ- prefix as Mali's letter-only civil registrations, and these were apparently allocated on a 'first come, first served' basis, not in batches assigned to specific types.

The first Soviet fighters supplied to the Mali Air Force were five Mikoyan/Gurevich MiG-17F *Fresco-Cs* delivered in the mid-1960s (some sources say in 1971). A single Mikoyan/Gurevich UTI-MiG-15 *Midget* was also delivered for pilot training. The MiGs equipped an *Escadrille de Chasse* (fighter squadron) based at Sénou AB near Bamako and were reportedly still in service in late 1991; the *Midget* is now a gate guard at the base, wearing a fake serial. There have also been unconfirmed reports in 1973 of six MiG-15 *Fagot* fighters being oper-

Mali Air Force MiG-17Fs, including TZ-365, on the dump at Bamako-Sénou AB.

Though long since retired, MiG-17F TZ-346 was still substantially intact when this picture was taken.

ated; possibly the truth is that they were ordered but never delivered – or that the MiG-17s were misidentified as MiG-15s. One of the MiG-17Fs crashed fatally in November 1974 during Mali's brief border conflict with Burkina-Faso over the Agacher Strip.

In 1974 the USSR supplied 12 Mikoyan MiG-21*bis* (*izdeliye* 75B) *Fishbed-L* fighters and two MiG-21UM (*izdeliye* 69A) *Mongol-B* trainers, and in 2005 two MiG-21MF (*izdeliye* 96) *Fishbed-Js* (Gor'kiy-built examples with MiG-21*bis*-style instrument panels) and another MiG-21UM were bought from the Czech Republic. There may have been other deliveries, as there are unconfirmed reports of two more *Fishbeds*. A USAF mission to Mali reported seeing five MiG-21*bis* fighters at Gao in 1991.

The remains of a Mali Air Force MiG-21*bis* shot down by Burkinabe troops.

As far as transport aircraft are concerned, The Mali Air Force had four or five Antonov An-26 *Curl* tactical transports. The type was introduced in 1976 and operated until at least 1995.

On 18th March 1961 the Soviet Union and the Republic of Mali signed a trade agreement which, among other things, provided for help in the setting up of a national air carrier, Air Mali. Two days later the airline ordered a sizeable number of aircraft, including three Il'yushin IL-18V *Coot* medium-haul airliners. A further example – this time a newer IL-18D – was delivered in 1975 to replace a crashed aircraft. Apart from scheduled international flights (the destinations included Paris, Marseilles and Rabat), the IL-18s were used by President Modibo Keita on his trips outside the country, acting as a surrogate government flight. Eventually, however, all surviving Air Mali IL-18s were returned when the relations with the USSR turned sour.

The first Soviet helicopters in Mali were two Mil' Mi-4 *Hounds* delivered in 1962. A number of Mil' Mi-8 *Hip* utility helicopters followed in 1976. As for the Mil' Mi-24D *Hind-D* attack helicopter, this did not come until 2008 when two ex-Bulgarian examples were acquired, followed by two more in 2009.

Mali Air Force MiG-21UM TZ-358, Bamako, 2005

Mali Air Force MiG-21*bis* TZ-381 and TZ 385 (note the different serial style) stored at Bamako.

TZ-358, one of the four known Mali Air Force MiG-21UMs.

Known Mali Air Force aircraft of Soviet origin				
Type	**Serial/registration**	**C/n**	**F/n**	**Notes**
An-26	TZ-347	3303		Crashed 31-3-1995
An-26	TZ-349	?		
An-26 (?)	TZ-353	?		Also reported as An-24
An-26	TZ-359	5306		Two-tone green camouflage plus red/yellow/blue cheatline
An-26	TZ-399	14007?		Ex-Aerocom ER-AWG, bought 2004???
IL-18V	TZ-ABD	181003303		Ex/to Aeroflot Soviet Airlines CCCP-75762, regd 8-1-965; returned 2-6-70
IL-18V	TZ-ABE	181003304		Ex-Aeroflot Soviet Airlines CCCP-75763, regd 8-1-1965. Crashed Lingomin, Upper Volta, 11-8-1974
IL-18V	TZ-ABY	182004505		Returned ?-1972, fate unknown
IL-18D	TZ-ADF	187010301		Ex/to CCCP-75459, leased from Aeroflot 2-4-1974. WFU/stored Bamako 3-1984; returned 31-3-1987
UTI-MiG-15	TZ-337	?		Derelict Sénou AB by 2-2005. Serial also reported in error as MiG-17F. Became, see next line
	'TZ-001'			Fake serial; gate guard Senou AB by 2008
UTI-MiG-15	TZ-366	?		Tan/green/brown camouflage. Derelict Mopti-Sévaré AB by 1999
MiG-17F	TZ-339	?		Green/dark green/brown camouflage. Derelict Mopti-Sévaré AB by 1999
MiG-17F	TZ-346	?		Green/brown camouflage. Derelict Sénou AB, last seen 2008
MiG-17F	TZ-348	?		Green/brown camouflage. Derelict Sénou AB, last seen 2008
MiG-17F	TZ-360	?		Fighter-bomber with extra wing pylons. Derelict Sénou AB, last seen 11-2003
MiG-17F	TZ-365	?		Fighter-bomber with extra wing pylons. Green/dark green/brown camouflage. Derelict Sénou AB, last seen 11-2003
MiG-17F	TZ-367	?		Fighter-bomber with extra wing pylons. Gate guard Sénou AB, last seen 11-2003
MiG-17F	TZ-368	?		Fighter-bomber with extra wing pylons. Green/dark green/brown camouflage. Derelict Mopti-Sévaré AB by 1999
MiG-21MFN	TZ-356	N96005508	96-04-...	Ex-Czech Air Force '5508'
MiG-21MFN	TZ-357	N96005512	96-04-...	Brown/green/dark green/dark earth camouflage, white serial. Ex-Czech Air Force '5512'
MiG-21*bis*	TZ-371	N750*****	?	Dark green/light green/dark earth camouflage. Noted 12-2005; stored Bamako 2007
MiG-21*bis*	TZ-372	N750*****	?	Noted 2007 stored Bamako
MiG-21*bis*	TZ-377	N750*****	?	Tan/green/dark earth camouflage. Noted 2007 stored Bamako
MiG-21*bis*	TZ-378	N750*****	?	Noted 2008 stored Bamako
MiG-21*bis*	TZ-381	N750*****	?	D/d 1974; tan/green/dark earth camouflage. Noted 1997
MiG-21*bis*	TZ-382	N750*****	?	Noted 2007 stored Bamako
MiG-21*bis*	TZ-383	N750*****	?	Tan/green/dark green camouflage. Stored Sénou AB
MiG-21*bis*	TZ-384	N750*****	?	
MiG-21*bis*	TZ 385	N750*****	?	Tan/green/dark earth camouflage, serial painted with no hyphen. Stored Sénou AB
MiG-21*bis*	TZ-389	N750*****	?	D/d 1974; noted 1997
MiG-21UM	'TZ-BA-100'	?	?	Fake serial; preserved Bamako by 2008
MiG-21UM	TZ-358	?	?	Tan/green/dark earth camouflage. Noted 12-2005; stored Bamako 2007
MiG-21UM	TZ-364	?	?	Tan/green/dark green/dark earth camouflage. Noted 2007 stored Bamako
MiG-21UM	TZ-375	516999341	?	Brown/green/dark green/dark earth camouflage. Ex-Czech Air Force '9341'
MiG-21UM	TZ-376	?	?	Tan/green/dark earth camouflage. Noted 12-2005
MiG-21UM	TZ-377	?	?	Probably became, see next line
	'TZ-002'			Fake serial; preserved Bamako/Sénou AB by 2008
Mi-24D	TZ-404	150152	?	Ex-Bulgarian Air Force '117 White', d/d 2007, sand/tan camouflage
Mi-24D	TZ-405	150154	?	Ex-Bulgarian Air Force '119 White', d/d 2007, sand/tan camouflage
Mi-24D	TZ-406	U5086	?	Ex-Bulgarian Air Force '105 White', d/d 2009
Mi-24D	TZ-407	U5088	?	Ex-Bulgarian Air Force '107 White', d/d 2009

Mauritania

While Mauritania is certainly not among the traditional buyers of Soviet hardware and its derivatives, a pair of Xian Y7H-200A transports (a Chinese derivative of the Antonov An-26 *Curl*) were delivered to the Mauritanian Air Force (*Force Aérienne Islamique du Mauritanie*). They were registered in the 5T-M** quasi-civil series and wore a two-tone camouflage; 5T-MAF was delivered in 1996 and crashed into the Atlantic Ocean off the Mauritanian coast on 5th April 1996, while 5T-MAG (only the last two letters were worn visibly; ex-B-3719, c/n 007H03) was delivered in 1997 and crashed on 12th May 1998.

A rare photo of an operational Royal Moroccan Air Force MiG-17 serialled 11-1FK; for some reason the last letter is missing. The aircraft in the background is 08-1FH.

Morocco

After attaining independence from France on 2nd March 1956 in 'the Revolution of the King and the Young People' Morocco sought help wherever possible to build up its nascent armed forces, purchasing military equipment from East and West alike. The spread of Arab nationalism led to a rapprochement with the Soviet Union, and in 1961 Morocco accepted a Soviet offer for military assistance. On 10th February 1961 the Soviet freighter S/S *Karaganda* brought 12 (some sources say 20) Mikoyan/Gurevich MiG-17 *sans suffixe Fresco-A* fighters built by the Kuibyshev aircraft factory and two Czechoslovak-built Mikoyan/Gurevich UTI-MiG-15 (Aero CS-102) *Midget* trainers for the Royal Moroccan Air Force (*al Quwwat al-Jawwiya al-Malakiya Marakishiya*, or *Aviation Royale Chérifienne*). Nearly 100 Soviet instructors and support personnel came to Morocco to help the RMAF master both types. The fighters wore an overall light grey finish. One publication stated the serials of the *Frescos* as 1-FA/01, 1-FB/02 etc. (French Air Force style), but actually they were applied as 01 (on the nose)/1FA (on the rear fuselage), 02/2FB etc. Later, in 1963, four MiG-17F *Fresco-C* fighters were added to the fleet.

Additionally, the RMAF operated two or four Il'yushin IL-28 *Beagle* tactical bombers supplied by the Soviet Union in 1963.

However, King Hassan II looked more to the West than to the East, and relations with Moscow were soured in 1963 when Morocco went to war with Algeria and the Soviet Union supported the latter. Therefore the Soviet jets had a brief career with the RMAF, being mothballed in 1966 and remaining in storage until the early 1980s when most of them were scrapped. After that, Morocco operated only Western aircraft (mostly of US and French origin).

Known Royal Moroccan Air Force aircraft of Soviet origin

Type	Serial	C/n	Notes
UTI-MiG-15	07	522516	Stored Meknes 11-1978; preserved Casablanca-Anfa by 4-2006
UTI-MiG-15	?	522570	Stored Meknes 11-1978
MiG-17	01/1FA	1407040	Preserved Marrakech (Menaral AB) by 1-2008 in non-authentic camouflage
MiG-17	02/1FB	1405043	Stored Meknes 11-1978
MiG-17	03/1FC	1407046	Stored Meknes 11-1978
MiG-17	04/1FD	?	
MiG-17	05/1FE	1406029	Preserved Meknes (Bassatine AB) by 3-2006
MiG-17	06/1FF	1406044	Stored Meknes 11-1978
MiG-17	07/1FG	?	
MiG-17	08/1FH	1406030	Stored Meknes 11-1978
MiG-17	09/1FI	?	
MiG-17	10/1FJ	1406016	Preserved Museum of Flight, Seattle (USA), by 8-2009 as 'Vietnamese People's Air Force 7469 Red' (later repainted back to 10/1FJ)
MiG-17	11/1FK	1407018	Preserved Casablanca-Anfa by 4-2006
MiG-17	12/1FL	?	

Mozambique

Mozambique, a former colony of Portugal, gained independence on 25th June 1975. At first the People's Air Force (FPA – *Força Popular Aérea*) was part of the army but in 1990 it became a separate armed service as the Mozambique Air Force (*Força Aérea da Moçambique*, FAM).

The FPA took delivery of at least 30 Mikoyan/Gurevich MiG-17 *sans suffixe* (*Fresco-A*) and MiG-17F *Fresco-C* (including Polish-built PZL Lim-5) fighter-bombers, some of them acquired from the East German Air Force in the 1980s (one batch was delivered in 1981 and another reportedly in 1985). The ex-EGAF examples were provided with wing pylons allowing them to carry rocket pods. The fighters were based at Nacala; the type stayed in service at least until 1991. On 8th July 1981 the pilot of an FPA MiG-17 defected to South Africa; surprisingly enough, the aircraft was later returned.

For training the MiG-17 pilots the FPA also received a number of Mikoyan/Gurevich UTI-MiG-15 *Midget* trainers (some sources again say 30 aircraft). At least two of them had been clandestinely imported from East Germany in December (some sources say August) 1981. Two UTI-MiG-15s were wrecked in post-reassembly checkout flights in 1979. Three *Midgets* were reportedly still operational in late 1991.

At least 38 Mikoyan MiG-21MF (*izdeliye* 96) *Fishbed-J* fighters were reportedly delivered to Mozambique. In 1982 they were augmented by 48 Mikoyan MiG-21*bis* (*izdeliye* 75A and 75B) *Fishbed-L/N* fighters, complete with Cuban pilots and ground staff; it is believed that they were used against the RENAMO insurgents (see below). By 1990 only 18 *Fishbeds* were still in use. The colour of the serials appeared to be left to the discretion of the units but was mostly red or blue. After the 1990 ceasefire and the departure of the Cuban personnel the *Fishbeds* were withdrawn from use. No MiG-21U/US/UM *Mongol* trainers were ever seen in Mozambique; perhaps none were imported because the Cubans knew the type well enough.

Mozambique Air Force UTI-MiG-15 '20 Yellow' seen in 1981.

The Mozambique People's Air Force operated at least two Antonov An-12 medium transports. The *Cubs* were apparently on lease from the Soviet Air Force (or were actually Soviet aircraft masquerading in Mozambican markings for appearance's sake) and were eventually returned to the USSR. The FPA's principal transport type, however, was the Antonov An-26 tactical transport, of which at least nine were delivered, starting in 1978. The *Curls* wore a two-tone green camouflage.

The late-style Mozambique Air Force roundel. The badge in the centre features a hoe and a Kalashnikov assault rifle symbolising readiness for both peace and war.

On 30th September 1980 the Mozambique government flight (some sources say the FPA) took delivery of a Tupolev Tu-134AK *Crusty* VIP jet in the radar-nosed export version. Originally it wore a five-digit serial and a blue/white livery with the tail adorned by an obscure emblem resembling the famous Sputnik-1 satellite on a blue/red/black/yellow rectangle. By 1981 the aircraft had gained a civil registration and an orange/white livery with the elaborate coat of arms of the Mozambique People's Republic on the tail. On 1st August

Mozambique Air Force UTI-MiG-15 (Aero CS-102) '20 Yellow'

Mozambique Air Force MiG-17F '104 Red' with old-style roundels

An air-to-air shot of a grey-painted Mozambique Air Force MiG-17 *sans suffixe* serialled '21 Red'.

Mozambique Air Force MiG-21*bis* '247 Red' seen at Beira-Sofala airport is definitely the worse for wear, probably having been cannibalised for spares.

1986 it was upgraded to Tu-134A-3 standard by fitting improved Solov'yov D-30 Srs III engines instead of Srs IIs.

The aircraft was operated by a Soviet flight crew seconded from Aeroflot Soviet Airlines' Leningrad Civil Aviation Directorate/1st Leningrad United Air Detachment/344th Flight. Unfortunately on 19th October 1986 the Tu-134AK crashed fatally near Nelspruit (Natal Province, South Africa) en route from M'bala (Zambia) to Maputo as a result of unlawful interference with the purpose of eliminating President Samora Moises Machel.

More than 25 Mil' Mi-8 helicopters of an unspecified version (probably armed Mi-8T *Hip-Cs*) were delivered to the Mozambique Air Force by 1985. The FPA also took delivery of 15 Mil' Mi-24 attack helicopters of an unspecified version (probably Mi-25 *Hind-Ds*). Of these, only four aircraft remained operational in early 1995.

Mozambique Civil War

After Mozambique became independent, the Mozambique Liberation Front (FRELIMO – *Frente de Libertação de Moçambique*) – the largest of the groups fighting for independence – took over the government of the country. FRELIMO was effectively communist and, in addition to economic mismanagement, supported guerrilla groups opposing the governments of South Africa and South Rhodesia. Unsurprisingly, these two countries supported the RENAMO (*Resistencia Nacional de Moçambique* – Mozambique Resistance Movement) insurgents when a civil war began in Mozambique in 1977. The USA also backed

Mozambique Air Force MiG-21*bis Fishbed-L* '208 Red' Beira-Sofala, 1997

Mozambique Air Force An-26 '024 Black'.

the rebels until disenchanted by their atrocities and the use of child soldiers. At first, the rebels thrived, the government forces being fully occupied with South African Defence Force and Rhodesian Army attacks and air raids; South African and Rhodesian insurgents had set up training camps and supply dumps in Mozambique. The situation for the communist government in Maputo began to improve when Southern Rhodesia became Zimbabwe in 1980; its new government headed by Robert Mugabe was formed from the rebels Mozambique had helped in the recent past.

More military aid came from the Soviet Union, China and Cuba. Initially, strike sorties were flown by MiG-17Fs (occasionally piloted by Soviet instructors), but the *Frescos* were of little use, being only able to strafe their targets as the wing hardpoints were occupied by drop tanks. An-26 transports equipped with four bomb shackles provided a welcome boost. The arrival of MiG-21*bis* fighters complete with Cuban personnel in early 1982 boosted resources further; they reportedly flew their first strikes against RENAMO assets in May 1982. (Some sources say the MiG-21s did not arrive until 1984.) Nonetheless, RENAMO was becoming very successful, making repeated attacks on the railway linking the seaport of Beira with Zimbabwe, but eventually they went too far by cutting the Beira-Mutare oil pipeline in January 1983. Zimbabwe reacted strongly by sending three army battalions to police vital communications in the so-called Beira Corridor. There are no reports of the involvement of the Zimbabwe Air Force, presumably because it was being reorganised.

By 1985, Mozambique was becoming desperate at its lack of success in smashing the insurgency and appealed to Zimbabwe and Zambia for more help. About 12,000 troops were sent to spearhead an attack on RENAMO's headquarters at Casa Banana; the force included 2,000 paratroops, who were dropped near the town to capture the airfield. MiG-21*bis* and MiG-17 fighter-bombers blasted the rebels, and Mi-25s provided close air support. No sooner had the Zimbabwean contingent been reduced in early 1986 than RENAMO began a counteroffensive, recapturing Gorongosa which had been one of its bases. Mozambique requested help not only from Zimbabwe but also from Zambia, Malawi and Tanzania. A combined force supported by 17 Mozambican MiG-17Fs, as well as Zambian MiG-21s and Zimbabwean aircraft, launched an attack against Gorongosa; however, the RENAMO guerrillas had learned of the operation in advance and vacated the base.

The RENAMO rebels made claims to have shot down several aircraft, including two MiG-17s on 16th April 1985, a further MiG-17 and a MiG-21*bis* near Casa Banana on 28th August 1985, two more MiG-17s near Majanga de Costa on 6th October 1985, a MiG-21*bis* near Luabo on 10th October 1985, one more near Gorongosa on 12th April 1986, a MiG-17 on 20th November 1988 and a MiG-21*bis* near Beira on 13th June 1989, but these claims were never confirmed. On the other hand, it *is* confirmed that a MiG-17 was lost on 9th February 1983 together with its pilot, Capt. O. A. Muthemba. One more MiG-17 crashed into the sea off Beira on 9th October 1985, killing Lt. M. Muiambo. On 14th June 1989 a MiG-17F crashed while attempting to land in a storm, killing the pilot; the exact location of the crash is unknown.

After fierce fighting, the area was pacified; the insurgents withdrew and regrouped into smaller sections, making detection more difficult. The war continued with Zambian and Tanzanian troops becoming involved and more sophisticated counter-insurgency tactics being employed with closer co-operation between ground troops and attacking aircraft.

Following the elimination of President Samora Machel and other senior government

Known Mozambique Air Force aircraft of Soviet origin				
Type	Serial/registration	C/n	F/n	Notes
An-12	011	?		Leased from the Soviet Air Force
An-12	012	?		Leased from the Soviet Air Force
An-26	022 Black	6402		
An-26	024 Black	6405		
An-26	026 Black	?		
An-26	028 Black	?		
An-26	030 Black	10805		
An-26	032 Black	10806?		
An-26	033 Black	?		Stored Maputo by 1-2004
An-26	036 Black	?		
An-26	038 Black	?		Wfu Quelimanein by 8-2004
An-26	040 Black	?		Derelict Nampula by 1995
An-26	042 Black	?		Crashed 30-3-1986
UTI-MiG-15	20 Yellow	?		Grey c/s
UTI-MiG-15	22 Yellow	922259		Czechoslovak-built (Aero CS-102); ex-East German Air Force 150 Black, ex-EGAF 04 Black
UTI-MiG-15	?	622084		Czechoslovak-built (Aero CS-102); ex-East German Air Force 103 Black, delivery unconfirmed
MiG-17	01 Black	?		
MiG-17	02 Black	?		Seen 2-1997
MiG-17	07 Black	?		Seen 1991
MiG-17	16 Black	?		Seen 1991
MiG-17	17 Black	?		Seen 1991
MiG-17	21 Red	5238		Built in Komsomol'sk-on-Amur? Defected to South Africa 8-7-1981
MiG-17	23 Black	?		Ex-East German Air Force? Derelict Beira
MiG-17	24 Black	?		Seen 1991
MiG-17	26 Black	?		Seen 1991
MiG-17F	101	?		Preserved Beira by 9-2003 as '01'
MiG-17F	104 Red	?		Green/brown camouflage
MiG-17F	129	0619		Ex-East German Air Force 129 Red*, ex-EGAF 211 Red
MiG-17F	132	0479		Ex-East German Air Force 132 Red*, ex-EGAF 393 Red
MiG-17F	138	7134		Ex-East German Air Force 138 Red*, ex-EGAF 910 Red
Lim-5	128	1C 0822		Ex-East German Air Force 128 Red*, ex-EGAF 210 Red
Lim-5	130	1C 0614		Ex-East German Air Force 130 Red*, ex-EGAF 314 Red
Lim-5	131	1C 0622		Ex-East German Air Force 131 Red*, ex-EGAF 322 Red
Lim-5	133	1C 0619		Ex-East German Air Force 133 Red*, ex-EGAF 519 Red
Lim-5	134	1C 0628		Ex-East German Air Force 134 Red*, ex-EGAF 528 Red
Lim-5	135	1C 0919		Ex-East German Air Force 135 Red*, ex-EGAF 569 Red
Lim-5	136	1C 0824		Ex-East German Air Force 136 Red*, ex-EGAF 757 Red
Lim-5	137	1C 0616		Ex-East German Air Force 137 Red*, ex-EGAF 886 Red
Lim-5	139	1C 0816		Ex-East German Air Force 139 Red*, ex-EGAF 911 Red
Lim-5	?	1C 0426		Ex-East German Air Force 915 Red
Lim-5	?	1C 0436		Ex-East German Air Force 785 Red
Lim-5	?	1C 0602		Ex-East German Air Force 202 Red
Lim-5	?	1C 0604		Ex-East German Air Force 435 Red
Lim-5	?	1C 0804		Ex-East German Air Force 914 Red
Lim-5	?	1C 0807		Ex-East German Air Force 668 Red
Lim-5	?	1C 0909		Ex-East German Air Force 945 Red
Lim-5	?	1C 0913		Ex-East German Air Force 513 Red
Lim-5	?	1C 0914		Ex-East German Air Force 955 Red
MiG-21MF	04	?	?	
MiG-21MF	06	?	?	

MiG-21MF	11	?	?	
MiG-21MF	26	?	?	
MiG-21*bis*	149	N75096782	75-69-...?	1982
MiG-21*bis*	201	N750*****	?	1982
MiG-21*bis*	202 Red	N75095084	75-68-...?	1982
MiG-21*bis*	203	N75095095	75-68-...?	1982
MiG-21*bis*	204	N75095103	75-68-...?	1982
MiG-21*bis*	205	N750*****	?	1982
MiG-21*bis*	206	N750*****	?	1982
MiG-21*bis*	207	N75095145	75-68-...?	1982
MiG-21*bis*	208 Red	N7509****	?	1982
MiG-21*bis*	209	N750*****	?	1982
MiG-21*bis*	210	N750*****	?	1982
MiG-21*bis*	211	N750*****	?	1982
MiG-21*bis*	212	N750*****	?	1982
MiG-21*bis*	213	N750*****	?	1982
MiG-21*bis*	214	N750*****	?	1982
MiG-21*bis*	215	N750*****	?	1982
MiG-21*bis*	216	N750*****	?	1982
MiG-21*bis*	217	N750*****	?	1982
MiG-21*bis*	218	N750*****	?	1982
MiG-21*bis*	219	N750*****	?	1982
MiG-21*bis*	220	N750*****	?	1982
MiG-21*bis*	221	N75096037	75-69-...?	1982
MiG-21*bis*	222	N75096055	75-69-...?	1982
MiG-21*bis*	223 Blue	N750*****	?	1982; noted Beira (Sofala airport) 1997
MiG-21*bis*	224	N750*****	?	1982
MiG-21*bis*	225 Blue	N750*****	?	1982; noted Beira 1997
MiG-21*bis*	226	N75096105	75-69-...?	1982
MiG-21*bis*	227	N75096124	75-69-...?	1982
MiG-21*bis*	228	N75096176	75-69-...?	1982
MiG-21*bis*	229	N75096182	75-69-...?	1982
MiG-21*bis*	230	N750*****	?	1982
MiG-21*bis*	231 Blue	N75096189	75-69-...?	1982; noted Beira 1997
MiG-21*bis*	232	N750*****	?	1982
MiG-21*bis*	233	N750*****	?	1982; noted Beira 1997
MiG-21*bis*	234	N750*****	?	1982
MiG-21*bis*	235	N750*****	?	1982
MiG-21*bis*	236	N750*****	?	1982
MiG-21*bis*	237	N750*****	?	1982
MiG-21*bis*	238	N750*****	?	1982
MiG-21*bis*	239	N750*****	?	1982
MiG-21*bis*	240	N750*****	?	1982
MiG-21*bis*	241 Blue	N75096524	75-69-...?	1982
MiG-21*bis*	242	N75096601	75-69-...?	1982
MiG-21*bis*	243	N750*****	?	1982
MiG-21*bis*	244	N75096608	75-69-...?	1982
MiG-21*bis*	245	N75096624	75-69-...?	1982
MiG-21*bis*	246	N750*****	?	1982
MiG-21*bis*	247 Red	N750*****	?	1982; noted derelict Beira 1997, used for spares?
MiG-21*bis*	248	N750*****	?	1982
Mi-8	050	?	?	
Mi-8	051	?	?	
Mi-8	052	?	?	
Mi-8	053	?	?	
Mi-8	054	?	?	

Mi-8	055	?	?	
Mi-8	056	?	?	
Mi-8	057	?	?	
Mi-8	058	?	?	
Mi-8	059	?	?	
Mi-8	060	?	?	
Mi-8	061	?	?	
Mi-8	062	?	?	
Mi-8	063	?	?	
Mi-8	064	?	?	
Mi-8	065	?	?	
Mi-8	066	?	?	
Mi-8	067	?	?	
Mi-8	068	?	?	
Mi-8	069	?	?	
Mi-8	070	?	?	
Mi-8	071	?	?	
Mi-8	072	?	?	
Mi-8	073	?	?	
Mi-8	074	?	?	
Mi-8	075	?	?	
Mi-24	102	?	?	
Mi-24	103	?	?	
Mi-24	104	?	?	
Mi-24	105	?	?	
Mi-24	106	?	?	
Mi-24	115	?	?	
Tu-134AK	63457	(03)63457	5909	D/D 30-9-1980, blue/white livery. Became, see next line
	C9-CAA			Reregistered 1981, orange/white livery. Crashed near Nelspruit, South Africa, 19-10-1986

The serials marked * were effectively bogus and were assigned for the delivery to Mozambique after the aircraft had been officially struck off charge by the EGAF and scrapped.

Mozambique Government Tu-134AK C9-CAA, with detail view of the Mozambican coat of arms on the fin

officials in the contrived air crash, a new regime instituted a series of economic reforms that led ultimately to a negotiated peace on 15th October 1992 with a UN peacekeeping force based in the country for the next two years.

Namibian Defence Force Air Wing fin flash.

Namibia

Namibia gained independence from South Africa on 21st March 1990 and created an Aviation Squadron of its small Defence Force in July 1992; this was subsequently known as the Namibian Defence Force Air Wing (NDFAW, or *Namibiese Lugmag* in Afrikaans). Subsequently after the acquisition of two (some sources say four) Mikoyan MiG-23MLA *Flogger-G* fighters from Angola, the Namibian Air Force was formed on 13th March 2005. The MiG-23s apparently sat idle at Grootfontein AB, seeing no operational use. At least two Gor'kiy-built Mikoyan MiG-21*bis* *Fishbed-L* fighters and one MiG-21UM *Mongol-B* trainer were purchased, to be refurbished and upgraded by IAI at Lod in 2002; eventually, however, none of these were deliv-

Namibian Defence Force Air Wing An-26 NAF-3-644; note the mis-spelling of the titles ('Airforce') and the c/n 14307 carried on the fin

An-26 NAF-3-644 seen immediately after landing.

Known Namibian Defence Force Air Wing aircraft of Soviet origin				
Type	**Serial**	**C/n**	**F/n**	**Notes**
An-26	NAF-3-644	14307		Ex-Komiinteravia RA-26239, bought 2002 via the Ukraine; green/grey camouflage
MiG-21*bis*	?	N75085304	?	Noted Lod, Israel 4-2002; not delivered
MiG-21*bis*	?	N75097455	75-70-...?	Noted Lod, Israel 4-2002; not delivered
MiG-21UM	?	516999499	?	Noted Lod, Israel 4-2002; not delivered, to Royal Cambodian AF as '7122'!
Chengdu F-7NM	0310	7NM0310?		D/D 2006; based Grootfontein
Chengdu F-7NM	0311*	7NM0311?		
Chengdu F-7NM	0312*	7NM0312?		
Chengdu F-7NM	0313	7NM0313?		D/D 2006; based Grootfontein
Chengdu F-7NM	0314*	7NM0314?		
Chengdu F-7NM	0315	7NM0315?		D/D 2006; based Grootfontein
Chengdu F-7NM	0316*	7NM0316?		
Chengdu F-7NM	0317*	7NM0317?		
Chengdu F-7NM	0318*	7NM0318?		
Chengdu F-7NM	0319	7NM0319?		D/D 2006; based Grootfontein
Chengdu F-7NM	0320*	7NM0320?		
Chengdu F-7NM	0321	7NM0321?		D/D 2006; based Grootfontein
Guizhou FT-7NG	0330	?		
Guizhou FT-7NG	0331	?		
Mi-8T	H-804	333...?	?	Ex-Libyan Arab Air Force '81xx Black'

* Existence not confirmed but likely

ered. Additionally, twelve Chengdu F-7NM fighters and two Guizhou FT-7NG trainers (Chinese derivatives of the MiG-21F-13 and the MiG-21UM respectively; NM is a customer designator standing for Namibia) arrived on 23rd February 2005, being airlifted to Grootfontein by chartered Antonov An-124-100 Ruslan (*Condor*) freighters.

Libya took an active part in stocking the Namibian Air Force, donating two Antonov An-26 *Curl* transports in 2001; a third An-26 was bought from the Russian airline Komiinteravia in 2002. Also in 2001, Libya donated two Mil' Mi-8T *Hip-C* utility helicopters and two Mil' Mi-24V *Hind-E* attack helicopters, and more may have been delivered.

Niger

The Niger Air Force roundel.

The air component of Niger's armed forces – *Escadrille Nationale du Niger* – received its first Soviet aircraft in June 1997 when Libya donated one of its Antonov An-26 *Curl* transports. By 2008 Niger obtained at least three Mil' Mi-17 *Hip-H* helicopters in unarmed utility configuration and an unknown number of Mil' Mi-24 *Hind* attack helicopters (two of the latter crashed on 27th June 2008 – presumably after colliding in mid-air). The Mi-17s wore a sand/tan camouflage scheme with 'Forces Armées Nigeriennes' titles and the codes 'HA' and 'HC'; since all known aircraft of *Escadrille Nationale du Niger* have quasi-civil registrations in the 5U-Mxx block, it stands to reason that the helicopters were allocated the registrations 5U-MHA and 5U-MHC, and that the intervening 5U-MHB is also a Mi-17.

Nigeria

Nigeria became independent of British rule on 1st October 1960, and in January 1964 the Federal Nigerian Air Force (FNAF) was created. The first Soviet aircraft to reach Nigeria were eight Mikoyan/Gurevich MiG-17 *sans suffixe Fresco-A* fighters and 28 MiG-17F *Fresco-Cs* supplied for use against the Ibo separatists in the eastern province (the so-called State of Biafra). The first batch of eight MiG-17Fs was delivered to Kano on 13th August 1967; four more followed on 27th April 1968 as attrition replacements, a further 16 between 12th October and 4th November 1968, and the final eight (the *Fresco-As*) on 13th and 18th October 1969. The MiG-17Fs mostly came from Egyptian Air Force stocks (according to some sources, via the USSR); some of the aircraft may have been PZL Lim-5P (Polish-built MiG-17PF *Fresco-D*) interceptors. The *Fresco-As*, on the other hand, were ex-East German Air Force aircraft. The fighters were delivered to Kano in northern Nigeria by Soviet Air Force

Federal Nigerian Air Force MiG-17F NAF 612 with a roundel on the fin and a crudely hand-painted serial

Federal Nigerian Air Force MiG-17F NAF 618 with a fin flash; note how the roundel is placed inside the serial

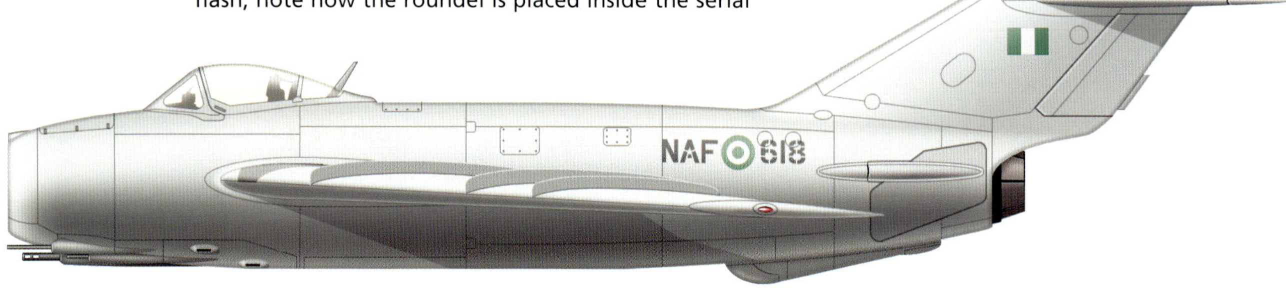

Antonov An-12 *Cub* transports. Another base used by the fighters was Kaduna, most of the MiGs ending their days there.

The fighters initially sported a natural metal finish, being supplied, as it were, before the Six-Day War and the introduction of camouflage in Egypt, but were eventually painted in various camouflage schemes. There was no clear standard of national insignia, three different styles being used at the same time. For example, MiG-17Fs serialled NAF 612 and NAF 614 had late-style markings with a green/white roundel on the fin and a crudely hand-painted serial; NAF 618, NAF 624 and NAF 625 had a stencilled serial and a Nigerian flag fin flash, while NAF 605 and NAF 615 also had a smaller roundel on the fuselage separating the two parts of the serial.

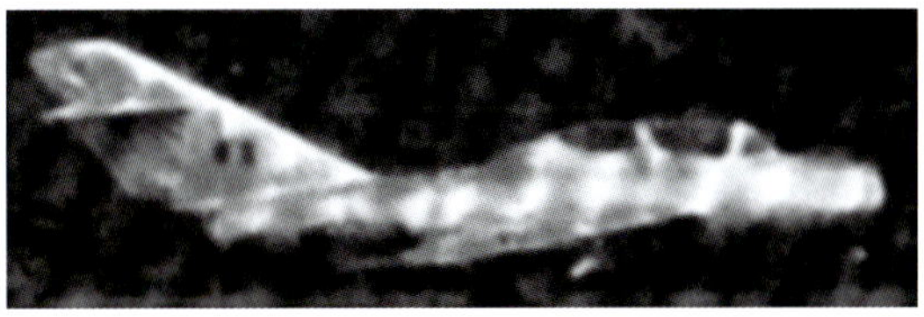

A poor but interesting photo of a Federal Nigerian Air Force UTI-MiG-15 wearing a fin flash.

The markings of Nigerian MiG-17Fs were applied haphazardly. NAF 612 here has a large roundel on the tail and a hand-painted serial...

...while NAF 615 has a small fin flash and a stencilled serial divided by an equally small roundel with no white colour.

Federal Nigerian Air Force MiG-21MF NAF 671

For training the MiG-17 pilots the FNAF also received a pair of Mikoyan/Gurevich UTI-MiG-15 *Midget* trainers (four, according to some sources) on 13th August 1967, and these were also based at Kaduna. According to some sources they were ex-Egyptian aircraft as well, which means they were Czechoslovak-built Aero CS-102s. Only two of the trainers were still operational by the mid-1980s.

The MiG-17 remained the FNAF's only jet fighter until 1975 when 25 Mikoyan MiG-21MF *Fishbed-J* supersonic fighters and six MiG-21UM *Mongol-B* trainers were delivered; they saw service with the 64th Air Defence Group at Makurdi. The single-seaters' serials were stencilled in red on the rear fuselage and the last two digits were repeated on the nose, also in red; the trainers probably had black serials. Accident attrition was heavy, and only 12 MiG-21MFs and five trainers remained by mid-October 1984. In 1985 the FNAF received an additional batch of 13 MiG-21*bis Fishbed-Ls* and three MiG-21UMs. In the 1990s lack of spares (and lack of cash with which to buy them) caused all Nigerian MiG-21s to be grounded and placed in storage. A US survey declared in 2000 that 22 *Fishbeds* were still there. Plans were put before the government in May 1990 to refurbish and upgrade them but they never reached fruition. However, increased Nigerian crude oil production and exports generated much-needed money, allowing the FNAF to order new Chinese-built *Fishbeds* – 12 Chengdu F-7NI fighters and three Guizhou FT-7NI trainers (once again, NI is a customer designator standing for Nigeria) – in 2005. These fighters are likewise operated by the 64th Air Defence Group. It took China a while to complete the order; the fighters were manufactured in 2009, but for some reason it was mid-2010 before they were finally delivered. The aircraft wear a tan/dark green/dark earth camouflage and green serials on the rear fuselage, with the last two digits again repeated on the nose (in black). On 22nd March 2011 one of the F-7NIs crashed at Kano, killing the pilot. On 11th May 2011 an FT-7NI crashed near Mbaniongu Mbazaam, Bebue State; this time the crew ejected safely.

Camouflaged Federal Nigerian Air Force MiG-21MFs and a silver-painted MiG-21US at Abuja.

A row of Nigerian MiG-21MFs, including NAF 671, shares the flight line with Aero L-29 Delfin trainers.

There have been unconfirmed reports that Nigeria had received 40 Mikoyan MiG-23BN *Flogger-H* fighter-bombers and ten MiG-23UB *Flogger-C* trainers to support their operations. However, the deal apparently did not materialise, as there have been no reports of Nigerian *Floggers* so far. Nor did Nigeria's intentions to obtain the Mikoyan MiG-29 fourth-generation fighter result in a delivery; the FNAF reportedly started negotiating the purchase in 1999, eventually ordering five MiG-29 (*izdeliye* 9.12B) *Fulcrum-A* single-seaters and one MiG-29UB *Fulcrum-B* trainer in 2001, but none have been reported.

In February 1968 Nigeria procured three second-hand Il'yushin IL-28 *Beagle* tactical bombers (two from Egypt and one from Algeria); some sources claim that the FNAF had at least six of the type and some of the bombers were purchased from Syria. The aircraft wore wraparound green/dark brown camouflage.

In March 2000 the FNAF ordered six Mil' Mi-35P *Hind-F* attack helicopters via Russia's Promexport arms export agency. The first two were delivered in mid-2000 after being refurbished by the Rostvertol plant in Rostov-on-Don (one of the two factories building the type). The *Hinds* are in service with the 97th Special Operations Group at Port Harcourt.

Another Mil' type operated by the FNAF is the rare Mi-34S *Hermit* light helicopter. Nine of these were delivered to the Training Command – specifically, the 305th Flying Training School at Enugu. The first three were taken on charge on 3rd April 2001, three more followed in January 2002 and the final three in August 2002.

The FNAF at war

The Nigerian MiG-17s and IL-28s saw action on home ground. In April 1967 a coup d'état occurred in Nigeria and the corrupt government was toppled by Gen. Irons, Commander-in-Chief of the Nigerian Armed Forces. A month later, however, Irons was killed in a new coup organised by Col. Ojukwu, the Governor of the Eastern province and one of the leaders of the Ibo tribe. The rebels declared their intention to secede, forming the so-called State of Biafra named after a bight in the Gulf of Guinea. This immediately sparked a bitter three-year civil war between the separatists and the federal government.

The FNAF originally used six impressed Nigeria Airways Douglas DC-3 airliners and 12 Aero L-29 Delfin advanced trainers against Ojukwu's rebels in the light bomber/paradrop and strike roles respectively. Pretty soon, however, it obtained real combat aircraft from Arab nations supporting the Islamic government in Lagos in its struggle against the Christian Ibo separatists. Egypt was the first to extend help, supplying 41 MiG-17Fs (misidentified as MiG-15s by some sources) and, together with Algeria, six IL-28s in 1969. The fighters were flown by British and South African mercenary pilots and based at Benin City. About 12 to 16 *Frescos* were operational at any one time, the rest remaining in storage. The bombers were likewise flown by mercenary crews (Egyptian ones this time). Operating from Enugu and Kalabar, the *Beagles* bore the brunt of the bombing missions but were reportedly found to be ineffective. They certainly did not venture near the Biafran capital, Uli, which was well protected by flak. Moreover, poor mission planning and the lack of a clearly-defined forward line of own troops (FLOT) sometimes resulted in own forces being bombed!

The war presented no great danger for FNAF fighter pilots and bomber crews, since the Biafran Air Force had no aircraft capable of air-to-air combat. At the start of the conflict the separatists had three bombers (a North American B-25 Mitchell and a pair of Douglas A-26 Invaders), plus an ex-Nigeria Airways Fokker F.27 Friendship airliner adapted for use as a bomber, but the B-25 and A-26s were soon destroyed on the ground by air raids on Enugu and Port Harcourt. After that, all the enemy could put up was Malmö MFI-9B primary trainers and two North American T-6G Texan primary trainers converted into makeshift attack aircraft. Flown by mercenary pilots led by the Swedish Count Carl Gustav von Rosen, the MFI-9s were known locally as Minicons – probably a corruption of 'mini-COIN' (counter-insurgency aircraft). They could only attack ground targets, which they did with a measure of success, shooting up a MiG-17 at Port Harcourt on 22nd May 1969 and destroying another at Benin City on 10th November 1969 (plus damaging several other aircraft, including an IL-28 at Enugu which was later repaired). The only combat loss of an FNAF MiG-17F was around July 1969; the aircraft was shot up so badly by flak while attacking a rebel gunboat on the Orashi River that it was declared a write-off. It was subsequently cannibalised for spares, thus saving the sister ship damaged at Port Harcourt.

On the other hand, the accident rate in the FNAF was rather high and more than 20 aircraft were totalled in the first year of the war. All the *Beagles* were soon grounded after being damaged in accidents. On 20th March 1969 one IL-28 struck trees during a low-level mission and suffered an engine failure, making a successful forced landing at Port Harcourt. Another *Beagle* veered off the runway at Port Harcourt on landing, burning out the brakes and tyres, but was later repaired.

The MiG-17s performed the usual fighter missions, managing to score a few 'kills'. On 9th November 1969 they shot down a Biafran T-6G over Binui Province; 20 days later they destroyed two Minicons, tailing them back to their jungle airstrip after a raid and shooting them up on the ground. Some *Frescos* flew

A Federal Nigerian Air Force IL-28. This *Beagle* retains the natural metal factory finish; most FNAF IL-28s wore various camouflage patterns.

ground attack missions, having been delivered with underwing launch rails for eight 76-mm Sakr rockets of the same type as used by the Egyptian Air Force. The MiGs also flew escort missions for the DC-3s on bombing/paradrop sorties, in which the IL-28s occasionally joined in, too. In February 1969 a group of DC-3s with such a mixed escort paradropped supplies and ammunition for a brigade of the 3rd Federal Army Division surrounded by rebel forces at Owerri. When attempts to unblock the brigade failed and the DC-3s were grounded for repairs after suffering flak damage, the government troops commandeered a Pan African Airways DC-4 which had landed at Port Harcourt, loading it with ammunition and ordering the captain to fly to Owerri, escorted by an IL-28. However, the captain cunningly simulated an engine malfunction, forcing a return to Port Harcourt – which probably saved both aircraft and crew.

A camouflaged Nigerian IL-28; unfortunately the serial is obscured.

There was also a tragic episode involving the shootdown of a civil airliner. The war had caused famine in Biafra, and the International Committee of the Red Cross (ICRC) organised the delivery of food and medicine to Uli – the sole Biafran airfield which remained intact, using chartered civil aircraft (for example, ex-USAF Boeing C-97s leased from the Swiss airline Balair). With typically bizarre wartime humour, the transports involved in the humanitarian airlift were dubbed 'powdermilk bombers' (possibly a play on the 'Berlin candy bombers' used in the 1948 Berlin airlift). However, the Nigerian government accused the ICRC of smuggling war materials to the separatists and the FNAF was ordered to shoot down any aircraft heading for Uli.

On 5th June 1969 Douglas DC-7B SE-ERP (c/n 45401, f/n 872) of Internord operated for the Swedish Red Cross was intercepted near Port Harcourt at dusk by a pair of MiG-17Fs on a routine patrol mission. Ignoring orders to change course and land at Port Harcourt, the airliner attempted to escape. Warning shots were also ignored. Next thing, one of the MiGs set one of the airliner's engines ablaze and the DC-7 exploded in mid-air, crashing near Eket and killing the crew – the American captain and three Swedish crewmen. Three days earlier, on 2nd June, a Flyghjälp ('Flight Aid')/ Swedish Red Cross Douglas DC-6 was strafed by a MiG-17F serialled NAF 617 while landing at Uli.

A camouflaged FNAF IL-28 shares the apron at Abuja with a MiG-17F.

Three more views of Nigerian IL-28s illustrating the varying (and rather crude) camouflage schemes.

A Nigerian IL-28 (the same as depicted above) taxies out at Abuja, showing what looks like a large identification letter (P) below the cockpit.

Federal Nigerian Air Force Mi-35P NAF 531

The Nigerian MiG-21MFs came too late to participate in the civil war, but they did get a piece of the action. In the early 1980s they delivered several strikes against targets in western Chad during security operations under the auspices of the African Union. A special unit, 'B Squadron' of the 64th Air Defence Group, was formed for this purpose at Maiduguri. In April 1983 FNAF MiG-21s attacked several fishermen's villages on the Chadian side of Lake Chad in a dispute over fishing, with a death toll of up to 90.

On 1st March 1978, just as MiG-21UM NAF 665 was practicing touch-and-go's at Kano airport, a Nigerian Airways Fokker F.28 Fellowship airliner registered 5N-ANA (c/n 11993) entered the landing pattern, inbound on a scheduled domestic flight from Sokoto. As the trainer climbed away to circuit altitude, the F.28 was already on the downwind leg, and the tower instructed the crew of the MiG to let the airliner land first. Apparently as a result of pilot error or miscommunication between the pilot and the tower the two jets collided at 1243 hrs local time when the airliner was on final approach to runway 05. Both aircraft crashed in flames 2 km (1.24 miles) from the runway; the pilots of the MiG-21 were killed, as were all five crew and 11 passengers on the F.28.

A striking picture of a shiny new Nigerian Mi-35P (NAF 531) in low-level flight.

A Federal Nigerian Air Force Mi-35P carrying drop tanks comes in to land.

A fine study of Mi-35P NAF 530.

Known Federal Nigerian Air Force aircraft of Soviet/Russian origin				
Type	**Serial**	**C/n**	**F/n**	**Notes**
IL-28	NAF 635	?		Photo exists
IL-28	NAF 801	?		
IL-28	NAF 802	?		
IL-28	NAF 803	?		
IL-28	NAF 804	?		

IL-28	NAF 805	?		Ex-Syrian AF '208'? Derelict Lagos by 1978; later preserved Nigerian War Museum, Umuahia
IL-28	NAF 806	?		
UTI-MiG-15	NAF 601	?		D/D 18-8-1967
UTI-MiG-15	NAF 602	?		D/D 18-8-1967
MiG-17?	NAF 630	?		D/D October-November 1968. Also reported as a MiG-17F! Preserved Kaduna (off base)
MiG-17	NAF 631	?		D/D 13-10-1969. Crashed at Makurdi 19-12-1969
MiG-17	NAF 632	?		D/D 13-10-1969
MiG-17	NAF 633	?		D/D 13-10-1969. Olive drab overall. Derelict Kaduna
MiG-17	NAF 634	?		D/D 13-10-1969. Preserved Kaduna
MiG-17	NAF 635	?		D/D 13-10-1969
MiG-17	NAF 636	?		D/D 18-10-1969
MiG-17	NAF 637	?		D/D 18-10-1969
MiG-17	NAF 638	?		D/D 18-10-1969
MiG-17F	NAF 603	?		D/D 13-8-1967
MiG-17F	NAF 604	?		D/D 13-8-1967
MiG-17F	NAF 605	?		D/D 13-8-1967, fitted with rocket launch rails. Gate guard at Kano
MiG-17F	NAF 606	?		D/D 13-8-1967
MiG-17F	NAF 607	?		D/D 13-8-1967
MiG-17F	NAF 608	?		D/D 13-8-1967, fitted with rocket launch rails
MiG-17F	NAF 609	0515317		D/D 13-8-1967
MiG-17F	NAF 610	?		D/D 13-8-1967
MiG-17F	NAF 611	?		D/D 27-4-1968; roundel on fin
MiG-17F	NAF 612	?		D/D 27-4-1968; roundel on fin
MiG-17F	NAF 613	?		D/D 27-4-1968; roundel on fin
MiG-17F	NAF 614	?		D/D 27-4-1968; roundel on fin
MiG-17F	NAF 615	?		D/D October-November 1968. Destroyed by a T-6G at Benin City 10-11-1969
MiG-17F	NAF 616	?		D/D October-November 1968. Crashed near Opobo 25-6-1969
MiG-17F	NAF 617	?		D/D October-November 1968. Crashed 12-1-1977
MiG-17F	NAF 618	?		D/D October-November 1968. Derelict Kaduna
MiG-17F	NAF 619	?		D/D October-November 1968. Derelict Kaduna
MiG-17F	NAF 620	?		D/D October-November 1968. Damaged beyond repair by MFI-9Bs at Port Harcourt 22-5-1969
MiG-17F	NAF 621	?		D/D October-November 1968
MiG-17F	NAF 622	?		D/D October-November 1968
MiG-17F	NAF 623	?		D/D October-November 1968. Crashed near Port Harcourt 19-7-1969, pilot killed
MiG-17F	NAF 624	?		D/D October-November 1968. Preserved Isingwu Amator in 1989; moved to Nigerian War Museum, Umuahia, by 2008
MiG-17F	NAF 625	?		D/D October-November 1968. Derelict Kaduna
MiG-17F	NAF 626	?		D/D October-November 1968
MiG-17F	NAF 627	?		D/D October-November 1968
MiG-17F?	NAF 628	?		D/D October-November 1968. Also reported as a UTI-MiG-15! Derelict Kaduna
MiG-17F?	NAF 629	?		D/D October-November 1968. Also reported as a UTI-MiG-15! Derelict Kaduna
MiG-21MF?	NAF 651	?	?	Also reported as MiG-21UM! Crashed 4-11-1981
MiG-21MF?	NAF 652	?	?	Also reported as MiG-21UM!
MiG-21MF?	NAF 653	?	?	Also reported as MiG-21UM! Gate guard at Makurdi AB
MiG-21MF?	NAF 654	?	?	Also reported as MiG-21UM!
MiG-21MF	NAF 656	?	?	
MiG-21MF	NAF 657	?	?	
MiG-21MF	NAF 658	?	?	
MiG-21MF	NAF 659	?	?	
MiG-21MF	NAF 660	?	?	Crashed 13-5-1986?
MiG-21MF	NAF 661	?	?	
MiG-21MF	NAF 662	?	?	Stored Makurdi AB 2009, for sale

MiG-21MF	NAF 663	?	?	Derelict Makurdi AB
MiG-21MF	NAF 664	?	?	
MiG-21MF	NAF 666	?	?	
MiG-21MF	NAF 667	?	?	
MiG-21MF	NAF 668	?	?	
MiG-21MF	NAF 669	?	?	
MiG-21MF	NAF 671	?	?	Tan/green camouflage; reported as delivered to 2nd Fighter Regiment
MiG-21MF	NAF 672	?	?	Crashed 19-6-1981
MiG-21MF	NAF 673	?	?	
MiG-21MF	NAF 674	?	?	
MiG-21MF	NAF 675	?	?	Crashed 1-10-1976
MiG-21MF	NAF 676	?	?	
MiG-21MF	NAF 677	?	?	
MiG-21MF	NAF 678	?	?	
MiG-21MF	NAF 679	?	?	
MiG-21*bis*	NAF681	N750*****	?	
MiG-21*bis*	NAF682	N750*****	?	Stored Makurdi AB 2009, for sale
MiG-21*bis*	NAF683	N750*****	?	Stored Makurdi AB 2009, for sale
MiG-21*bis*	NAF684	N750*****	?	Stored Makurdi AB 2009, for sale
MiG-21*bis*	NAF685	N750*****	?	
MiG-21*bis*	NAF686	N750*****	?	
MiG-21*bis*	NAF687	N750*****	?	
MiG-21*bis*	NAF688	N750*****	?	
MiG-21*bis*	NAF689	N750*****	?	Stored Makurdi AB 2009, for sale
MiG-21*bis*	NAF690	N750*****	?	Stored Makurdi AB 2009, for sale
MiG-21*bis*	NAF691	N750*****	?	
MiG-21*bis*	NAF692	N750*****	?	Stored Makurdi AB 2009, for sale
MiG-21*bis*	NAF693	N750*****	?	Gate guard at Abuja airbase
MiG-21*bis*	NAF694	N750*****	?	
MiG-21UM	NAF 650	?	?	
MiG-21UM	NAF 655	?	?	
MiG-21UM	NAF 665	?	?	D/D 1975. Crashed near Kano 1-3-1978 after colliding with an airliner
MiG-21UM	NAF 670	?	?	D/D 1985. Stored Makurdi AB 2009, for sale
MiG-21UM	NAF 680	?	?	D/D 1985
MiG-21UM	NAF 685	?	?	D/D 1985
Chengdu F-7NI	NAF 800	?		
Chengdu F-7NI	NAF 801	?		
Chengdu F-7NI	NAF 802	?		
Chengdu F-7NI	NAF 803	?		
Chengdu F-7NI	NAF 804	?		
Chengdu F-7NI	NAF 805	?		
Chengdu F-7NI	NAF 806	?		
Chengdu F-7NI	NAF 807	?		
Chengdu F-7NI	NAF 808	?		
Chengdu F-7NI	NAF 809	?		
Chengdu F-7NI	NAF 810	?		
Chengdu F-7NI	NAF 811	?		
Guizhou FT-7NI	NAF 812	?		Seen 4-2010
Guizhou FT-7NI	NAF 813	?		Seen 5-2010
Guizhou FT-7NI	NAF 814	?		Seen 5-2010
Mi-35P	NAF 525	?	?	Seen 8-2006
Mi-35P	NAF 526	?	?	Seen 8-2006
Mi-35P	NAF 527	?	?	Seen 8-2006

Mi-35P	NAF 528	?	?	Seen 8-2006
Mi-35P	NAF 529	?	?	Seen 8-2006
Mi-35P	NAF 530	?	?	Seen 10-2009; sand/green/dark green camouflage
Mi-35P	NAF 531	?	?	Seen Port Harcourt 8-2009 in damaged condition; sand/green/dark green camouflage
Mi-35P	NAF 532	?	?	Wfu Port Harcourt by 8-2009
Mi-35P	NAF 533	?	?	Damaged 2-2008 but repaired by 8-2008; sand/green/dark green camouflage
Mi-35P	NAF 534	?	?	Seen Port Harcourt 8-2009 in damaged condition
Mi-35P	NAF 535	?	?	Wfu Port Harcourt by 9-2009
Mi-24P	NAF 536	?	?	Ex-Ukrainian Air Force, seen 5-2010; tan/olive drab camouflage
Mi-24V	NAF 537	?	?	Ex-Ukrainian Air Force, seen 5-2010; tan/olive drab camouflage
Mi-24V	NAF 538	?	?	Ex-Ukrainian Air Force, seen 5-2010; tan/olive drab camouflage
Mi-34S	NAF 550?	978305*******		Existence not confirmed but likely
Mi-34S	NAF 551	978305*******		
Mi-34S	NAF 552	978305*******		
Mi-34S	NAF 553?	978305*******		Existence not confirmed but likely
Mi-34S	NAF 554?	978305*******		Existence not confirmed but likely
Mi-34S	NAF 555	978305*******		
Mi-34S	NAF 556	978305*******		
Mi-34S	NAF 557	978305*******		
Mi-34S	NAF 558?	978305*******		Existence not confirmed but likely

Rwanda

The Rwandan Air Force (*Force Aérienne Rwandaise*) did not operate Russian aircraft until the late 1990s. On 24th-27th October 1999 it took delivery of three Mil' Mi-17MD (Mi-17V-5) utility helicopters. Two more Kazan'-built Mi-17s, including a Mi-172 presidential VVIP aircraft, were added later.

Rwanda also acquired two Mil' Mi-24 *Hind* attack helicopters of an unspecified version from Belarus in 1997. These were the first aircraft delivered to the Rwandan Air Force since the 1994 civil war during which all available aircraft had been destroyed. In early 1999 Rwanda signed a contract for the delivery of further *Hinds* – this time cannon-armed Mi-24P *Hind-Fs* – with the Consolidated Sales Corporation (CSC), a company registered in the British Virgin Islands.

There is evidence of a Rwandan-registered Antonov An-72 *Coaler-A* STOL transport, 9XR-CV (c/n unknown), operated by Regional International Air Services in 2003. This aircraft may have had some military applications.

Rwandan Air Force Mi-17 RAF-1905

A fine early morning shot of Rwandan Mi-17 RAF-0407.

Opposite page: Two Rwandan Air Force Mi-17V-5s (RAF-0210 and RAF-1707) on a pre-delivery test flight from Kazan'.

Valuables being shifted to and from Mi-17 RAF-1905 sporting a rather gaudy colour scheme.

Known Rwandan Air Force aircraft of Soviet/Russian origin				
Type	**Serial**	**C/n**	**F/n**	**Notes**
Mi-17V-5	RAF-0210	?	?	Sand/brown camouflage
Mi-17-1V	RAF-0407	?	?	Sand/brown camouflage
Mi-17V-5	RAF-1510	?	?	Seen 8-2008
Mi-17-1V	RAF-1705	?	?	Seen 5-2006
Mi-17V-5	RAF-1707	?	?	Sand/brown camouflage. Seen 2-2002
Mi-17-1V	RAF-1905	?	?	Seen 7-2009; stepped nose but clamshell doors, sand/green/brown c/s
Mi-172	RAF-1907	?	?	Seen 11-2000; stepped nose, sand/brown camouflage
Mi-17-1V	?	646M05	?	Seen 5-2009
Mi-17-1V	?	646M09	?	Seen 5-2009
Mi-17-1V	?	646M10	?	Seen 5-2009
Mi-24V	RAF-0102	?	?	Seen 7-2007, sand/brown camouflage
Mi-24V	RAF-0110	?	?	Seen 12-2004, sand/brown camouflage
Mi-24P	RAF-0802	?	?	Seen 4-2004
Mi-24P	RAF-2102	?	?	Seen 8-2009
Mi-24P	RAF-2308	?	?	Seen 8-2009

Two Rwandan Air Force Mi-35s, with RAF-1001 nearest to camera.

Senegal

The Senegal Air Force (*Armée de l'Air du Senegal*) took delivery of a small number of Russian helicopters after the turn of the century. Like all other Senegalese military aircraft, they are quasi-civil and registered in the 6W-S** block. Known examples include two Mil' Mi-2 *Hoplite* utility helicopters first noted in December 2006.

Two Kazan'-built Mil' Mi-17 *Hip-H* utility helicopters were likewise first noted in December 2006; the construction number of one example has been quoted in error as 96862 but a Mi-8 with this c/n did not yet exist then. Confusingly, with the Mi-8/Mi-17 family, the same specific designation is sometimes used for helicopters with considerable structural differences. Though listed as Mi-17V-5s,

Senegal Air Force Mi-17 6W-SHT/6862 carrying four UB-32A pods. Note the non-standard radome with a bulged bottom.

The Senegal Air Force 'stars and bars' insignia, which is not carried by the Mil' helicopters.

Senegal Air Force Mi-17 6W-SHT/6862

Another view of 6W-SHT/6862, with sister ship 6W-SHU/6861 visible beyond. Though listed as Mi-17V-5s, they are unique in having the old glazed nose (plus a non-standard radar) instead of a stepped 'dolphin nose', and were possibly built to order in this configuration.

Senegal Air Force Mi-35P 6W-SHZ seen after lift-off.

This view of 6W-SHZ shows the additional serial 07 worn on the port side only.

The other Senegalese Mi-35P, 6W-SHY/04, ground-runs its engines.

and showing the main features of this version (a double-width port side entry door, a normal-width entry door to starboard and a cargo ramp instead of clamshell doors), the Senegalese examples have the rounded glazed nose typical of earlier versions of the *Hip* (plus a weather radar in a small radome of non-standard shape) instead of the usual stepped nose.

The Mil' Mi-35P *Hind-E* attack helicopter was added to the inventory by 2008. Two examples have been seen so far. The helicopters are based at Dakar-Ouakam airfield.

Known Senegal Air Force aircraft of Soviet/Russian origin				
Type	Registration/serial	C/n	F/n	Notes
Mi-2	6W-SHR	?		Seen 12-2006
Mi-2	6W-SHS/601	?		Seen 12-2006
Mi-17V-5	6W-SHT/6862 White	?	?	Seen 12-2006; non-standard glass nose, grey c/s. Not c/n 96862!
Mi-17V-5	6W-SHU/6861 White	?	?	Seen 12-2006; non-standard glass nose, grey c/s. Not c/n 96861!
Mi-35P	6W-SHY/04 Black	?	?	Two-tone green camouflage, code on port side only
Mi-35P	6W-SHZ/07 Black	?	?	Two-tone green camouflage, code on port side only

Sierra Leone Air Force Mi-24V SL202 flying over the Atlantic coast. Note the two Mi-17s in the top left-hand corner.

Sierra Leone

In Sierra Leone, where a civil war had been going on since 1991, the ill-trained government forces were steadily losing ground to the Revolutionary United Front (RUF) – in effect, a bunch of thugs led by Alfred Foday Sankoh, a corporal with Libyan affiliations, anti-Western ideas and a determination to seize power. In an attempt to hold its own against the insurgents, in the early 1990s the Republic of Sierra Leone Military Force (RSLMF) purchased a pair of second-hand Mil' Mi-24V *Hip-E* attack helicopters – reportedly for US$9 million. Only one of them was serviceable, the other one apparently being used as a spares source. The helicopter was based at Freetown-Lungi airport and flown by Belorussian contract pilots.

Another in-flight study of a Sierra Leonean Mi-24V carrying two drop tanks.

Known Sierra Leonean military helicopters of Soviet origin				
Type	**Serial/registration**	**C/n**	**F/n**	**Notes**
Mi-17	9L-LBD/1 Black	520M15	?	Ex-G-BVXN
Mi-17	9L-LBE/2 Black	520M16	?	Ex-G-BVXO
Mi-24V	AF-0010	3532421622258	?	Ex-Belorussian Air Force. Became, see next line
	SLAF-001			SLAF = Sierra Leone Air Force
Mi-24V	SL-201	?	?	Ex-Ukrainian Air Force. DBR 19-10-2001; wfu Cockerill Barracks
Mi-24V	SL-202	?	?	Ex-Ukrainian Air Force

Additionally, several Sierra Leonean Antonov An-12 *Cub* and Il'yushin IL-76 *Candid* freighters may have been used to perform a military airlift role, since the RSLMF has no proper transports.				
Type	**Registration**	**C/n**	**F/n**	**Notes**
An-12B (+)	9L-LCR	4341801		Showa Air; ex-CCCP-12166, acquired by 7-2003. Damaged beyond repair Goma, Zaïre, 27-5-2003
An-12B (+)	9L-LDW	2340804?		Showa Air; ex-Avial'-NV RA-11324?, acquired by 7-2003
An-12BP (+)	9L-LEA	5343408		Ex-Avia-Pusk UN 11001, operated in late 2003. Sold to Air Armenia 12-2003 as EK-11001
An-12A (+)	9L-LFQ	2400502		Ex-RA-48970, operated mid-2006
IL-76TD 'Falsie'*	9L-LBK			Air Sultan Ltd.; bought 10-1996, ex-RA-76383. Sold to Atlas Air 2-1998 as EP-ALB
IL-76TD	9L-LBO	1023408265	8207	Intalair; ex-Beriyev Design Bureau RA-76355, bought 1997. Sold to Atlas Air as EP-ALD

* The aircraft retained the tail gunner's station of the IL-76MD UR-86909 from which it was converted

The RSLMF also operated a single Mil' Mi-17-1V *Hip-H* armed utility helicopter, likewise flown by a Belorussian contract crew and presumably acquired in Belorussia. While greatly intimidating the rebels, the helicopters had difficulty finding and attacking them in the dense jungle.

In 1995 the rebels were getting dangerously close to the capital, Freetown. In desperation Sierra Leone's ruling military junta, the National Provisional Ruling Council (NPRC), hired the aforementioned proactive security organisation Executive Outcomes to conduct operations against the RUF and provide tactical planning and personnel training. Only then did the situation begin to improve. Within a week (not counting training and mission planning/preparation time), Executive Outcomes accomplished more than the RSLMF had in four years of bush warfare, and with far fewer casualties. Soon EO and the government troops had recaptured the diamond-rich Kono district, and the first of the 30,000-plus refugees who had fled the area began to return. Eventually the RUF was forced to call a ceasefire (though mainly in order to regroup and re-arm, as intelligence reports plainly indicated) and begin negotiations with the government. Thus, ironically, the 'dogs of war' had succeeded in securing peace – something everybody else involved had failed to!

The *Hind* was used mostly as a troopship and an escort aircraft, providing top cover for the Mi-17-1V gunship. In addition to the latter, EO brought its own Mi-17s in unarmed utility configuration, which were purchased in the UK via the Corporate Management company and nominally operated by the local airline Ibis Air. The aircraft operated by EO were based at Freetown-Lungi, but *ad hoc* helipads were also established near Koidu. The Mi-17s inserted and extracted commando groups and mortar teams and carried supplies. Close air support (CAS) missions were also flown occasionally. For these, the Mi-24 carried UB-32A rocket pods with thirty-two 57-mm S-5 rockets each, and sometimes GUV gun pods; depending on the target, the pods could be configured with either three Gatling machine-guns or a 30-mm AGS-17 grenade launcher (which was found to be very effective against area targets). The flexible nose-mounted YakB-12,7 machine gun was used for strafing and described as 'deadly accurate'. It should be noted that the Belorussian pilots were Afghan War veterans and used their Afghan experience to advantage. Drop tanks were rarely carried because they only left two wing stations free for weapons. The problem of the Mi-24's limited range was tackled by airlifting fuel drums to the helipads from which it operated in the Mi-17s.

The main problem was that neither the Belorussians nor the South Africans spoke each other's language. Hence a Sierra Leonean officer who spoke Russian had to be carried in the cabin as a tactical interpreter between the two; this practice complicated operations and increased the risk of 'friendly fire' casualties during CAS missions.

Even though all EO pilots had taken conversion training in the Mi-24, the Belorussians would not let them fly it on actual operations, clearly regarding the South Africans as competitors. Eventually, however, Executive Outcomes convinced the Sierra Leonean government that the helicopter's potential was not being used fully due to the language barrier. Before long, South African crews were allowed to fly the *Hind* – and were impressed; to quote EO's chief pilot Charlie Tait, *'it's like flying a fighter! [...] For an aircraft of 13 tons she is exceptionally agile and manoeuvrable'*. The cockpit ergonomics also earned some praise, as did the air conditioner; in Tait's words, *'designed for an NBC environment, it also just happens to make operations in the tropics quite enjoyable'*.

The EO crews grew quite fond of the Mi-24, bestowing nicknames on all three helicopters. The Mi-24V was dubbed 'Renoster' because its rather ugly and menacing looks, with the protruding machine-gun in the nose, gave it a certain resemblance to a rhinoceros. Mi-17 9L-LBD was christened 'Bokkie' because it was as sprightly as a young springbok antelope, while sister ship 9L-LBE was dubbed 'Daisy' because it behaved a bit like a cow.

On 24th April 1999 Sierra Leone purchased an additional two Mi-24s (one of which was reportedly a Mi-24V and the other an older Mi-24D *Hind-D*) from Ukrainian surplus stocks via the abovementioned Consolidated Sales Corporation. The helicopters would be used both against the RUF guerrillas and for border protection.

Somalia

The Somali Republic was created in 1960 from former British and Italian colonies, and the Somalian Aeronautical Corps (SAC, or *Dayuuradaha Xooga Dalka Somaliyeed* in Somalian) was established that same year with the primary task of supporting the army. In December 1960 it was officially renamed the Somalian Air Force (CCS, or *Ciidanka Cirka Somaliyeed*). Its aircraft fleet was initially Western but from 1963 onwards, in exchange for the use of three Somalian seaports by the Soviet Navy, the Soviet Union supplied military hardware in large amounts and sent military advisors to Somalia.

The aircraft supplied to the SAC included up to 20 Mikoyan/Gurevich MiG-15*bis Fagot-B* fighter-bombers – enough to equip one squadron based at Hargeisa, plus seven UTI-MiG-15 *Midget* trainers. Two of the latter were reportedly still operational in the 1980s

In 1963-67 the Somalian Aeronautical Corps took delivery of at least 12 late-production Mikoyan/Gurevich MiG-17 *sans suffixe Fresco-As* (the total number has been reported as 20 or even 54). Most of them were based at Mogadishu airport, but a MiG detachment was also established at Hargeisa in the former British Somaliland. The MiG-17s wore two-tone green or green/brown camouflage and had serials in the 100 block, though some of these were allocated to MiG-15s. Many of the *Fresco-As* were lost during Somalia's war with Ethiopia in the summer of 1977. The survivors quickly became unserviceable in the 1980s after Somalia terminated its friendship treaty with the USSR in November 1977. There have also been claims that the SAC received about 40 MiG-17F *Fresco-Cs*, only ten of which were flyable. Yet, there is no reliable evidence of this – no Somalian MiG-17Fs have been positively identified.

In 1974 Somalia received the first ten Mikoyan MiG-21MF (*izdeliye* 96F) *Fishbed-J* fighters and four MiG-21UM (*izdeliye* 69)

Below: Still in natural metal finish and wearing no serial, a Somalian Aeronautical Corps UTI-MiG-15 'burns rubber' on landing. The MiG-15*bis*-style airbrakes suggest it is an ex-Polish Air Force SBLim-2.

Bottom: A camouflaged Somalian MiG-17. The aircraft appears to be tied down with chains.

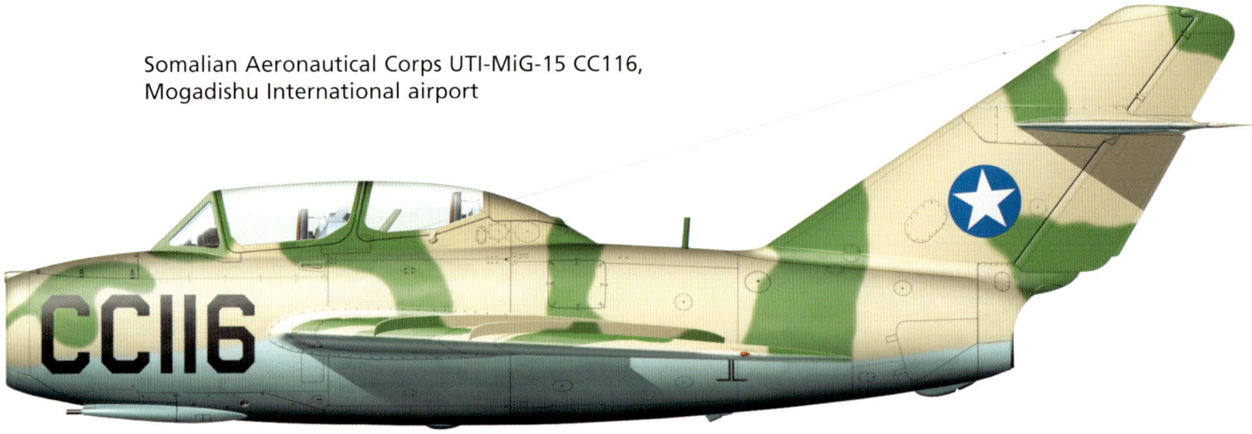

Somalian Aeronautical Corps UTI-MiG-15 CC116, Mogadishu International airport

'Kilroy was here'. Covered in graffiti, the hulk of UTI-MiG-15 CC116 at Mogadishu airport has suffered the attentions of vandals. Wrecked SAC MiG-21s are visible beyond.

Mongol-B trainers. It is not certain how many ultimately arrived but most sources suggest a maximum of 45 fighters and ten trainers. By 1993 the surviving examples had been withdrawn from use.

In November 1977 Somalia terminated its friendship treaty with the USSR because the latter had backed Ethiopia in the Ogaden War of 1977-78. Hence Soviet military aid was cut off, forcing the Somalian Aeronautical Corps to re-equip with Western types and Chinese copies of Soviet aircraft. These included up to 50 Shenyang F-6C fighters (the export version of the J-6C, an improved Chinese version of the Mikoyan/Gurevich MiG-19S *Farmer-C*) and eight Shenyang FT-6C trainers (the export version of the JJ-6C, which had no direct Soviet equivalent). Deliveries began in 1980, the

Somalian Aeronautical Corps F-6C 460, Mogadishu

Somalian Aeronautical Corps MiG-17 CC125

Its cockpit covered by a tarpaulin, this weathered but intact SAC F-6C is examined by US Marines in a hangar at Mogadishu. Another F-6C is visible beyond. Note the 'No smoking' sign in Italian (Vietato fumare), a left-over from colonial days.

Seen from under the wing of a SAC An-24, a Somalian FT-6C trainer languishes on the dump at Mogadishu. Note the AFEOD (Air Force – Explosive Ordnance Disposal) titles scrawled on the nose by US servicemen to indicate that the aircraft is 'safe', with no ammo that may detonate accidentally.

With a Somalian Aeritalia G.222 (M.M.60214/AM94) as a backdrop, MiG-21MF '224' sits forlorn at Mogadishu airport in February 1993. Note the damage to the fuselage nose.

fighters being based at Mogadishu and Hargeisa; Somalian F-6Cs originally wore an overall white finish but were subsequently repainted in a haphazardly applied three-tone camouflage – mustard yellow, olive drab and a rather yucky swampy green with bright blue undersurfaces. By 1993 all of them, too, had fallen into dereliction.

In 1972 the Somalian Aeronautical Corps reportedly operated ten Il'yushin IL-28 *Beagle* tactical bombers. Unfortunately no details are known.

The transport element of the SAC included at least three Antonov An-2 *Colt* utility biplanes delivered in the 1960s; it is not known whether they were Soviet-built or Polish-built. Heavier aircraft included three Antonov An-24B *Coke* airliners (in VIP configuration) bought in 1976 and two Antonov An-26 *Curl* tactical transports. The turboprops had five-digit serials in the 60*** block with the Italian Air Force-style prefix M.M. (*matricola militare* – military registration) plus alphanumeric codes. (Some sources, however, suggest this was not '60' but in fact 6O-, the Somalian nationality prefix.) All other aircraft of the Somalian Air Force had two- or three-digit serials, occasionally prefixed CC (for *Ciidanka Cirka*).

A number of Mil' Mi-8T *Hip-C* helicopters in armed utility configuration were also supplied.

Wrecked Somalian military aircraft at Mogadishu International Airport, including MiG-21MFs and a MiG-21UM. An American soldier examines the non-airworthy fighters.

Ogaden War and Somalian Civil War

Ethiopia was in a chaotic state for several years after the overthrow of Emperor Haile Selassie I in September 1974, and the Somalian dictator Gen. Siad Barré decided to take advantage of this to annexe the Ogaden province bordering on Somalia. Because of the large Islamic Somali population, Somalia regarded the Ogaden as its own and supported the Western Somali Liberation Front (WSLF) and Somali-Abo Liberation Front (SALF) insurgency in the province with the purpose of incorporating it into a 'Greater Somalia'. Without declaring war, the Somali Army invaded the Ogaden on 13th July 1977. It took the Addis Ababa government two weeks to realise that an aggression had taken place and take countermeasures, and by the end of the month 60% of the province was occupied

The Somalian MiG-15s, MiG-17s and MiG-21MFs, some of them flown by Iraqi and Syrian pilots, took part in the invasion. Their Ethiopian opponents used Northrop F-5A Freedom Fighters, mostly flown by Israeli volunteers. These aircraft were inferior in performance to the MiG-21MF but their AIM-9J Sidewinder missiles outclassed the Somali's R-13s.

The Soviet Union, while supplying arms to both sides, persisted in negotiating a ceasefire. At the pinnacle of his army's success Siad Barré, incensed that his troops were being halted in their advance by an enemy with Soviet-supplied weapons, unilaterally terminated the Soviet-Somalian Friendship Treaty in November 1977 and expelled 6,000 Soviet advisors, demanding that they leave by 13th November. As a result, Soviet aid to Somali was discontinued and that to Ethiopia appre-

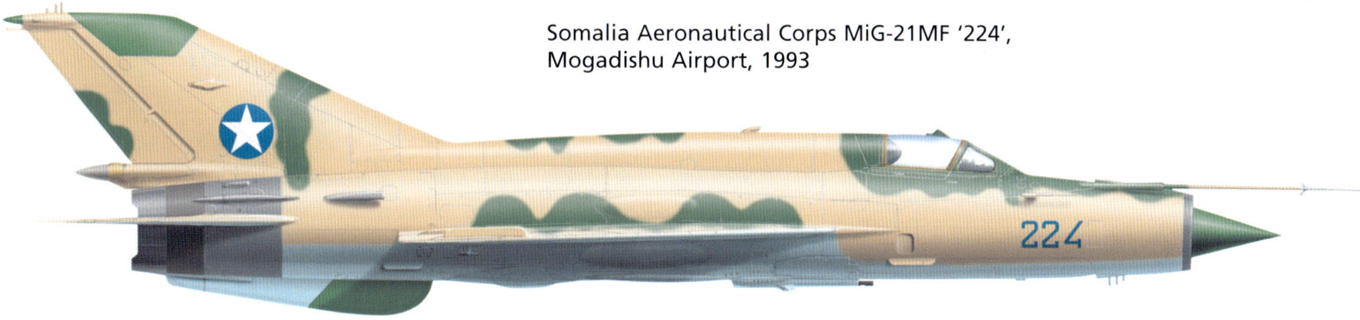

Somalia Aeronautical Corps MiG-21MF '224', Mogadishu Airport, 1993

SAC An-26 M.M.60207/AM88 in its better days; the colour scheme echoes one of the pre-1973 liveries of Aeroflot Soviet Airlines' An-24Bs. Note the 'flying cheetah' emblem and the letter F below it.

Camouflaged Somalian An-26 60204 (ex-Libyan Arab Air Force '8204') parked at Nairobi. Note the 'LAAF' tail titles amended to 'SAF'.

ciably increased. Additionally, a Cuban-manned squadron of MiG-17Fs was from Yemen. An urgently organised massive airlift of arms came just at the right time for the Ethiopians as the Somalis launched their final assault on 28th November, and on 3rd February the Ethiopians launched their counterattack. The Somalis were relentlessly driven back, and on 9th March 1978 Barré ordered his forces to pull out of Ethiopia, the last unit leaving on 15th March. This last action signalled the end of the Ogaden War.

As far as is known, the Somalian MiG-21MFs claimed eight victories – four MiG-21MFs, one English Electric Canberra bomber and three Douglas DC-3 transports flying resupply missions to besieged Ethiopian garrisons (though one of the DC-3s is also attributed to a MiG-17). The Ethiopians claimed 11 aerial 'kills' (ten MiG-21MFs and one MiG-17), all scored by F-5As with Israeli pilots. (According to Ethiopian sources, though, more than 23 Somalian MiGs were destroyed in the first six months of the war.) The first 'kill' – a MiG-21MF – was scored on 24th July. The following day four SAC MiG-21MFs loaded with bombs (and thus carrying no missiles for self-defence) were jumped by two F-5As. The Ethiopian fighters destroyed two of the MiGs, one with cannon fire and the other with a missile. The F-5 pilots watched in amazement as the two remaining MiGs collided and crashed while trying to avoid another missile. (Another account of the battle says that the collision came first, and that another MiG-21 and two of the MiG-17 fighter-bombers which the *Fishbeds* were escorting were shot down afterwards; also, there were three F-5s, not two.) That same day a MiG-17 was shot down by Ethiopian AA artillery over the Karamara Pass. On 29th July the Ethiopian F-5s downed another Somalian MiG-21MF. On 2nd August four Freedom Fighters raided Hargeisa, destroying one MiG-21MF, one IL-28 and one An-26 on the ground. Two more MiG-21MFs were shot down by F-5s during the battle near the town of Jijjiga, which was ultimately taken by the Somalians.

After the war, Somalian Shenyang F-6Cs patrolled the Sudanese and Ethiopian borders. In 1988-90 the Somalian Aeronautical Corps was put into action on home ground in an attempt to stomp out rebellion in the northern regions of Somalia. Flown by local and mercenary pilots, the remaining MiG-21MFs and F-6Cs operated out of Hargeisa against Somali National Movement insurgents until the latter overran Hargeisa in June 1988; after that, all operational aircraft were concentrated at Mogadishu. Two F-6Cs were shot down in 1988 during these operations.

However, when things got too hot for comfort, Siad Barré fled the country in 1991 and in the ensuing anarchy the Somalian Aeronautical Corps effectively ceased to exist. By the time the United Nations launched its first (unsuccessful) operation to restore law and order in Somalia (UNOSOM I) in 1992, the airfields had been destroyed and all remaining SAC aircraft were derelict due to lack of spares.

Known Somalian Aeronautic Corps aircraft of Soviet origin

Type	Serial/code	C/n	F/n	Notes
An-24B	MM60204/AM 85	67302705		Pre-1973 Aeroflot-style livery, 'flying cheetah' nose art
An-24B	MM60205/AM 86	67302708		Damaged. Pre-1973 Aeroflot-style livery, 'flying cheetah' nose art
An-24B	MM60206/AM 87	67302709		Pre-1973 Aeroflot-style livery, 'flying cheetah' nose art. Later recoded to, see next line
	MM60206/GD 87			Returned to the USSR, became CCCP-02709
An-26	60204	?		Ex-Libyan Arab Air Force 8204, transferred 1990; typical Libyan camouflage, 'LAAF' titles on tail amended to 'SAF'. Stored Nairobi-Jomo Kenyatta airport since 1993
An-26	MM60207/AM 88/F	?		Pre-1973 Aeroflot-style livery, 'flying cheetah' nose art
An-26	MM60208	?		
UTI-MiG-15	CC114	?		
UTI-MiG-15	CC115	?		Derelict Mogadishu
UTI-MiG-15	CC116	?		Czechoslovak-built (Aero CS-102)? Derelict Mogadishu, green/sand camouflage
MiG-17	CC125	1025		Derelict Mogadishu, green/brown camouflage, 'flying cheetah' nose art
MiG-17	CC126	1026?		
MiG-17	CC128	1028?		

MiG-17	CC129	1029	Derelict Kismayu
MiG-17	CC130	1030?	
MiG-17	CC131	1031?	
MiG-17	CC132	1032?	
MiG-17	CC133	1033?	
MiG-17	CC134	1034	Derelict Mogadishu
MiG-17	CC135	1035?	
MiG-17	CC136	1036	Built in Komsomol'sk-on-Amur; derelict Mogadishu
MiG-17	no serial	?	Ex-CC120? Derelict Hargeisa in relatively good condition
Shenyang F-6C	426	46-10519	Derelict Mogadishu
Shenyang F-6C	428	46-10***	Derelict Mogadishu; this is one of the aircraft listed below
Shenyang F-6C	430	46-10***	This is one of the aircraft listed below
Shenyang F-6C	432	46-10***	This is one of the aircraft listed below
Shenyang F-6C	434	46-10***	This is one of the aircraft listed below
Shenyang F-6C	436	46-10***	This is one of the aircraft listed below
Shenyang F-6C	438	46-10***	This is one of the aircraft listed below
Shenyang F-6C	440	46-10***	This is one of the aircraft listed below
Shenyang F-6C	442	46-10***	This is one of the aircraft listed below
Shenyang F-6C	444	46-10***	This is one of the aircraft listed below
Shenyang F-6C	446	46-10***	This is one of the aircraft listed below
Shenyang F-6C	448	46-10***	Derelict Hargeisa*; this is one of the aircraft listed below
Shenyang F-6C	450	46-10***	This is one of the aircraft listed below
Shenyang F-6C	452	46-10***	This is one of the aircraft listed below
Shenyang F-6C	454	46-10***	Derelict Hargeisa*; this is one of the aircraft listed below
Shenyang F-6C	456	46-10514	Derelict Mogadishu
Shenyang F-6C	458	46-10***	This is one of the aircraft listed below
Shenyang F-6C	460	46-10511	Derelict Mogadishu, sand/green/brown camouflage
Shenyang F-6C	?	46-10505	
Shenyang F-6C	?	46-10506	
Shenyang F-6C	?	46-10507	
Shenyang F-6C	?	46-10508	
Shenyang F-6C	?	46-10509	
Shenyang F-6C	?	46-10510	
Shenyang F-6C	?	46-10512	
Shenyang F-6C	?	46-10513	
Shenyang F-6C	?	46-10515	
Shenyang F-6C	?	46-10516	
Shenyang F-6C	?	46-10517	
Shenyang F-6C	?	46-10518	
Shenyang F-6C	?	46-10520	
Shenyang F-6C	?	46-10521	
Shenyang F-6C	?	46-10522	
Shenyang F-6C	?	46-10523	
Shenyang F-6C	?	46-10524	
Shenyang F-6C	?	46-10525	
Shenyang F-6C	?	46-10526	
Shenyang F-6C	?	46-10527	
Shenyang F-6C	?	46-10528	
Shenyang F-6C	?	46-10529	
Shenyang F-6C	?	46-10530	
Shenyang F-6C	?	46-10531	
Shenyang F-6C	?	46-10532	
Shenyang F-6C	?	46-10533	
Shenyang F-6C	?	46-10534	
Shenyang F-6C	?	46-10535	
Shenyang F-6C	?	46-10536	
Shenyang F-6C	?	46-10537	

Shenyang F-6C	?	46-10538	
Shenyang F-6C	?	46-10539	
Shenyang F-6C	?	46-10540	
Shenyang F-6C	?	46-10601	
Shenyang F-6C	?	46-10607	
Shenyang F-6C	?	46-10608	
Shenyang F-6C	?	46-10613	
Shenyang F-6C	?	46-10614	
Shenyang F-6C	?	46-10615	
Shenyang F-6C	?	46-10616	
Shenyang F-6C	?	46-10617	
Shenyang F-6C	?	46-10618	
Shenyang F-6C	?	46-10619	
Shenyang F-6C	?	46-10620	
Shenyang F-6C	?	46-10621	
Shenyang F-6C	?	46-10622	
Shenyang F-6C	?	46-10623	
Shenyang F-6C	?	46-10624	
Shenyang F-6C	?	46-10625	
Shenyang F-6C	?	46-10626	
Shenyang F-6C	?	46-10634	
Shenyang F-6C	?	46-10635	
Shenyang F-6C	?	46-10636	
Shenyang F-6C	?	46-10637	
Shenyang F-6C	?	46-10638	
Shenyang F-6C	?	46-10639	
Shenyang F-6C	?	46-10640	
Shenyang FT-6C	405	48-1021*	Abandoned Mogadishu, intact; this is one of the eight aircraft listed below
Shenyang FT-6C	?	48-10211	
Shenyang FT-6C	?	48-10212	
Shenyang FT-6C	?	48-10213	
Shenyang FT-6C	?	48-10214	
Shenyang FT-6C	?	48-10215	
Shenyang FT-6C	?	48-10216	
Shenyang FT-6C	?	48-10217	
Shenyang FT-6C	?	48-10218	
MiG-21MF	220	96****	
MiG-21MF	222	96****	
MiG-21MF	224	968711	Later incorporated parts of MiG-21MF c/n 968709, wfu Mogadishu
MiG-21MF	226	968708	Later incorporated parts of MiG-21MF c/n 968710, wfu Mogadishu
MiG-21MF	228	96****	
MiG-21MF	230	96****	
MiG-21MF	232	96****	
MiG-21MF	234	96****	Sometimes reported as c/n 968711, see 224!
MiG-21MF	236	96****	Sometimes reported as c/n 968708, see 226!
MiG-21MF	238	96****	
MiG-21MF	240	96****	
MiG-21MF	242	96****	
MiG-21MF	244	96****	
MiG-21MF	246	96****	
MiG-21MF	248	968811	Later incorporated parts of MiG-21MF c/n 968714, wfu Mogadishu
MiG-21MF	250	96****	
MiG-21MF	252	968901	Wfu Mogadishu
MiG-21MF	254	968902	
MiG-21MF	256	968903	Wfu Mogadishu
MiG-21MF	258	968904	Wfu Mogadishu
MiG-21MF	?	968709	Cannibalised for spares

MiG-21MF	?	968710		Cannibalised for spares
MiG-21MF	?	968714		Cannibalised for spares
MiG-21UM	205	516963026	?	Wfu Mogadishu
MiG-21UM	208	516909001	?	Wfu Mogadishu
Mi-8T	CC-62	?	?	Wfu Kismayu
Mi-8T	CC-69	?	?	Wfu Hargeisa
Mi-8T	C.C.-72	?	?	Wfu Mogadishu, green overall

*Two more F-6Cs lacking serials were abandoned at Hargeisa in relatively good condition.

South African Republic

In 1998 the South African company Advanced Technologies and Engineering (ATE) teamed up with the local arms manufacturer Denel to develop an upgrade package for the Mil' Mi-24 attack helicopter designated SuperHind Mk III. ATE obtained an Arsen'yev-built Mi-24V Hind-E from the Algerian Air Force for prototype conversion in 1998. Surprisingly, the c/n has been reported as 3532421809222 (that is, a 13-digit number typical of Soviet examples rather than a six-digit c/n usually found on export *Hinds*).

A large upswept boxy structure was grafted onto the extreme nose ahead of the weapons systems operator's cockpit, replacing the standard USPU-24 barbette mounting a YakB-12,7 four-barrel Gatling machine-gun; it blended into large boxy fairings on both sides of the forward fuselage. The structure carried a 20-mm (.78 calibre) Denel G12 single-barrel cannon on a barbette, which was slaved to a Denel Argus 550 gyrostabilised optoelectronic targeting system in a compact 'ball turret' mounted higher and further forward. The downward-pointing stub wing endplates with mounting frames for twin 9M114 Shtoorm (AT-6 Spiral) anti-tank guided missiles were removed and replaced by ventral fittings, allowing neat four-packs of Kentron ZT35 Ingwe tube-launched ATGMs (also of South African design) to be carried under the wingtips; these packs could be tilted nose-down to facilitate engagement. The armament was thus similar to that of the Atlas CSH-2 Rooivalk attack helicopter. The existing steel armour in the crew section was replaced by Kevlar armour to save weight. A large programmable IRCM flare dispenser was fitted on the underside of the tailboom.

Mi-24V SuperHind Mk III ZU-BOI in interim configuration (with new intake filters). This view shows clearly the 'proboscis' carrying the Denel G12 cannon and Argus 550 opto-electronic 'ball turret', as well as the the starboard lateral fairing.

This view of ZU-BOI taxying at Midrand shows the four-packs of Ingwe ATGMs and the flare dispenser under the tail-boom.

Wearing a sinister overall black colour scheme (though this was seemingly changed to dark grey at a later stage), the SuperHind Mk III demonstrator made its first flight at Midrand-Grand Central Airport on 15th February 1999. Nine days later it was registered ZU-BOI (the ZU- prefix is used for non-type-certified aircraft in South Africa). By 2008 the standard air intake dust/debris filters of quasi-spherical shape had been replaced by elongated filters similar to those fitted to the Rooivalk; this was the SuperHind Mk III. Still

Mi-24V SuperHind Mk III ZU-BOI in original post-conversion guise takes off; the landing gear is just beginning to retract. The black colour makes an interesting comparison with the later slate-grey finish.

For every 'boi', there is a 'gal'! The second prototype of the ATE SuperHind Mk III conversion was registered ZU-GAL and painted pale grey overall.

This view of ZU-GAL shows the extreme elevation angle of the cannon allowing the SuperHind Mk III to strafe targets in level flight.

Here, ZU-BOI is shown at Midrand-Grand Central airport following conversion to the ultimate SuperHind Mk V configuration. The new cockpit section strongly resembles that of the Bell AH-1 Huey Cobra.

later, ZU-BOI underwent an even more radical conversion: the standard vertically staggered tandem cockpits for the pilot and WSO were replaced by a new nose section resembling that of the Bell AH-1 Cobra attack helicopter, the designation changing to SuperHind Mk V.

Later, in 2002, ATE obtained a second Mi-24V (SB-21, c/n 070129) that had been upgraded to SuperHind Mk III configuration for the Algerian Air Force. To match the first one the helicopter was playfully registered ZU-GAL and painted light grey overall.

Sudan

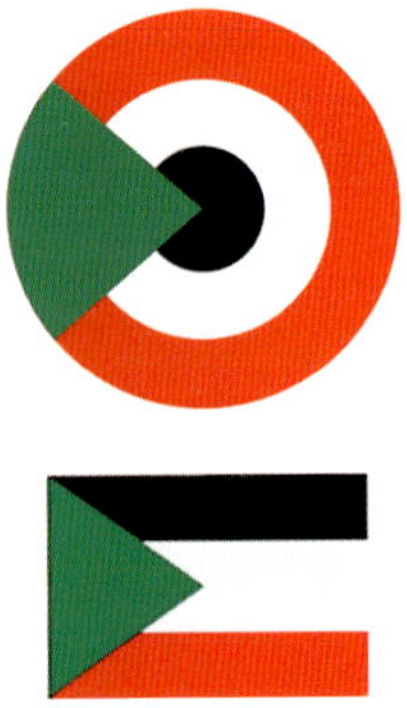

Sudan, which had been a British/Egyptian co-dominion since 1899, achieved independence in 1956, and the Sudanese Air Force (SuAF, or *al Quwwat al-Jawwiya as-Sudaniya*; also referred to as *Silakh al-Jawwiya as-Sudaniya*) was created with British help in 1957. In 1961 the Soviet Union signed a long-term trade agreement and a technical/economic co-operation agreement with Sudan. In 1964 the military junta running the country since 1958 was toppled, but a new reactionary government took over just a year later. Widespread discontent with this government's policy led to a new military coup in May 1969; the new Sudanese government signed a renewed co-operation treaty with the USSR. One of the results of this deal was the delivery of an arms package, which included Soviet aircraft for the SuAF.

The first Soviet jets supplied to Sudan were 18 Moscow-built Mikoyan MiG-21PF (*izdeliye* 76) *Fishbed-D* fighters and two MiG-21U (*izdeliye* 66-600) *Mongol-A* trainers in 1970; they were operated by No.2 Sqn at Wadi Sayyidna AB near Omdurman. A second delivery of 18 more advanced MiG-21M (*izdeliye* 96) *Fishbed-Js* and four MiG-21US (*izdeliye* 68) *Mongol-B* trainers followed in 1971. The fighters were in their natural metal factory finish and carried the serial in both European and Arab numerals. The number of these aircraft had reduced to seven and two respectively by 1992, with only about half of these totals operational. From 1988 they were progressively replaced by more modern aircraft; none remain in service today.

In 1987 the Sudanese Air Force received ten Mikoyan MiG-23MS *Flogger-E* fighters (also reported as MiG-23MLAE *Flogger-Gs*!) and two MiG-23UB *Flogger-C* trainers from Libya. The civil war has undoubtedly taken its toll and most experts suggest that only about six were still serviceable in 2004.

In 2001 the Sudanese Air Force ordered an initial ten MiG-29SE (*izdeliye* 9.13SE) *Fulcrum-C* fighters and a single MiG-29UB *Fulcrum-B* combat trainers from Russia for delivery in 2002. Similar orders were placed with the Ukraine and Belorussia. The 12 ex-Ukrainian

Sudanese Air Force MiG-21US '303' hangared at Wadi Sayyidna.

'345', one of the Sudanese Air Force MiG-21Ms. Note the dual presentation of the serial, which is not common practice in the Arab nations.

Air Force machines were the first to arrive in December 2003/June 2004. The Russian fighters and trainers followed in July 2004, being delivered to Khartoum by Antonov An-124 Ruslan freighters. 12 ex-Belorussian examples arrived in 2008. The *Fulcrums* are in service with the No.1 and No.2 Fighter Squadrons at Wadi Sayyidna; they wear a sand/green/brown camouflage scheme.

The Soviet Union was not the only supplier of fighters to the Sudanese Air Force. After the 1969 revolution Sudan was on good terms with China, too. Hence in 1969 the SuAF received 20 Shenyang F-5 (Mikoyan/Gurevich MiG-17F *Fresco-C*) fighters and at least 12 Shenyang FT-5 trainers. (Some sources state the single-seaters were F-5As, the Chinese-built version of the radar-equipped MiG-17PF *Fresco-D* all-weather interceptor.) The aircraft were probably based in Khartoum. 17 F-5s/FT-5s were reportedly in service in June 1976; by late 1991 this number had decreased to ten. According to some sources, the Sudan Air Force operated nine or ten Shenyang F-6 and FT-6 (Mikoyan/Gurevich MiG-19S *Farmer-C* derivative) delivered in 1980.

After President Jaafar Nimeiry assumed power, the USSR fell out of favour with the Sudanese government. China stepped into the breach in 1987 by supplying 15 Chengdu F-7B fighters (a MiG-21F-13 *Fishbed-C* derivative; another source says nine in 1987 and six in 1995), plus two Guizhou FT-7 trainers (a MiG-21UM derivative). Further supplies came from Iran in 1996 or 1997 when 22 Chengdu F-7Ns were reportedly delivered. A final batch of about six new F-7Ms was delivered in 1998. No details are known of serials or construction numbers for any of the above aircraft. By 2007 only 12 F-7s were still active with 1st Fighter Sqn and the remainder were in storage, having been replaced in 2nd Fighter Sqn by MiG-29s.

A small number of strike aircraft was delivered as well. First, Libya supplied 12 Mikoyan MiG-23BN *Flogger-H* fighter-bombers in 1987. It is not known if any were still operational by 2004, what with the climate, lack of spares and combat attrition during the Second Sudan Civil War. Later, in 2008, Belarus supplied Sudan with 13 single-seat Sukhoi Su-25 *Frogfoot-A* attack aircraft and one Su-25UB *Frogfoot-B* two-seater; two more Su-25UBs were added in 2010. They are in service with the No.2 Fighter-Bomber Sqn at Wadi Sayyidna.

The transport element of the Sudan Air Force included six Soviet-built Antonov An-12 *Cub* medium transports, two of which were quasi-civil, and one quasi-civil Shaanxi Y8D – a Chinese version of the An-12BP. (Some sources have claimed that the SuAF had An-12s serialled 700 (which was actually an An-24RT), 711, 722, 733, 744 and 755; however, recent reports fail to confirm this.) Additionally, the Khartoum-based airline AZZA Transport Co. Ltd. (which, though nominally a civilian enterprise, was chiefly tasked with transporting military personnel and materiel) had four second-hand Soviet-built Antonov An-12s and a further Shaanxi Y8 – this time a Y8F-100.

Soviet tactical transports were initially represented in Sudanese Air Force service by the Antonov An-24T *Coke*. Nine of these were delivered, starting in 1979, and the last one was an An-24RT – the version on which the two Ivchenko AI-24VT turboprops were augmented by a Tumanskiy RU19A-300 turbojet booster in the starboard nacelle. One of the An-24Ts was transferred to South Yemen by January 1985 and two others were lost in crashes on 20th March 1997 and 25th April 2003, the rest being stored at Khartoum.

Later, in the 1990s, the Sudanese Air Force obtained a number of second-hand Antonov An-26 *Curl* transports. These were sourced from far and wide, serving with the No.1 and No.3 Transport Sqns at Khartoum; some came from Libya, while others were purchased from Russian and Ukrainian airlines. Many of the SuAF An-26s were quasi-civil and wore airline-style colour schemes, or sported Air Force serials along with civil registrations.

A real oddity in the aircraft fleet of the Sudanese Air Force is the Antonov An-30 *Clank* photo survey/reconnaissance aircraft. Three such aircraft are operated since 2007. Two of them have military markings and may well retain the camera suite, but the third machine wearing a civil registration and a smart white colour scheme with a red/black cheatline is an An-30A-100 19-seat VIP conversion operated by Green Flag Aviation and was purchased as such from Aircraft Repair Plant No.410 in Kiev, the Ukraine, which handles the An-24/-26/-30/-32 family.

One more Antonov tactical transport type operated by the Sudanese military is the An-32B *Cline*. Six aircraft were reportedly ordered but only three An-32Bs are known to have been delivered, and even these are second-hand aircraft (the stockpile of unfinished An-32 airframes at the Aviant plant in Kiev in the early 2000s has been 'eaten up' by other

Sudanese Air Force MiG-21M '345'
No.2 Sqn, Wadi Sayyidna

customers). Two of them were delivered to the Sudanese Air Force and, again, wear both military serials and civil registrations; more will be said about the third aircraft later.

Also in 2007, Sudan became an operator of the Antonov An-72/An-74 *Coaler-A/B* twin-turbofan STOL transport family. The government flight purchased a single An-72 modified to An-72-100D extended-range executive configuration, as well as three regular An-74s and two An-74-200s; the registration for one more An-74 was reserved in the Sudanese register but the aircraft was not delivered. All of these are second-hand aircraft. True, in 2005 Sudan placed an order for four new-build An-74TK-200 convertible cargo/passenger jets and two An-74TK-300s (the version with underwing engines), but no deliveries of these have been reported.

The backbone of the Sudanese Air Force's transport element, however, is the Il'yushin IL-76TD. Before acquiring at least one such aircraft of its own in 2010 (the aircraft wears an anonymous all-white scheme without titles and apparently without a visibly applied serial) the SuAF had resorted to leasing two IL-76TDs from Russian airlines. Also, the three *Candid-As* of AZZA Transport Company Ltd. are used for carrying military materiel and even performing airborne assault missions (paradropping personnel).

The 'first-generation' Mil' Mi-8T *Hip-C* utility helicopter (usually in armed configuration) and Mi-8PS VIP helicopter entered service with the Sudanese Air Force in 1971. At least 15 have been identified by their serials, although there may be a good deal more if the serials were allocated in sequence. The 'second-generation' *Hip-H* versions – the Kazan'-built Mi-17, Mi-17V-5 and Mi-172, the Ulan-Ude-built Mi-8AMT and Mi-171 – were delivered from the mid-2000s onwards; these included both brand-new and second-hand helicopters, and the serials of earlier Mi-8Ts were occasionally reused. Sudanese Mi-8/-17s lacked the KO-50 cabin heater on the starboard side, as it was superfluous in the hot climate.

More than 50 Mil' Mi-24V/Mi-35 *Hind-E* and Mi-24P/Mi-35P *Hind-F* attack helicopters have been delivered to the Sudanese Air Force. Of these, at least two are ex-Libyan Arab Air Force Mi-35s delivered before 1989; Belarus supplied six Mi-24s in 1996 and Russia furnished another 45 between 2001 and 2007.

Sudanese Air Force MiG-21M '344' taxying at Wadi Sayyidna. Note how the serial is carried on the forward and rear fuselage.

Sudanese Air Force MiG-21M '344'
No.2 Sqn, Wadi Sayyidna

The *Hinds* are operated by the No.1 Helicopter Sqn at Khartoum.

Special mention should be made of the Police Air Wing, an outfit of the Ministry of the Interior. Its fleet includes a single An-32B transport, a single An-72-100D executive jet (operated in 2008 but later transferred to another government agency and reregistered) and at least two *Hip-H* helicopters with extra long-range tanks (a Mi-171 and a Mi-17V-5), all wearing a distinctive pale blue/dark blue/ white livery. They are used for airlifting per-

A superb pre-delivery shot of Sudanese Air Force MiG-29SE '623' at Lookhovitsy-Tret'yakovo.

Sudanese Air Force MiG-29SE *Fulcrum-C* '623', No.2 Fighter Sqn

Sudanese Air Force Su-25 '205' seen before delivery.

Su-25s '204' and '210' parked at Wadi Sayyidna, with B-8M1 FFAR pods on the wing pylons.

sonnel and police equipment and conducting special operations.

Like some other operators of Soviet hardware, Sudan took an 'if-you-want-a-thing-well-done-do-it-yourself' approach to maintenance issues, establishing an overhaul facility called SAFAT Aviation Complex (not to be confused with the Italian artillery systems manufacturer, *Società Anonima Fabbricazione Armi Torino*) in 2004. The facility is located 20 km (12.4 miles) from Khartoum. It is capable of performing heavy maintenance on such types as the An-2, An-24, An-26, An-30, An-32, An-74, MiG-21, MiG-29, F-6/FT-6 and Mi-8, and is set to handle the IL-76 as well.

Sudan Civil Wars

Sudan was administratively divided into north and south regions. There had long been strife between the two, the north being predominantly Muslim whereas the south was Christian and Animist; this had resulted in civil war many times over the centuries, the Christian and other non-Islamic people resisting the imposition of strict Islamic Law in the South. In 1955, shortly before the independence of Sudan, the conflict flared up again, both sides fighting for predominance in what became known as the First Sudan Civil War.

In the course of the war the freshly delivered Shenyang F-5s and MiG-21PFs stationed

All-white Sudanese Air Force An-26 wearing no markings other than the serial '7705'.

This An-26 was hit by ground fire during a bombing sortie in the Second Sudan Civil War, crash-landing near Gogrial and subsequently being pillaged. The serial appears to be in Arabic only, but in fact a skin panel with 'SAF 7711' painted on it has been stolen by metal scavengers.

at Wadi Sayyidna were tasked with attacking not only the Land Freedom Army insurgents (better known by the name Anya-Nya), which had been active in the south since 1955, but also airfields in Ethiopia in 1971 with the purpose of stopping the airlift of arms to the rebels in southern Sudan. Most of the MiG-21 pilots were Sudanese (persistent rumours that Soviet and Egyptian pilots were involved remain unconfirmed) and their lack of experience, together with the desert conditions, led to huge attrition over the next five years, particularly after the new president Jaafar Nimeiry took an anti-Soviet stand, expelling the Soviet technical support staff, and the USSR cut off spares supplies for the fighters. President Nimeiry had survived a communist coup in 1971 and subsequently decided to align Sudan with the West.

The first use of the MiG-21s against the Anya-Nya was on 24th September 1970 when three MiG-21s deployed to Juba (supported by one An-12) attacked the rebels' headquarters in Morta, Equatoria State, with bombs and unguided rockets. More attacks followed in the next few days; despite putting up fierce resistance, the rebels were dispersed by a decisive attack on 1st October. The SuAF lost one MiG-21 in an accident at Juba, and another (plus a Mi-8 helicopter) were reportedly shot down by the rebels. More raids against Anya-Nya camps near Morta took place in February 1971; the rebels claimed that the government forces resorted to random bombing, with numerous civilian casualties as a result – a claim which Khartoum strongly denied.

On 20th September 1972, five Libyan Lockheed C-130H Hercules transports airlifting about 400 troops to Uganda were intercepted by Sudanese MiG-21s, which forced them to divert and land at Khartoum. Permission was given for them to return to Libya, but after taking off they flew unimpeded to Uganda.

Sustained efforts by the World Council of Churches to mediate a ceasefire finally yielded results in 1972. The Addis Ababa Agreement was signed that year, giving the South a measure of regional autonomy.

Sudanese Government An-30A-100 ST-GFD (c/n 0605) with a red/white/green cheatline; the logo next to the c/n is that of the Kiev Aircraft Repair Plant No.410

Although there was peace in the Sudan, the African continent suffered a succession of droughts, which caused the terrible African famine of the 1970s. The people in the Darfur Region in western Sudan accused the government of neglecting their needs, and as a result the area of unrest expanded.

Further problems arose in 1981 after Libya invaded Chad and bombed Sudanese towns and villages, where Chadian soldiers had supposedly taken refuge. The onslaught culminated with an attack on Omdurman where many civilians were killed and their houses destroyed. Immediately, Egypt responded by sending a squadron of McDonnell Douglas F-4E Phantom II fighter-bombers to an airfield near its border with Sudan and the US Air Force transferred eight McDonnell Douglas F-15C Eagle fighters and a Boeing E-3A Sentry AWACS to Egypt to discourage Libya from invading Sudan. One of the reasons for the external response was that the SuAF lacked the means to oppose a possible invasion; only eleven Sudanese MiG-21 fighters and three MiG-21U trainers out of the 42 delivered had survived the First Sudanese Civil War, and only five or six of them were airworthy.

In June 1976 Sudan extended aid to the rebellious Eritrea province which was seeking independence from Ethiopia. Thus, two different kinds of F-5 met in combat over Eritrea – Ethiopian Air Force Northrop F-5A/E Tigers were up against Sudanese Shenyang F-5s (MiG-17Fs). Several F-5s (the Chinese kind) were shot down or damaged by the Ethiopian fighters. A ceasefire between Ethiopia and Eritrea was agreed in March 1978.

Open rebellion broke out again in 1983 after the Muslim government in Khartoum tried to reduce the South's autonomy; the Sudan People's Liberation Army (SPLA) was formed in May 1983 and the Second Sudan Civil War began. The fighting became fiercer when Khartoum totally abrogated the Addis Ababa Agreement and imposed Sharia Law on the Christian South. Ethiopian transport aircraft were still able to continue supplying the rebel army, the Sudanese radar cover being insufficient to monitor the whole area.

Despite the delivery of ten BAC Strikemaster Mk 90s, ten Northrop F-5Es and two F-5Fs to Sudan in the early 1980s, the remaining MiG-21Ms were extensively used as fighter-bombers against the SPLA; the latter had no aircraft but did have 9K32 Strela (SA-7 *Grail*) MANPADS. The SuAF suffered heavy losses, which were partly compensated by the arrival of ten Chengdu F-7A fighters in to help defeat the SPLA rebels. By 1992 there were still seven MiG-21Ms and two MiG-21USs in service but another two MiG-21Ms were lost in 1996; six of the MiG-23s were reportedly shot down as well.

Libya lent a hand in the conflict, supplying Libyan-flown MiG-23MS fighters and two MiG-25RBK reconnaissance aircraft; the latter were on loan and operated from Khartoum in 1988, monitoring the war zone. A MiG-23MS was shot down 500 km (310 miles) south of Khartoum on 18th December 1987; one more was downed a year later, on 13th December 1988.

Additionally, An-26 transports were extensively used as makeshift bombers against the SPLA, becoming an essential element of the SuAF's COIN tactics; this even gave rise to the colloquial term 'Antonov bomber'. The An-26 had a bombsight fitted as standard (for accurate delivery of items during paradrop operations) and four detachable bomb shackles, and could carry more bombs in the capacious cabin. As a rule, the An-26s flew their bombing missions after Sudanese fighter-bombers had 'softened up' the rebels' air defences, and

Here, ST-GFD is shown in a different guise, the lower stripe of the cheatline being black instead of green

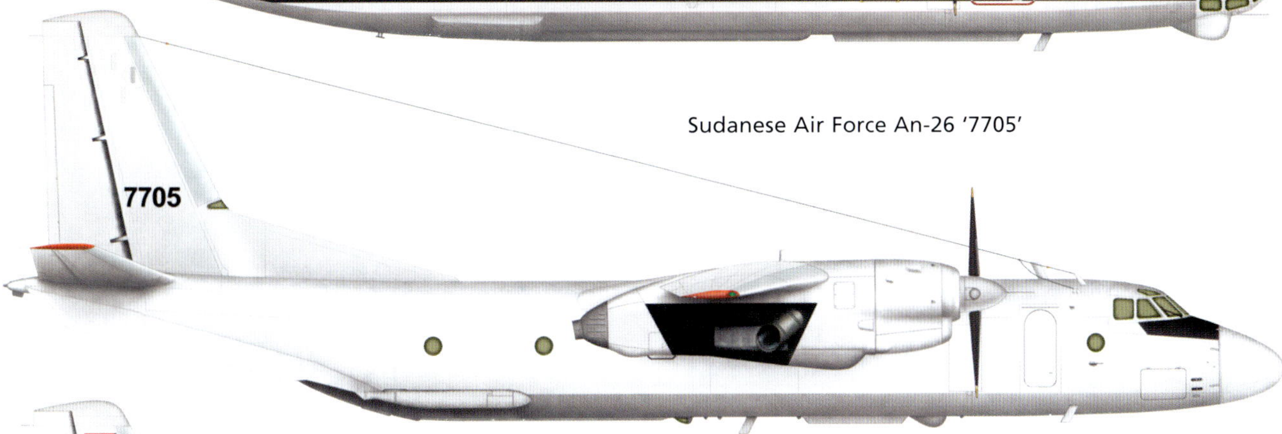

Sudanese Air Force An-26 '7705'

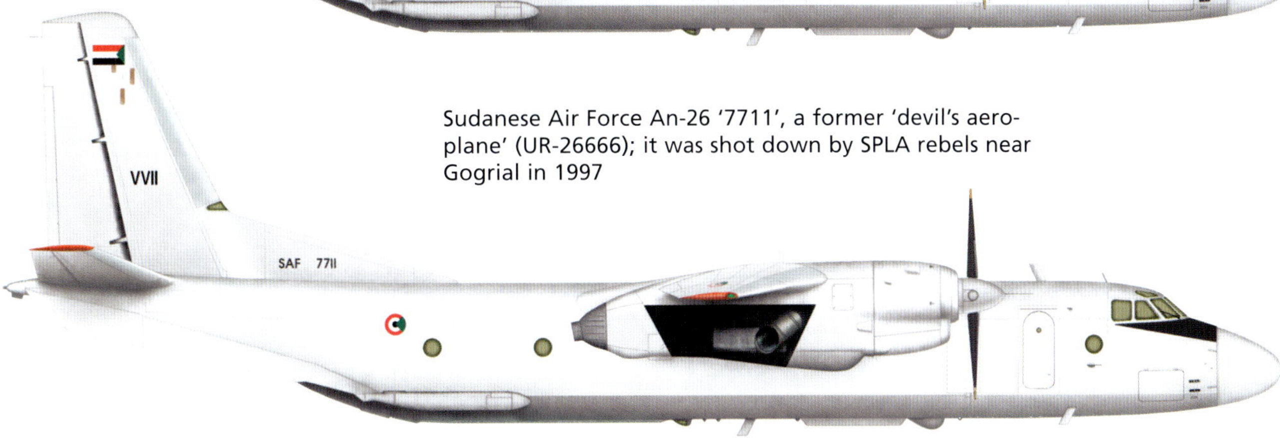

Sudanese Air Force An-26 '7711', a former 'devil's aeroplane' (UR-26666); it was shot down by SPLA rebels near Gogrial in 1997

were always followed by Mi-24 gunship helicopters. About 260 such missions were flown in Southern Sudan in 1997-2000 alone, often with large civilian casualties as a result.

At one stage the government gave arms to local tribesmen opposing the rebels; in Darfur (western Sudan), where open warfare began in 2003, it even recruited a private army known as the Janjaweed which, assisted by the Sudanese Air Force, aimed to bring the area under control. The result was worldwide notoriety and condemnation for the atrocities committed by the Janjaweed against civilians.

Neither side was able to win a clear-cut victory and the war was further complicated by the development of oil production from the Bentiu Fields in 1998. The refinery and its pipelines provided a new and easy target for rebels, as well as an extended area for government forces to patrol. However, the government now had more hard currency available from oil exports to buy up-to-date equipment and began to replace F-7 fighter-bombers with MiG-29s in 2004.

Worried by the famine, the fighting and no doubt the possible loss of oil supplies, the superpowers put pressure on the belligerents to negotiate a ceasefire. The outcome of it was the Nairobi Comprehensive Peace Agreement signed on 9th January 2005 granting Southern Sudan autonomy for six years, to be followed by a referendum on its independence. The result of the referendum was a unanimous vote in favour of secession, and a

An-12B ST-JUA wears the smart livery of Juba Air Cargo. Despite its civil looks, it was very probably used in the interests of the military during the Second Sudanese Civil War.

new state called South Sudan came into existence on 9th July 2011.

Meanwhile, the fighting in Darfur continued, and the Sudanese Air Force MiG-29s were actively involved in spite of the Khartoum government's pledge not to use them against the Sudan Liberation Army (SLA; formerly DLF – Darfur Liberation Front) and the Justice and Equality Movement (JEM) insurgents. On 10th May 2008 the SLA guerrillas managed to shoot down one of the MiG-29s during the battle for Omdurman; the pilot ejected but lost his life in so doing. Some sources reported that the mercenary pilot was a former Russian Air Force officer, a claim which the Russian Ministry of Defence strongly denied.

In early May 2008 the conflict in Darfur became international when Chad lent support to the rebels – probably in retaliation for Khartoum's support of Chadian insurgents that had attacked N'djamena. On 15th-16th May 2008 Chadian aircraft and helicopters (including Su-25s) launched three attacks against insurgents that were massing on the Sudanese side of the border, preparing for a new raid on N'djamena. Following an exchange of diplomatic protests, Sudan declared the attacks to be an act of war and threatened Chad with a full-scale military response.

A ceasefire was agreed in November 2008, but this was soon broken by the JEM in

Sudanese Air Force An-12B '988' with a cheatline in national flag colours

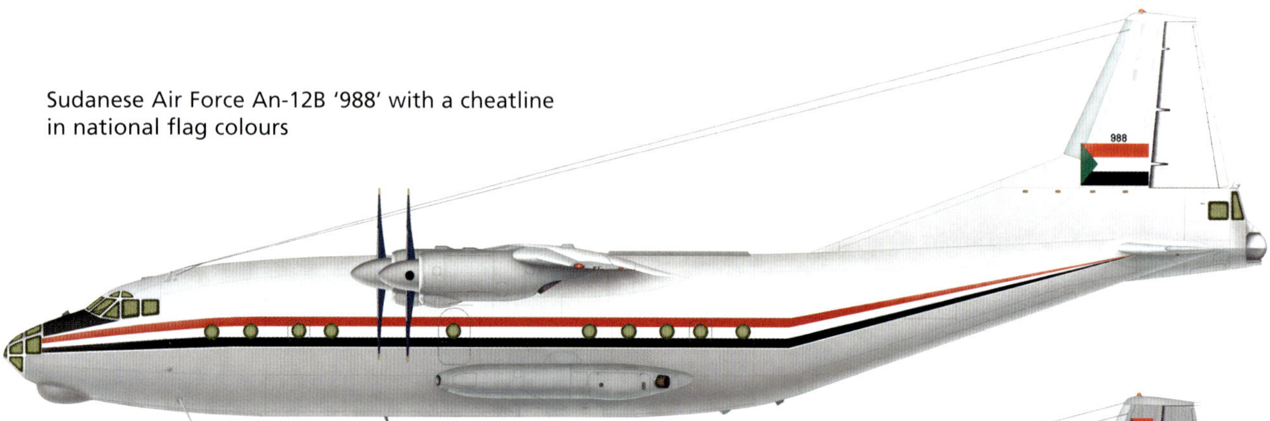

An-12B '831' wears a colour scheme akin to that of its Indian Air Force sister ships, with a white flight deck roof to minimise heating by the sun

An-74 ST-PRB stored at Moscow-Ostaf'yevo in 2005, resplendent in Sudanese government flight colours. The intake covers still bear the aircraft's previous identity, RA-74024.

Similarly painted An-74 ST-GFF of Green Flag Aviation.

Smart-looking An-72-100D ST-PRM is operated for the government.

In contrast, An-72-100 ST-BDK wears an anonymous plain white colour scheme.

Another view of An-74 ST-GFF.

Sudanese Air Force Mi-171 '534'. Note that the paint shop has made a mistake, applying the insignia back to front on the starboard side (the green sector on the roundel and the fin flash should face front).

Sudanese Air Force Mi-171 '534' (note the wide cabin door to starboard)

Mi-171 '100 White' (with extra fuel tanks) in the blue and white colours of the Sudanese Police Air Wing.

Something seems to be wrong with the loading ramp of Mi-17V-5 SAF531, causing the personnel to study it closely. Note the lack of the KO-50 cabin heater.

Sudanese Air Force Mi-35P '911'

Sudanese Air Force Mi-17V-5 '533' and Mi-35 '933' at Khartoum. Note the 'dead crocodile' (Mi-35 '912') in the background.

Sudanese Mi-35P '924' armed with B-8V20 FFAR pods. Note the high gloss finish.

Mi-35P '921' with B-8V20 pods in low flight.

January 2009 by attacking the town of Muhajaria in south-western Darfur. In response, SuAF Nanchang A-5C *Fantan* attack aircraft and An-26s (used as bombers) flew sorties from Nyala against JEM positions near Muhajaria on 13th-14th January and against various villages held by JEM guerrillas on 20th January. By March the JEM had withdrawn to Chad again, but by no means given up the struggle – in fact, they became stronger, recruiting new fighters and attacking across the border at will. When two insurgent units crossed the border and captured the garrison at Kornoy, the SuAF intensified its bombing raids against JEM positions in Kornoy and their base camp at Furawiya. On 30th May 2009 Sudanese Su-25s were used against the insurgents for the first time, attacking a JEM camp at Goz Beida 70 km (443 miles) inside Chadian territory.

A Sudanese Air Force Mi-2 wearing no serial. The paint job suggests the helicopter has been freshly overhauled.

Known Sudanese Air Force aircraft of Soviet/Russian origin				
Type	**Serial/registration**	**C/n**	**F/n**	**Notes**
An-12BP (+)	ST-ALV (1)	8345909		Ex-Iraqi Air Force YI-AEP. Registration later to Y8F-100 (see below)
An-12BK	ST-AZM	9346808		Previous identity unknown. Also reported as ST-AZN/Sudanese Air Force '956'!
An-12...	977 (2)			Reported 8-2008; see An-24T!
An-12B (+)	831	?		Grey overall, white top to flight deck (Indian Air Force style)
An-12BP (+)	988	8345908		Ex-Iraqi Air Force YI-AER, bought 2007; white with cheatline in national flag colours
An-12...	999	?		Reported 7-2008
An-12...	9933	?		Ex-ST-ZNN
An-12	ST-AQG	401907		AZZA Transport Co. Ltd.; ex-RA-11234
An-12B (+)	ST-ASA	402010		AZZA Transport Co. Ltd.; ex-RA-11374, bought by 2007. Crashed near Khartoum 7-10-2012 due to engine failure
An-12BK	ST-ARV	8345310		AZZA Transport Co. Ltd.; ex-Dvin Avia EK-11028, bought 2006
An-12BK	ST-DAS	7345209		AZZA Transport Co. Ltd.; ex-Valan ER-AXC, bought 2002
Shaanxi Y8D	ST-ALU (1)	070804		Derelict Wau, Sudan. Registration passed to An-26 c/n 10810 in 1999; see also An-26 ST-ALU/7711!
Shaanxi Y8F-100	ST-ALV (2)	?		AZZA Transport Co. Ltd.
An-24T	700 Black	102*0**?		Crashed 25-4-2003
An-24RT	900 Black	1025023?		
An-24T	911 Black	1025021?		
An-24T	922 Black	1025020?		
An-24T	933 Black	1025019?		
An-24T	944 Black	1025018?		
An-24T	955 Black	1022017?		
An-24T	966 Black	1022016?		To South Yemen Air Force by 1-1985 as 2016?
An-24T	977 Black	1022015?		
An-26	26563	3506		Ex-Sibir' Airlines RA-26563; AZZA Transport Co. Ltd., all-white c/s. Became, see next line
	7705 Black			All-white, no titles or insignia
An-26	7711 Black	8109		Ex-UR-26666. Serial in Arabic digits. Crash-landed near Gogrial, southern Sudan, 1997
An-26B	7715 Black	13307		Ex-UR-CFZ
An-26B	7716 Black	13405		Ex-UR-CGA
An-26	7722 Black	?		
An-26	ST-ALU/7755 Black	11805		Ex-Libyan Arab Air Force 8201. Wfu Raga, southern Sudan

An-26	ST-AQO/7777 Black	9905		
An-26	ST-JAC	10203		AZZA Transport Co. Ltd.; ex-EX-003, leased from Aerovista
An-26	ST-ZZZ (1)	10407		Ex-Komiinteravia RA-26236. Crashed 7-8-2006; registration reused for, see next line
An-26	ST-ZZZ (2)	10404		Ex-Komiinteravia RA-26240. Became, see next line
	7706			
An-30	7704	?		
An-30	7708	1201?		Ex-Rzhevka Air RA-30062?
An-30A-100	ST-GFD	0605		Government Flying Service; ex-ARP 410 Airlines UR-30030
An-32B	ST-ATL/7709	?		
An-32B	ST-ATM/7710	3409		Ex-Antonov Design Bureau UR-48023
An-32B	ST-PAW/203 White	3209		Police Air Wing; ex-Armenian Air Lines EK-48026
An-72-100	ST-BDK	365.720.**.***	?	All-white, no titles
An-72-100D	ST-PAM	365.720.96.914	1503	Police Air Wing; ex-Enimex ES-NOI, D/D 2008; white with trim in national flag colours. To Government Flying Service (?) by 2010 as, see next line
	ST-PRM			
An-74-200	ST-BDA	365.470.98.968	1807	Ex-Bashkirian Airlines RA-74014
An-74TK-200	ST-BDS	365.470.70.655	0701	Ex-Bashkirian Airlines RA-74001; all-white c/s, no titles, bought by 2010
An-74	ST-BDT	365.470.97.935	1606	Ex-Bashkirian Airlines RA-74046
An-74-200	ST-BDX	365.470.96.924	1509	Ex-Aviacor RA-74041
An-74	ST-GFF	365.470.97.932	1604	Ex-BIO Air T9-ABE
An-74	ST-PRB	365.470.96.918	1505	Ex-Kupol-Avia RA-74024
An-74	[ST-PRK]	365.470.98.956	1708	Ex-4L-EAS; reservation only (not taken up, became EX-74201)
IL-76TD	ST-AIR	1033414474	8709	Ex/to RA-76367, leased from C-Air in 1997-98
IL-76TD	ST-AIY	1023410344	8406	Ex/to RA-76350, leased from C-Air in 1997-98
IL-76TD	ST-APS	1023409316	8309	AZZA Transport Co. Ltd.; ex-Belair EW-76837, bought 1996
IL-76TD	ST-AQB	0023442218	3105	AZZA Transport Co. Ltd.; ex-EP-ALC, formerly operated by Trans ATTICO (African Transport, Trading & Investment Co.)
IL-76TD	ST-AZZ	1023408265	8207	AZZA Transport Co. Ltd.; ex-Faso Airlines XT-FCB, all-white c/s. Possibly same aircraft as, see next line!
IL-76TD	'SAF 1106'	?	?	ATC callsign only??? Seen 2010, all-white c/s
MiG-21PF	201	767401		
MiG-21PF	202	767402		
MiG-21PF	203	767403		
MiG-21PF	204	767404		
MiG-21PF	205	767405		
MiG-21PF	206	767406		
MiG-21PF	207	767407		
MiG-21PF	208	767408		
MiG-21PF	209	767409		Crashed
MiG-21PF	210	767410		
MiG-21PF	211	767411		
MiG-21PF	212	767412		
MiG-21PF	213	767413		
MiG-21PF	214	767414		
MiG-21PF	215	767415		Crashed
MiG-21PF	216	767416		
MiG-21PF	217	767417		
MiG-21PF	218	767418		
MiG-21M	331	?		Silver overall
MiG-21M	344	?		Silver overall
MiG-21M	345	?		Silver overall
MiG-21*bis*	1130	N750*****	?	Ex-Ethiopian Air Force (EtAF serial!); flown to Sudan by defector 6-6-1989. Used as ground instructional airframe at Wadi Sayyindna. EtAF sand/green camouflage, SuAF insignia

MiG-21U	219	?		
MiG-21U	220	?		
MiG-21US	300	?		
MiG-21US	221	516806208?	?	C/n as reported but doubtful! Reserialled as, see next line
	301			301 was also reported as a MiG-21M!
MiG-21US	222	516806209?	?	C/n as reported but doubtful! Reserialled as, see next line
	302			
MiG-21US	223	516806210?	?	C/n as reported but doubtful! Reserialled as, see next line
	303			Silver overall
MiG-21US	224	516806211?	?	C/n as reported but doubtful! Reserialled as, see next line
	304			304 was also reported as a MiG-21M!
MiG-23MS	06916	0391206916	?	Ex-Libyan Arab Air Force; derelict Juba 2007, LAAF insignia
MiG-23MS	06918	0391206918	?	Ex-Libyan Arab Air Force; crashed, date unknown
MiG-23MS	09055	0391209055	?	Ex-Libyan Arab Air Force; derelict
MiG-25RBK	1035	N020*****	?	Loaned from Libyan Arab Air Force 1988
MiG-29SE	606 Black	29607*****	?	
MiG-29SE	612 Black	29607*****	?	
MiG-29SE	614 Black	29607*****	?	
MiG-29SE	620 Black	29607*****	?	
MiG-29SE	621 Black	29607*****	?	
MiG-29SE	622 Black	29607*****	?	
MiG-29SE	623 Black	29607*****	?	
MiG-29SE	624 Black	29607*****	?	
MiG-29SE	625 Black	29607*****	?	
MiG-29SE	626 Black	29607*****	?	
MiG-29SE	627 Black	29607*****	?	
MiG-29SE	628 Black	29607*****	?	
MiG-29SE	629 Black	29607*****	?	
MiG-29UB	602 Black	N509030*****	?	
MiG-29UB	630 Black	N509030*****	?	
MiG-29UB	631 Black	N509030*****	?	
Mi-8T	432	335**?	?	
Mi-8T	451	335**?	?	
Mi-8PS	502	10598?	?	Ex-Helisota LY-HBP, ex-Luftwaffe 93+55?; basic Luftwaffe government flight livery
Mi-8T	506 White	335**?	?	Seen 1981
Mi-8T	507	335**?	?	Seen 9-2002
Mi-8T	511	335**?	?	Seen 10-1981
Mi-8T	515	33515	?	Seen 10-1996
Mi-8T	520	335**?	?	Seen 2-1985
Mi-8T	521	335**?	?	Seen
Mi-8T	524	335**?	?	Seen 2-1985
Mi-8T	528 (1)	335**?	?	Seen 2-1985
Mi-8T	531 (1)	335**?	?	Seen 1973
Mi-8T	534	335**?	?	Seen 3-2009, sand/green camouflage
Mi-8T	536	335**?	?	Seen 4-2009
Mi-8T	658	335**?	?	Seen 4-2009
Mi-8T	866	335**?	?	Seen 8-1988
Mi-8AMT	100 White	59489605704	03**	Police Air Wing; ex-RA-27115
Mi-17V-5	101 White	736M07	?	Police Air Wing
Mi-172	?	736C08	?	
Mi-17	525	?	?	Seen 2009
Mi-171	528 (2)	?	?	Seen 2-2007
Mi-17V-5?	529	?	?	Possibly Mi-171Sh! Seen 10-2006

Mi-17V-5	531 (2)	?	?	White/silver c/s, serial repeated on nose as SAF 531, and main serial on tailboom later painted out; still later two-tone green camouflage. Crashed 18-4-2011
Mi-17V-5	533	?	?	Seen 3-2007; all-white c/s, later two-tone green camouflage
Mi-17	534	?	?	Seen 10-2010; sand/green camouflage
Mi-17	537	?	?	Seen 2009
Mi-171Sh	546	?	?	'Glass nose'/'flatass'
Mi-17	575	?	?	
Mi-17	ST-GFA	202M27	?	Ex-Rostvertol CCCP-70958, bought 1996
Mi-17	ST-GFB	202M25	?	Ex-Rostvertol CCCP-70956, bought 1996
Mi-172	ST-GFK	?	?	Stepped nose, armoured cabin; white with cheatline in national flag colours
Mi-172	ST-PRN/257 Black	?	?	Stepped nose; white with cheatline in national flag colours
Mi-24V	901 Black	?	?	
Mi-24...	902 Black	?	?	
Mi-24...	905 Black	?	?	
Mi-24V	909 Black	?	?	
Mi-24...	911 Black	?	?	
Mi-24V	912 Black	?	?	
Mi-24V	913 Black	?	?	
Mi-24V	915 Black	?	?	Derelict Khartoum
Mi-24...	916 Black	?	?	
Mi-24P?	918 Black	?	?	
Mi-24...	920 Black	?	?	
Mi-24P?	921 Black	?	?	Cannon removed?
Mi-24...	922 Black	?	?	
Mi-24...	923 Black	?	?	
Mi-24P	924 Black	?	?	
Mi-24V	925 Black	?	?	
Mi-24...	926 Black	?	?	
Mi-24...	927 Black	?	?	
Mi-24V	928 Black	?	?	
Mi-24...	929 Black	?	?	
Mi-24...	930 Black	?	?	
Mi-24V?	931 Black	?	?	Derelict Khartoum
Mi-24...	932 Black	?	?	
Mi-24V	933 Black	?	?	
Mi-24...	935 Black	?	?	
Mi-24...	936 Black	?	?	Derelict Khartoum
Mi-24V	937 Black	?	?	
Mi-24V	938 Black	?	?	
Mi-24V	939 Black	?	?	
Mi-24...	940 Black	?	?	Reported in error as 540?
Mi-24P	941 Black	?	?	
Mi-35P	942 Black	?	?	
Mi-35P	943 Black	?	?	
Mi-35	944 Black	?	?	
Mi-35	945 Black	?	?	
Mi-35	946 Black	?	?	
Mi-35	947 Black	?	?	
Mi-24P	949 Black	35324341*****	?	
Shenyang F-5?	158	?		Seen 4-1972; also reported as an FT-5!
Shenyang F-5	160	?		Seen 4-1972
Shenyang F-5	161	?		Seen 4-1972
Shenyang F-5	163	?		Seen 4-1972
Shenyang F-5	164	?		Seen 4-1972
Shenyang F-5	170	?		Seen 4-1972

Shenyang F-5	701	?		Seen 4-1975
Shenyang F-5	702	?		Seen 12-1974
Shenyang F-5	703	?		Seen 2-1975
Shenyang F-5	704	?		Seen 1-1975
Shenyang F-5	705	?		Seen 1-1975
Shenyang F-5	706	?		Seen 1-1975
Shenyang F-5	707	?		Seen 1-1975
Shenyang F-5	708	?		Seen 11-1974
Shenyang F-5	709	?		Seen 1-1975
Shenyang F-5	710	?		Seen 5-1974
Shenyang F-5	711	?		Seen 1-1975
Shenyang F-5	712	?		Seen 3-1975
Shenyang FT-5	168	55-****?		White overall. Seen 4-1972
Shenyang FT-5	180	55-****?		Seen 4-1972
Shenyang FT-5	181	55-****?		Seen 4-1972
Shenyang FT-5	182	55-****?		Seen 4-1972
Shenyang FT-5	183	55-****?		Seen 4-1972
Shenyang FT-5	184	55-****?		Seen 4-1972
Shenyang FT-5	186	55-****?		Seen 4-1972
Shenyang FT-5	187	55-****?		Seen 4-1972
Shenyang FT-5	188	55-****?		Seen 4-1972
Shenyang FT-5	190	55-****?		Seen 4-1972
Shenyang FT-5	713	55-****?		Seen 11-1974
Shenyang FT-5	714	55-****?		Seen 12-1974
Shenyang FT-5	745	55-****?		White overall. Derelict Wadi Sayyidna 2010
Shenyang F-6	701	46-*****		
Shenyang F-6	702	46-*****		
Shenyang F-6	703	46-*****		
Shenyang F-6	704	46-*****		
Shenyang F-6	705	46-*****		
Shenyang F-6	706	46-*****		
Shenyang F-6	707	46-*****		
Shenyang F-6	708	46-*****		
Shenyang F-6	709	46-*****		
Shenyang F-6	710	46-*****		
Shenyang F-6	711	46-*****		
Shenyang F-6	712	46-*****		
Shenyang F-6	748	46-*****		White overall. Derelict Juba 2005
Shenyang F-6	764	46-*****		Sand/green/brown camouflage
Shenyang FT-6	713	48-*****		
Shenyang FT-6	714	48-*****		White overall
Su-25	201 Black	255081*****	?	Seen 12-2008
Su-25	203 Black	255081*****	?	Seen 8-2009
Su-25	204 Black	255081*****	?	Seen 8-2009
Su-25	205 Black	255081*****	?	
Su-25	206 Black	255081*****	?	Seen 8-2009
Su-25	207 Black	255081*****	?	
Su-25	208 Black	255081*****	?	
Su-25	209 Black	255081*****	?	
Su-25	210 Black	255081*****	?	Seen 8-2009
Su-25	211 Black	255081*****	?	Seen 8-2009
Su-25	212 Black	255081*****	?	Seen 8-2009
Su-25	213 Black	255081*****	?	Seen 8-2009
Su-25	214 Black	255081*****	?	Seen 8-2009
Su-25UB	202 Black	382201*****	?	Seen 12-2008
Su-25UB	215 Black	382201*****	?	D/D 2010
Su-25UB	217 Black	382201*****	?	D/D 4-2010

South Sudan

The southern region of Sudan has had a long history of separatism. There were several reasons for this. As mentioned earlier, one of them was religious differences between the Muslim north and the predominantly Christian south; an attempt by Khartoum to enforce Shari'a law in the south led to dissent. Another issue was the control over natural resources – southern Sudan was rich in oil. Barely a year after Sudan had achieved independence in 1956, the First Sudanese Civil War broke out, lasting until 1972 when the Southern Sudan Autonomous Region was formed. This existed until 1983 when the Second Sudanese Civil War began, which lasted more than 20 years and ended with the Comprehensive Peace Agreement of 2005. Later that year, southern autonomy was restored when an Autonomous Government of Southern Sudan was formed.

South Sudan became a sovereign state on 9th July 2011 after the population almost unanimously voted in favour of independence. However, even before the referendum on this issue – that is, in 2009 – the self-proclaimed state of South Sudan ordered ten brand-new Mi-17s (nine Mi-17V-5 utility helicopters and a single Mi-172 for government VIP use) from the Kazan' Helicopters factory. The helicopters were delivered to the nascent Sudanese People's Air Force (SPAF) in 2010 and, for a while, sat parked at the airport of the South Sudanese capital Juba, 'doing little more than looking pretty', as one commentator put it. One of the helicopters made its first public appearance during the independence day celebrations, trailing a large national flag.

Originally the Mi-17s were anonymous but soon gained serials – which, incidentally, give the impression there are more aircraft than the SPAF actually has. Serials reported to date are SPAF-101, SPAF-102, SPAF-104, SPAF-106, SPAF-108, SPAF-109, SPAF-110 and SPAF-112.

Tanzania

Tanzania, the former British colony of Tanganyika and Zanzibar, became an independent republic on 9th December 1961 and the Tanzanian People's Defence Force Air Wing (TPDFAW, or JWTZ – *Jeshi la Wananchi la Tanzania* in Swahili) was formed in September 1964.

The TPDFAW was largely equipped with aircraft of Chinese origin. These included an order for 48 Shenyang F-5 fighters (the export

Tanzanian People's Defence Force Shenyang FT-2 '52-212', Mikumi AB, Dar es Salaam; note the galloping zebra nose art

A trio of white-painted Tanzanian FT-5s (9118, 9120 and 9121) makes a formation flypast in 2004.

version of the J-5, a licence-built Mikoyan/ Gurevich MiG-17F *Fresco-C*). The first 12 aircraft arrived in 1973 (some sources say 1970), and the remainder were never delivered, as Tanzania ordered more capable fighters. Training was undertaken on two Shenyang FT-2 trainers (the export version of the JJ-2, a Mikoyan/Gurevich UTI-MiG-15 *Midget* clone), also delivered in 1970, and more than 20 Shenyang FT-5 trainers (the export version of the JJ-5 – a unique Chinese version looking like a cross-breed between the UTI-MiG-15 and the MiG-17F) delivered in 1973. The F-5s and FT-2s were based at Mikumi AB 130 km (80 miles) north of Dar-es-Salaam and were still operational in the mid-1980s. The FT-5s, on the other hand, were based at Ngerengere. On 29th June 2010 one of them suffered serious technical problems and the crew attempted an emergency landing on a straight stretch of highway near Sagera. Unfortunately the trainer collided with a bus that happened to be in the wrong place at the wrong time and rolled inverted, killing both pilots.

An 'open house' at Ngerengere AB; civilians swarm like ants over two camouflaged F-6s (9205 and 9211). Note the JWTZ (Tanzanian People's Defence Force Air Wing) titles.

F-6C '9236' taxies with a full load of two drop tanks and two small FFAR pods. Note the squadron badge on the tail.

In contrast, FT-6C '9124' retains the white factory finish. Again, the aircraft carries drop tanks and FFAR pods.

More Chinese fighters followed in the spring of 1974 when the TPDFAW took delivery of 12 Shenyang F-6 (Mikoyan/Gurevich MiG-19S *Farmer-C*) fighters, augmented by 12 improved Shenyang F-6C and four Shenyang FT-6C trainers in 1982. These aircraft, too, were based at Ngerengere (some sources say Mikumi); according to *Flight International*, ten were still operational in late 1991. Originally painted white overall, the F-6s later wore a sand/green/brown camouflage scheme with the serial on the nose and 'JWTZ' titles on the rear fuselage.

Also in 1974, the Soviet Union delivered 14 Mikoyan MiG-21MF *Fishbed-J* fighters and two MiG-21U *Mongol-A* trainers to Tanzania. They were based at Mwanza; no serials or tactical codes are known. Several *Fishbeds* were lost in accidents before Tanzania went to war with Uganda in 1978 (see Ugandan section). One MiG-21MF was shot down by Ugandan troops and another fell victim to 'friendly' anti-aircraft fire in the course of this conflict; on the other hand, seven captured Ugandan MiG-21MFs and one MiG-21U were flown to Mwanza and put into service by the TPDFAW as war booty. (At least the Tanzanians tried to do it; there are reports that two MiG-21s crashed en route to Mwanza, probably due to poor maintenance.) Later, in 1998, the Ukraine supplied a further four MiG-21MFs from surplus stocks. Unfortunately none have been positively identified so far.

In 1980 the Tanzanian *Fishbed* fleet was bolstered by Chinese 'cousins' – ten Chengdu F-7B (or F-7M Airguard) fighters and two Guizhou FT-7 trainer were ordered new from China, followed by the arrival of two second-hand Chengdu F-7Ns from Iran in 1997. A further shipment of new F-7s – this time in the F-7TG version – arrived in August-September 2009. Again, the F-7s are based at Mwanza; the single-seaters wear an overall grey colour scheme and TAFC nose titles (standing for Tanzanian Air Force Command), while the trainers wear a sand/turquoise/olive drab camouflage and no titles.

In November 2003 the TPDFAW took delivery of two grey-painted Shaanxi Y8 transports (a Chinese derivative of the Antonov An-12BP *Cub*). Oddly, they have been described as Y8F-200s, though it would be more logical to suppose they are export Y8D-IIs with Western avionics. The following year two second-hand Antonov An-28 *Cash* STOL feederliners were purchased for utility duties. The transports are based at Mwanza.

The TPDFAW in action

The Tanzanian F-6s saw action in the war between Tanzania and Uganda in 1978-79. However, even before that, in 1976, two were lost in a 'friendly fire' incident when AA artillery teams at the newly constructed Mwanza AB mistook the fighters arriving from Ngerengere for Ugandan MiGs and opened fire. One F-6 was shot down, killing the pilot, Ayekuwa Akiirusha. The other fighter slipped through the barrage but, with no other airfield close at hand, its pilot was forced to eject after running out of fuel. The investigation revealed that no one had warned the local units of the impending redeployment.

The war started when Uganda, proceeding from intelligence reports of an impending Tanzanian invasion, decided to deal a pre-emptive strike. In September 1978 Ugandan fighters made a series of raids in the Bukoba area, with lots of civilian casualties, whereupon about 2,000 Ugandan troops crossed the border and annexed the Kagera Salient. In return, Tanzania launched a counter-invasion. Little is known of TPDFAW activities in the war, except that initially the Tanzanian F-6s and MiG-21s flew combat air patrol missions from Mwanza and rarely ventured into Ugandan airspace until 1979 (unlike the Ugandans, who repeatedly attacked Mwanza but to little avail). Despite the initial success, the Ugandan Army was unable to exploit it, lacking the means to cross the Kyaka River, and was eventually forced to pull out of the Kagera Salient. Conversely, the TPDF advanced north across the border, taking several towns and constructing an *ad hoc* airstrip near Mutukula that allowed TPDFAW transports to bring in supplies.

Mid-March 1979 was marked by the only case when Ugandan and Tanzanian MiG-21s met in combat. The Ugandan pilot pulled the short straw – his aircraft was shot down near Byesika village 5 km (3.1 miles) from the road between Masaka and Mubende, the pilot losing his life. It is not known who scored the kill and when exactly.

Libya had intervened in mid-February, supporting Uganda both on the ground and in the air. Among other things, on 2nd April 1979 one of the Libyan Arab Air Force Tupolev Tu-22B bombers was tasked with destroying the fuel dump at Mwanza, but the bombs missed their target. The Tanzanians retaliated by sending at least four MiG-21s which attacked Entebbe airport, cratering the runway. Also on 2nd April, TPDFAW MiG-21s

A Tanzanian People's Defence Force Air Wing F-7A

attacked fuel dumps at Jinja and Kampala. Later they supported the Tanzanian ground forces' onslaught on Entebbe airport, which was captured by 8th April. Two days later Tanzanian troops entered Kampala from three sides, purposefully leaving a narrow corridor for the Libyans to retreat to Kenya and Ethiopia. The war ended in total defeat for Uganda, which remained under Tanzanian occupation until 1982.

Known Tanzanian People's Defence Force Air Wing aircraft of Soviet origin			
Type	**Serial**	**C/n**	**Notes**
An-28	JW9031*	1AJ 0**-**	Reported in error as ex-Rodina RA-28927 (c/n 1AJ 008-14), which actually is now RF-38387
An-28	JW9032	1AJ 0**-**	Reported in error as ex-Rodina RA-28937 (c/n 1AJ 009-03), which actually is now RF-38388
Shaanxi Y8F-200	JW-9034	110804	D/D 3-11-2003
Shaanxi Y8F-200	JW-9035	120801	D/D 3-11-2003
Shenyang FT-2	52-212	52-212?	
Shenyang FT-5	9101	?	
Shenyang FT-5	9102	?	Natural metal finish, no insignia except serial; active 2004
Shenyang FT-5	9103	?	
Shenyang FT-5	9104	?	
Shenyang FT-5	9105	?	
Shenyang FT-5	9106	?	
Shenyang FT-5	9107	?	
Shenyang FT-5	9108	?	
Shenyang FT-5	9109	?	
Shenyang FT-5	9110	?	
Shenyang FT-5	9111	?	
Shenyang FT-5	9112	?	
Shenyang FT-5	9113	?	
Shenyang FT-5	9114	?	
Shenyang FT-5	9115	?	
Shenyang FT-5	9116	?	
Shenyang FT-5	9117	?	
Shenyang FT-5	9118	?	White overall; active 2004
Shenyang FT-5	9119	?	White overall, no insignia except serial. Crashed near Sagera 29-6-2010
Shenyang FT-5	9120	?	White overall, no insignia except serial; active 2004
Shenyang FT-5	9121	?	White overall, no insignia except serial; active 2004
Shenyang F-6	9201	?	
Shenyang F-6	9202	?	
Shenyang F-6	9203	?	
Shenyang F-6	9204	?	
Shenyang F-6	9205	?	
Shenyang F-6	9206	?	

Shenyang F-6	9207	?	
Shenyang F-6	9208	?	
Shenyang F-6	9209	?	
Shenyang F-6	9210	?	
Shenyang F-6	9211	?	
Shenyang F-6	9212	?	
Shenyang F-6C	9226	?	
Shenyang F-6C	9227	?	
Shenyang F-6C	9228	?	
Shenyang F-6C	9229	?	
Shenyang F-6C	9230	?	
Shenyang F-6C	9231	?	
Shenyang F-6C	9232	?	
Shenyang F-6C	9233	?	
Shenyang F-6C	9234	?	
Shenyang F-6C	9235	?	
Shenyang F-6C	9236	?	
Shenyang FT-6C	9122	?	
Shenyang FT-6C	9123	?	
Shenyang FT-6C	9124	?	White overall; active 2004
Chengdu F-7TG	JW 9242	1140	
Guizhou FT-7	JW 9125	?	
Guizhou FT-7	JW 9126	?	

* The JW serial prefix stands for Jeshi la Wananchi

Uganda

On 9th October 1962 the British Protectorate of Uganda became an independent state and the Uganda Army Air Force (UAAF) was formed that same year. In 1965 Uganda signed a co-operation treaty with the Soviet Union that made deliveries of military equipment possible. In 1971 a coup d'état occurred in which the government of President Milton Obote was overthrown, resulting in a military dictatorship headed by Gen. Idi Amin Dada. The latter continued a policy of co-operation with the Soviet Union.

According to press reports, the UAAF received seven Mikoyan/Gurevich MiG-17F *Fresco-C* fighters from the USSR in 1966; five more aircraft followed later. The MiGs equipped Uganda's first combat squadron known simply as 'The Combat Unit' and based at Entebbe International airport. By 1976 the *Fresco* had been supplanted by the MiG-21 as the UAAF's main fighter and relegated to the ground attack role. Four of the MiG-17Fs were destroyed on the ground at Entebbe Airport in 1976 during the widely publicised hijack drama (see below).

Two Mikoyan/Gurevich UTI-MiG-15 *Midget* trainers were supplied along with the MiG-17Fs in 1966 to support their operations. The aircraft were based at Entebbe.

The first deliveries of Mikoyan MiG-21s to Uganda took place in the early 1970s, when 18 MiG-21MF *Fishbed-J* fighters and three MiG-21U *Mongol-A* trainers arrived. Seven were destroyed in the Israeli raid on Entebbe in 1976 and the rest were captured and taken home as war booty by the Tanzanians when their army supporting the National Liberation Front of Uganda (NLFU) rebels overthrew Idi Amin and seized Entebbe in 1979 (see below). After that, the UAAF was reorganised as the Uganda People's Defence Force/Air Wing (UPDF/AW) in the 1980s, and later renamed the Uganda People's Defence Air Force (UPDAF).

1999 saw the purchase of six MiG-21*bis* (*izdeliye* 75A) *Fishbed-L* fighters and two MiG-2UM *Mongol-B* trainers from Poland. Four of the jets were reportedly overhauled and upgraded to MiG-21-2000 standard by IAI at Lod, Israel, at a price of US$20-25 million and airlifted to Uganda by chartered Il'yushin IL-76TD freighters, with the rest presumably to follow. Unlike earlier Soviet-supplied MiG-21s, which had a natural metal finish with red or yellow alphanumeric serials, these aircraft

The original (1964-standard) insignia of the Ugandan Army Air Force showing the head of the crested crane, the national symbol.

Three camouflaged Uganda Army Air Force MiG-17Fs (including U603) share the apron at Entebbe with two UAAF C-47s and a civilian Piper PA-31 Navajo.

wore a grey (the single-seaters) or green/tan camouflage and red numeric serials (the original Polish ones); interestingly, two of them – a MiG-21*bis* and a MiG-21UM – sported the same serial until the single-seater was reserialled to avoid confusion. Trouble arose immediately with these newcomers; it appeared that two of the jets had in fact not been upgraded, and pilots were unwilling to fly them because the aircraft suffered from serious mechanical problems. One MiG-21*bis* crashed into Lake Victoria on 15th July 2003, killing the Russian contract pilot, and the others had to be returned to IAI for servicing. There they sat for at least a year, pending payment for the work already done; eventually they returned to serve with 'The Combat Unit' at Entebbe. One more MiG-21 – a MiG-21UM trainer – crashed into Lake Victoria near Buwu Island on 17th June 2006; this time the crew consisting of a Russian contract pilot and an Israeli pilot working for IAI ejected safely.

The latest Russian fixed-wing aircraft to enter service with the UPDAF is the Sukhoi

Ugandan Army Air Force MiG-21MF U906 with 1964-style insignia
Destroyed at Entebbe International Airport on 3rd July 1976

Ugandan Army Air Force MiG-21MF U-907
Destroyed at Entebbe International Airport on 3rd July 1976

Su-30MK2 multi-role fighter. The eight aircraft were built by the Komsomol'sk-on-Amur Aircraft Production Association and match the *Flanker-F* standard of the Chinese (Su-30MKK) and Venezuelan (Su-30MKV) examples, with tall squared-off fin caps and no canard foreplanes. Uganda denied buying Su-30s at first when the Russian press broke the news in April 2010; the cost of the deal was estimated at US$ 200 million. The first two fighters were delivered to Entebbe by a Russian Antonov An-124 Ruslan heavy-lift aircraft and reassembled in early July 2011; the first engine runs took place on 8th July and the first Su-30 made its first flight from Entebbe on 11th July, the other machine following suit a day later. Two more aircraft were delivered in November 2011 and the final four were to follow in 2012. The Su-30MK2s wear a green/brown camouflage, new-style UPDAF insignia and serials prefixed AF (instead of the usual U); incidentally, the serials were not allocated in sequence. The *Flankers'* primary mission is to protect the oil fields near Lake Albert.

The first Soviet helicopters operated by the UAAF were two Mil' Mi-8T *Hip-C* utility helicopters delivered by December 1975. There are reports that one more 'first-generation' Mi-8 was delivered in January 1988 and a further two in 1998. Additionally, various versions of the 'second-generation' Mi-17 *Hip-H* (including the 'glass-nosed' Mi-17-1V, the radar-nosed Mi-17V-5 with a cargo ramp and the VIP-configured Mi-172) were supplied to Uganda, starting in the late 1990s; known

This grey-painted Ugandan MiG-21*bis* is a former Polish machine overhauled by IAI in 1999.

An Ugandan People's Defence Air Force MiG-21*bis* prepares for take-off.

A pair of UPDAF *Fishbed-Ls* resting between missions at Entebbe airport. Note the late-style roundels and the 'land rocket' GPUs.

The Ugandan Air Force changed the style of its roundels several times. This is the 1999 version.

UPDAF MiG-21UM '9307 Red' fires up at Entebbe. Note the raised instructor's periscope and the non-standard drop tank with a rounded nose.

serials suggest that 30 or more have been delivered. On 30th July 2005 one of the Mi-172s crashed fatally in northern Uganda, colliding with Mt. New Cush in clouds due to pilot error; the victims included Ugandan Vice-President John Garang.

Another Mil' type operated by the UAAF is the Mi-24 attack helicopter. Its first appearance on Ugandan soil triggered a major corruption scandal. In July 1996 the government of President Yoweri Museveni took a decision to acquire four Mi-24s from Belarus, together with appropriate ground support equipment, spares and ammunition. In addition, several Belorussian experts would be hired to train local pilots in the operation of the helicopters. In April 1997 the government signed a purchasing contract for the helicopters, to be supplied via the Consolidated Sales Corporation (CSC) – allegedly incorporated in the British Virgin Islands. The price tag of the choppers was quoted as US$ 1.5 million apiece. However, when the initial two helicopters arrived in early 1998, it turned out they had not been refurbished prior to delivery, as provided by the contract, and were not airworthy (when the case was dragged into public view, the media labelled it 'the case of the junk helicopters'). The scandal was further fuelled by claims that the choppers' record cards had been forged to create the impression they were lower-time airframes.

On 15th March 1999 the UPDAF announced it was going to return the *Hinds* because of their poor condition. The CSC was required to supply serviceable replacement helicopters within three weeks, otherwise the contract would be cancelled and Uganda would demand liquidated damages. CSC argued that the contract had no provisions for cancellation. The big problem was who would do the refurbishment. The Ugandan government estimated the cost of the work at US$ 1.2 million. Dismissing this figure as exaggerated, CSC proposed a company approved by the OEM (the Mil' Design Bureau) that would do the job for US$ 400,000 but this was rejected by the UPDAF.

As the *ad hoc* board of inquiry proceeded to investigate the deal, the case got 'curiouser and curiouser', to put it in terms of Lewis Carroll. It turned out that CSC *did not exist* at the time when the contract for the helicopters was signed with the Ministry of Defence; it was registered in Uganda several months *after* the contract was signed. In fact, CSC was a mere broker with no direct links to the supplier of the goods – a profiteering company that

Ugandan People's Defence Air Force MiG-21UM
Entebbe International Airport, 2007

A camouflaged Ugandan MiG-21UM with no serial. Again, it wears 1999-style roundels with a white centre showing a complete crested crane rather than just the head.

had purchased the helicopters from Belarus through a series of middlemen, and resold them for a hefty profit. Furthermore, it was revealed that Uganda's Minister of Defence, Maj.-Gen. Salim Saleh (who just happens to be President Museveni's younger brother), had received a US$ 800,000 commission (read: bribe) from CSC's lobbyist, Mr. Emmanuel Katto, for supporting the award of the helicopter contract. Salim Saleh later confessed to the President that he had received the bribe, and the president advised him to give the money to the army for special operations against the Lord's Resistance Army militant group in the north. The board of inquiry established that, all things considered, the fraudulent transaction had cost Uganda a pretty penny – close to US$ 13 million.

The commission recommended the prosecution of Maj.-Gen. Saleh and the other individuals involved in the corruption scandal. The Cabinet endorsed this recommendation in a Government White Paper, stating that *'all officials of the Ministry of Defence, the Uganda People's Defence Forces, the Bank of Uganda or any other person implicated should be held accountable for causing financial loss to government or corruption'*. The Cabinet also ordered the Attorney General to lift the CSC corporate veil and proceed personally against Emmanuel Katto and other persons who offered the bribes to the military officials. Eventually, however, no charges were brought against anyone except Katto due to – guess what? – *lack of evidence*; nor did the Minister of Defence face court martial. Having friends and relatives in such high places certainly helps!

Eventually Uganda did receive a number of Mi-24V *Hind-Es* and Mi-24P *Hind-Fs* in 2003 (reportedly three refurbished ex-Belorussian aircraft and three new ones), including at least one night-capable Mi-35PN. Known serials suggest that more than 20 *Hinds* have been delivered, but that may be a ploy to create the impression there are more than there actually are. The helicopters are used to provide security for the presidential VIP aircraft, patrolling the airport area before the jet takes off and lands; they have also been used against the LRA guerrillas. In August 2012 three Mi-24s escorted by a Mi-172 were dispatched to Somalia for peacekeeping operations under the auspices of the African Union.

Ugandan People's Defence Air Force Su-30MK2 AF 011
Entebbe International Airport, July 2011

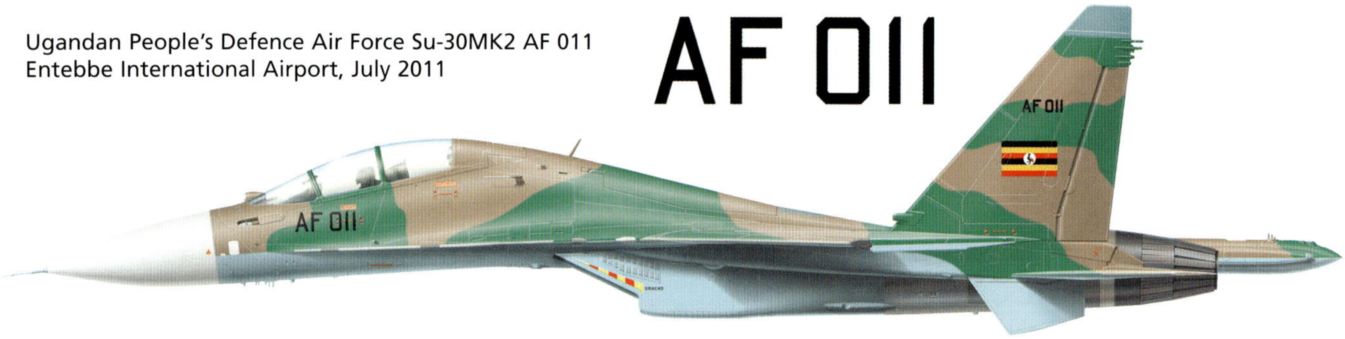

Unfortunately the *Hinds* never got there; on 12th August all three Mi-24s crashed into Mt. Kenya between Nanyuki airbase and Garissa, two of the crews being killed and the third crew surviving with injuries. The cause is not clear but may be controlled flight into terrain.

Ugandan wars

After seizing power, Idi Amin imposed a reign of terror on his political opponents, with the result that 20,000 refugees fled to neighbouring Tanzania. In September 1972 Amin expelled most of the 50,000 people of Asian origin living in Uganda, confiscating their property, the greater part of which filled the pockets of the army. Meanwhile, ex-President Milton Obote was mustering a resistance movement in Uganda and obtaining arms. Falling back on the refugees in Tanzania, the new NLFU resistance group invaded Uganda on 17th September 1972 when Obote believed the time was right.

However, the rebels' assumption that the Ugandans would rise up against the dictator proved to be wrong. There was no uprising and, at Amin's request, Libya sent 3,000 troops and reportedly Dassault Mirage 5D fighters to quell the insurgency. Bolstered by this show of strength, the Ugandan Army, supported by MiG-17F and MiG-21MF fighter-bombers, attacked the insurgents on 20th September, forcing them to retreat across the border. Appreciative of the help received from Libya, Amin declared Uganda to be an 'Islamic Republic', even though 85% of its population was Christian.

Shortly afterwards, the Palestine Liberation Organisation (PLO) was encouraged to build a base in Uganda. When Palestinian and West German terrorists hijacked an Air France Airbus A300B-2 airliner (F-BVGG) bound from Tel Aviv to Paris on 27th June 1976, taking 268 passengers (including many Israeli nationals) and crew hostage, Idi Amin gave refuge to the hijackers. When the Airbus had landed at Entebbe, the PLO demanded that the Israeli government release 53 PLO members from Israeli prisons in return for freeing the

UPDAF Su-30MK2 is towed for the official unveiling ceremony attended by President Yoweri Museveni. Note the fin flash used instead of a roundel.

Opposite page: Su-30MK2 AF 011 shows off the camouflage pattern on its upper surfaces.

Opposite page, below: AF 011 flies over the Ugandan countryside, seen from sister ship AF 015.

UPDAF Mi-17-1V AF-601 disgorges troops. This machine crashed fatally in 1998.

One of the newer *Hips* in Ugandan service is this Mi-17V-5, AF-611. Note the exhaust/air mixers.

Another UPDAF Mi-17V-5, AF-603. For some reason the Ugandan Mi-17s never wore roundels.

Ugandan People's Defence Air Force Mi-35 AF-803

hostages. Pretending to negotiate with the hijackers in order to buy some time for organising a rescue, the Israelis launched Operation *Thunderbolt*. On 3rd July six Israeli Defence Force/Air Force aircraft (four Lockheed C-130H Hercules transports and two Boeing 707s converted to tankers, one of which also acted as relay station) took off from Israeli-controlled Sinai. Landing at Entebbe airport hot on the heels of a British Airways Boeing 707-436, the first Hercules disgorged an Israeli commando team, plus a black Mercedes-Benz saloon and several Land Rover jeeps, which formed a convoy simulating Amin's entourage. The ruse was successful; the hostages were freed, the hijackers killed or captured, and 53 minutes after landing the C-130s departed for Israel. Before that, to prevent pursuit (and, no doubt, as a punitive action for siding with the bad guys), all Ugandan Air Force aircraft on site – four MiG-17Fs and seven MiG-21MFs – were destroyed by explosives.

Encouraged by this setback for the government forces, more recruits joined the NLFU, which was based in Tanzania, and in 1978, fearing a new invasion (with good reason), Idi Amin ordered a pre-emptive strike on Tanzania. To prepare the way, MiG-17Fs and MiG-21MF fighter-bombers attacked the city of Bukoba in northern Tanzania on 10th and 11th October 1978, followed by a new raid on 27th October when three waves of aircraft attacked a hospital and other buildings, causing appreciable damage.

On 28th September the Ugandan Army invaded Tanzania and soon occupied the Kagere Salient, slaughtering nearly 10,000 civilians in the process. Once again, air support was given by MiG-21MFs, the Tanzanian air defences later claimed to have shot down several of these. Uganda acknowledged the loss of a single MiG-17F that was shot down on 1st October, the pilot ejecting and rejoining the Ugandan forces. Unfortunately, one of the Tanzanian MiG-21MFs was also destroyed by 'friendly fire'. This allowed the Ugandan to raise a ballyhoo, claiming that 'a Tanzanian air attack headed for major Ugandan cities had been routed'.

In November a pair of UAAF MiG-21s attacked Mwanza, the Tanzanian base closest to the border. They inflicted no serious damage; moreover, one of the fighters flown by Capt. Kiiza (who would eventually become

Two Ugandan *Hinds*. The one in the foreground (undergoing maintenance) is a Mi-35; the other one is a night-capable Mi-35PN with clipped wings, fixed landing gear and an LLLTV/laser system.

Mi-35 AF-803 comes in to land past a Moldovan-registered An-12B.

An unmarked Ugandan Mi-35P wearing a different camouflage; note the absence of the cannon.

Commander of the UAAF) was damaged by a 9K32 *Strela* (Arrow; SA-7 *Grail*) MANPADS.

Soon, however, the Ugandan offensive stalled, being unable to cross the Kyaka River that delimited the Kagere Salient from the south. After a general mobilisation and careful planning, the Tanzanians and the NLFU counter-attacked, crossing the river and hitting the exposed flank of the Ugandans. The latter called in air support but the MiG-17s attacking the Tanzanian pontoon bridge were met with heavy AA fire, two of them being shot down by MANPADS. On 20th January 1979 the Tanzanians and their allies forced the Ugandan Army back across the border and continued the pursuit into Uganda. UAAF MiG-21s flew strike missions against Tanzanian troops that had captured the Simba Hills and the airstrip at Lukoma, and on 27th February a large formation of MiG-21s attacked the Tanzanian for-

ward airstrip at Mutukula, losing three aircraft to MANPADS, but to no avail.

In the meantime, a Ugandan Army unit at Tororo in western Uganda (near the Kenyan border) revolted. Reinforcements had to be brought in from Jinja to suppress the mutiny, supported by air strikes by UAAF fighter-bombers.

The Tanzanians continued their advance, capturing Masaka on 25th February and approaching Entebbe by the end of March. In mid-March a Ugandan MiG-21 was shot down by a Tanzanian sister ship. On 2nd April, a Libyan Tupolev Tu-22B *Blinder-As* dropped twenty 250-kg (551-lb) high-explosive bombs on M'wanza, missing military targets; there was no interception because the Tanzanian MiG-21MFs were all occupied with ground attack sorties near Kampala. On 7th April the

Known UAAF and UPDAF aircraft of Soviet/Russian origin

Serial	Version	C/n	F/n	Notes
U605	UTI-MiG-15	?		Brown/slate grey camouflage, yellow serial
U606	UTI-MiG-15	?		
U-618	UTI-MiG-15	?		Sand/green camouflage, black serial. Gate guard at Entebbe IAP
U601	MiG-17F	?		
U602	MiG-17F	?		Brown/slate grey camouflage, yellow serial
U603	MiG-17F	?		
U604	MiG-17F	?		
U607	MiG-17F	?		
U614	MiG-17F	?		Brown/slate grey camouflage, yellow serial. Gate guard at Entebbe IAP
U-904	MiG-21MF	?	?	
U-905	MiG-21MF	?	?	
U-906	MiG-21MF	?	?	Destroyed at Entebbe 3-7-1976, wreck still present 2003
U-907	MiG-21MF	?	?	Destroyed at Entebbe 3-7-1976, wreck still present 1999
U-908	MiG-21MF	?	?	Destroyed at Entebbe 3-7-1976, wreck still present 1999
U-909	MiG-21MF	?	?	
U-910	MiG-21MF	?	?	
U-911	MiG-21MF	?	?	
U-912	MiG-21MF	?	?	
U-913	MiG-21MF	?	?	
U-914	MiG-21MF	?	?	
U-915	MiG-21MF	?	?	
U-916	MiG-21MF	?	?	Destroyed at Entebbe 7-1979, wreck still present 1999
U-917	MiG-21MF	?	?	
U-918	MiG-21MF	?	?	
U-919	MiG-21MF	?	?	Destroyed at Entebbe 7-1979, wreck still present 1999
9211 Red	MiG-21*bis*	N75089211	75-64-**?	Overhauled IAI and delivered to Uganda
9307 Red (1)	MiG-21*bis*	N75089307	75-64-**?	Ex-Polish Air Force 9307 Red (1); see MiG-21UM below! Became, see next line
AF-973				
9799 Red	MiG-21*bis*	N75089799	75-64-**?	Ex-Polish Air Force 9799
9801 Red	MiG-21*bis*	N75089801	75-64-**?	Ex-Polish Air Force 9801
9802 Red	MiG-21*bis*	N750*****	?	Reported on overhaul by IAI at Lod in 2004, origin unknown
9811 Red	MiG-21*bis*	N75089811	75-64-**?	Ex-Polish Air Force 9811; overhauled by IAI
9818 Red	MiG-21*bis*	N75089818	75-64-**?	Ex-Polish Air Force 9818; overhauled by IAI
U-307	MiG-21UM	?	?	Overhauled by IAI, seen Lod 10-2003
U-900	MiG-21U...	?	?	Exact version unknown. Derelict Entebbe 1995
U-901	MiG-21UM	?	?	Derelict Entebbe 1996
U-902	MiG-21U...	?	?	Exact version unknown. Derelict Entebbe 1996
U-903	MiG-21U...	?	?	Exact version unknown; reported in error as MiG-21MF. Destroyed in Israeli raid 7-1979, wreckage noted Entebbe 1995
9307 Red (2)	MiG-21UM	516999307	?	Ex-Polish Air Force 9307 Red (2); noted operational in 2007. Possibly crashed 15-6-2006
no serial?	MiG-21UM	516999306?	?	Overhauled IAI and delivered to Uganda. Ex-Polish Air Force 9306 Red?
no serial	MiG-21UM	516999393	?	Purchased via Tbilisi Aircraft Plant (TAM). Camouflaged
AF-001	Mi-172	?	?	Seen 1997

AF-601 (1)	Mi-17-1V	336M01	?	Crashed near Gulu, Uganda, 4-7-1998
AF-601 (2)	Mi-17V-5	?	?	
AF-602	Mi-17	?	?	
AF-603	Mi-17V-5	?	?	Seen 1998
AF-604	Mi-17	?	?	Seen 5-2008
AF-605 (1)	Mi-17V-5	?	?	Seen 3-1998; crashed? See next line!
AF-605 (2)	Mi-172	?	?	Seen 5-2008!
AF-606	Mi-17	?	?	Seen 1997
AF-607	Mi-17	?	?	Seen 1998
AF-611	Mi-17V-5	?	?	Seen 10-2007
AF-613	Mi-17V-5	?	?	Seen 4-2008
AF-615 (2)?	Mi-17	627C01	?	Possibly same aircraft as AF-615 above! Crashed 30-7-2005
AF-627	Mi-17	?	?	Seen 2012
AF-6...	Mi-17	223M28	?	
AF-802	Mi-24V	?	?	
AF-803	Mi-24V	?	?	Crashed in Kenya 12-8-2012
AF-804	Mi-24	?	?	
AF-806	Mi-24	?	?	
AF-807	Mi-24P	?	?	Crashed in Kenya 12-8-2012
AF-808	Mi-24P	?	?	Crashed in Kenya 12-8-2012
AF-821	Mi-24V	3532421521412	?	Seen 11-1998
AF-823	Mi-24V	3532421319261	?	Seen 11-1998
AF-8...	Mi-24P	3532432622957	?	Seen 8-2001
AF 011	Su-30MK2	798103*****	?	D/D 7-2011, first flight from Entebbe 12-7-2011
AF 015	Su-30MK2	798103*****	?	D/D 7-2011, first flight from Entebbe 11-7-2011

Tanzanians stormed Entebbe IAP, Idi Amin managing to flee to Libya in the nick of time aboard a Libyan C-130. Kampala fell on 11th April and by 3rd June Amin's junta had been toppled, Godfrey Binaisa becoming the new president. (This was by no means the end of the row – appetite comes when eating. In May 1980 Binaisa was unseated by Obote, who thus restored his position as president until he was deposed in a new military coup on 27th July 1985 and fled the country. Next, Tito Okello, the leader of the Military Council, became the new president. In late January 1986 the National Resistance Council was formed as the ruling body; its chairman Yoweri K. Museveni was elected president, lasting longer than most of his predecessors.)

In early May 2011 the UPDAF MiG-21s were pressed into action again during Operation *Iron Fist* undertaken jointly with Sudan. This time the enemy was the infamous Lord's Resistance Army (LRA) – a particularly violent insurgent group that had been active in northern Uganda since 1987 and had committed great atrocities against the civilian population. The MiG-21s delivered strikes against five LRA camps in Sudan, followed by an assault by Ugandan ground forces which seized more than US$ 2 million worth of weapons. However, the UPDF suffered heavy losses during the operation while the insurgents scattered, and eventually the Ugandans left Sudan by 18th May as arranged.

Attempts to persuade the LRA to stop the armed struggle failed, and in 2006 the UPDAF began a new series of raids against the insurgents. When the LRA leaders refused to sign the Juba Peace Accord prepared after two years of negotiations, it was decided to wipe out the insurgents that had holed up in the Garamba National Park in the north-east of the Democratic Republic of Congo. On 14th December 2008 Uganda, the DRC and South Sudan began a joint military operation called *Lightning Thunder*. According to plans the UPDAF MiG-21s deployed to Nakasongola AB in north-western Uganda were to deliver a strike against a hut at Suke, DRC, where the LRA leaders were staying, and the AAA positions protecting it, whereupon Mi-24s carrying an assault party would come in and ground forces would finish off the job. However, bad weather foiled this plan; the MiGs were unable to take off, and the Mi-24s that reached the target were met with dense AAA fire because the gun emplacements had not been destroyed. When the helicopters left the scene in order to refuel, the insurgents wasted no time moving out, abandoning the camp; the Ugandan Army began a pursuit supported by air strikes, but this eventually failed.

Operation *Lightning Thunder* was declared officially completed in late February 2009 but the LRA was still active, the majority of the guerrillas remaining at large. Moreover, the LRA ignited insurgency in northern Uganda itself. Thus, operations against the LRA jointly with the Congolese Army continued throughout 2009 and 2010, the Ugandan MiG-21s and Mi-24s operating from tactical airstrips in northern Uganda and the DRC.

As if that weren't enough, in 2009 Uganda had a border conflict with Kenya. The apple of discord was Migingo Island on Lake Victoria, which was claimed by Kenya. On 20th February several Ugandan Mi-17s and Mi-24s inserted a commando team which secured the island and arrested the Kenyan police officers stationed there.

Zambia

Zambia was created in 1963 when the former British colony of Nyasaland separated from the Central African Federation. Since the mid-1970s the Zambian Air Force and Air Defence Command has been purchasing military equipment from the East and the West alike. Thus, in 1976 Zambia took delivery of 14 Mikoyan MiG-21*bis* (*izdeliye* 75A) *Fishbed-L* fighters and two MiG-21UM (*izdeliye* 69) *Mongol-B* trainers. Initially based at Lusaka International airport, they later moved to Livingstone airbase. Known serials suggest that either more MiG-21s were delivered later or the existing aircraft were subsequently reserialled. Two trainers and eight (some sources suggest 11) surviving fighters were sent to Israel in 1997-98 for upgrading to MiG-21-2000 standard by IAI at Lod; however, some of these were not returned. (Some sources suggest that more than two MiG-21UMs were delivered and that eight single-seaters and five trainers were sent to Israel.)

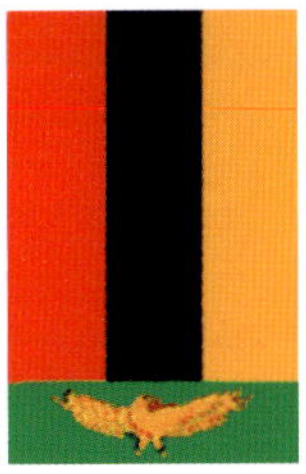

Zambia also established ties with China as an arms supplier, resulting in the delivery of 12 Shenyang F-5 (Mikoyan/Gurevich MiG-17F *Fresco-C*) fighters and FT-5 trainers based on the F-5, plus at least 12 Shenyang F-6C (modified Mikoyan/Gurevich MiG-19S *Farmer-C*) fighters, between 1976 and 1978. They were to serve as a deterrent against incursions by

Known Zambian Air Force aircraft of Soviet origin

Type	Serial	C/n	F/n	Notes
MiG-21*bis*	AF603	?	?	
MiG-21*bis*	AF604	?	?	
MiG-21*bis*	AF605	?	?	
MiG-21*bis*	AF606	?	?	
MiG-21*bis*	AF607	?	?	
MiG-21*bis*	AF608	?	?	
MiG-21*bis*	AF609	?	?	
MiG-21*bis*	AF610	?	?	
MiG-21*bis*	AF611	?	?	
MiG-21*bis*	AF612	?	?	Also reported as AF912
MiG-21*bis*	AF613	?	?	
MiG-21*bis*	AF614	?	?	
MiG-21*bis*	AF615	?	?	
MiG-21*bis*	AF616	?	?	
MiG-21*bis*	AF-921	N750*****	?	Sand/green/dark green/brown camouflage
MiG-21*bis*	AF-926	N750*****	?	Sand/green camouflage
MiG-21*bis*	AF-927	N750*****	?	Sand/green/dark green/brown camouflage
MiG-21*bis*	AF-930	N75086175	?	Sand/green camouflage. Refurbished by IAI
MiG-21UM	AF601	?	?	
MiG-21UM	AF602	?	?	
MiG-21UM	AF-923	516999318	?	Stored Lod, Israel, 2010
MiG-21UM	AF-924	516999319	?	Stored Lod, Israel, 2011; sand/green/brown camouflage. MiG-21UM AF 824 was reported in 1986 but may be a mis-sighting for this aircraft

Mi-8T	AF751	?	?
Mi-8T	AF752	?	?
Mi-8T	AF753	?	?
Mi-8T	AF754	?	?
Mi-8T	AF755	?	?
Mi-8T	AF756	?	?
Mi-8T	AF757	?	?
Mi-8T	AF758	?	?
Yak-40 'Salon'	AF604	?	
Yak-40 'Salon 1st Class'	AF605	9532042	
Yak-40 'Salon 1st Class'	AF606	9530843	

Zambian Air Force MiG-21*bis Fishbed-L* AF612

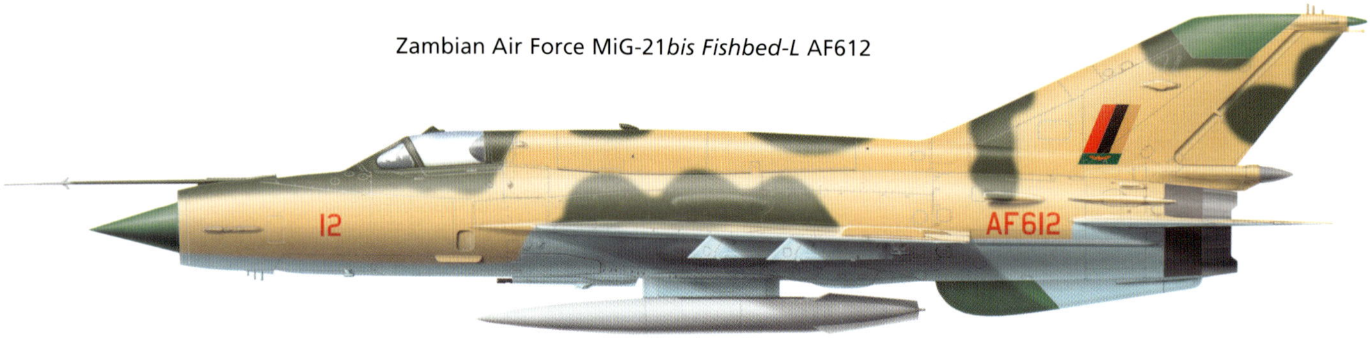

Zambian Air Force MiG-21UM AF601

Rhodesian Air Force aircraft; however, the Zambian Air Force's experience with them was chiefly negative. It is not known whether they were used operationally against the RENAMO guerrillas during the Zambian deployment in Mozambique. One of these F-6Cs became a gate guard at Lusaka airport.

VIP transport duties were performed by three Yakovlev Yak-40 *Codling* trijets, at least two of which were in the Yak-40 'Salon 1st Class' version. The type was reportedly in Zambian service since 1977 but the construction numbers of two aircraft indicate they were built in 1975.

Eight Mil' Mi-8T *Hip-C* helicopters were delivered to the Zambian Air Force in 1975. Most of them were derelict at Lusaka by 2000.

Zimbabwe Air Force Chengdu F-7 IIN '708', 5 Sqn 'Arrows', Thornhill-Gweru

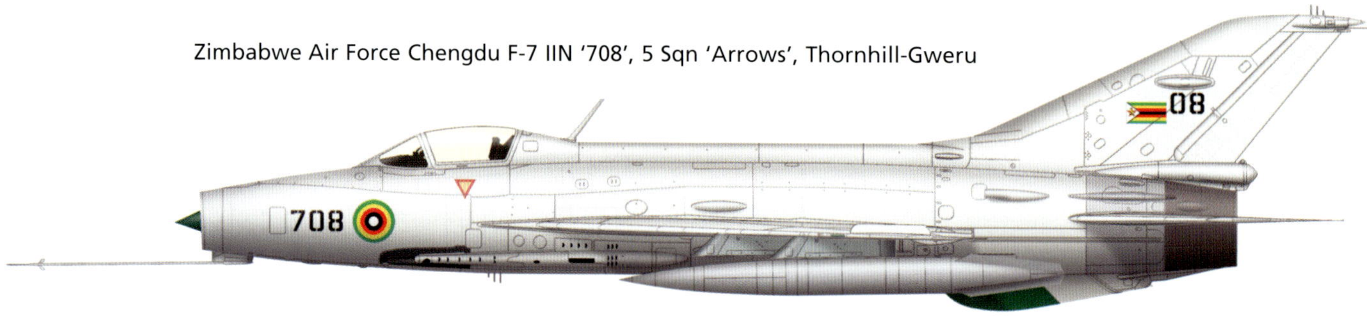

796, a Zimbabwe Air Force FT-5. Note the unpainted air-brakes and the emblem on the fin.

Zimbabwe

The Air Force of Zimbabwe originated as the South Rhodesian Air Force on 19th September 1939. It reformed after the war on 28th November 1947, becoming the Royal Rhodesian Air Force in 1954 after South Rhodesia had federated with Nyasaland. In 1963 Nyasaland split off and renamed itself Zambia. In 1965, after a Unilateral Declaration of Independence, the Royal title was removed, followed in 1979 by a change of the country's name to Zimbabwe; the air force received its current name on 18th April 1980.

In 1986 the Zimbabwe Air Force reportedly

Two Zimbabwean F-7 IINs, 704 and 707, make a formation flypast.

Two more ZAF F-7 IINs at Thornhill AB. Note the absence of roundels.

730, the first of two Zimbabwean FT-7BZ trainers.

A Zimbabwe Air Force Mi-35 at close quarters.

Opposite page: A Zimbabwe Air Force Mi-35P leads a Eurocopter Super Puma and a Bell 214.

received two Shenyang FT-5 trainers (based on the F-5, the Chinese-built Mikoyan/Gurevich MiG-17F *Fresco-C*). They were reportedly retired in late 1991. Four Chengdu F-7 II and eight F-7 IIN fighters (with two and four wing pylons respectively) were delivered to the No.5 Sqn 'Arrows' at Thornhill AB, Gweru, between 1986 and September 1987, and two Guizhou FT-7BZ trainers followed in 1991. Some of the fighters, known locally as F-7 IIs, mount two wing pylons; the rest, designated F-7 IIN, have four. Three fighters were lost in January 2001, crashing while en route to the Democratic Republic of Congo to join a flypast at the funeral of the Congolese President Laurent Desiré Kabila. Reports of a further delivery of 12 have not been confirmed, neither has that of an order for the more advanced F-7MG allegedly placed in 2004. The sorry state of the Zimbabwean economy and the strong denials by the Chengdu Aircraft Corporation make these deliveries unlikely. For the first ten years the F-7s wore an overall while colour scheme; later, in the course of an overhaul, this gave place to and off-white/light grey camouflage scheme.

In the year 2000 the Air Force of Zimbabwe purchased three MiG-23s (MiG-23MS *Flogger-E* fighters and MiG-23UB *Flogger-C* trainers) from the Libyan Arab Air Force. More MiG-23s were obtained later, as four were reportedly still airworthy in 2008 and you cannot have four out of three!

On 8th April 1999 the Zimbabwean independent newspaper *Financial Gazette* reported that the Air Force of Zimbabwe planned to purchase ten Mil' Mi-24 attack helicopters from Russia as part of a US$ 26.25 million weapons package for the COIN role – specifically, for fighting in the Democratic Republic of Congo where Zimbabwe supported the Kinshasa government. Six (some sources say as many as 12) are known to have been delivered to date, including two Mi-35 *Hind-Es* and two Mi-35P *Hind-Fs*; they are operated by the No.7 Sqn 'Scorpions' or No.8 Sqn 'Spiders' at Manyame AB near Harare. The helicopters wear a wraparound dark green/dark earth camouflage.

In the spring of 2000 four of the six Mi-35s and Mi-35Ps delivered to Zimbabwe were upgraded by the manufacturer (the Rostvertol Helicopter Plant) to give them day/night all-weather capability. They now feature Russian-made night vision goggles, a new defensive countermeasures system, a Garmin GPS 115 global positioning system with VPS-200 interface, and an IRTV-445MGII thermal imaging system allowing the crew to detect and identify objects at a distance of up to 4 km (2.5 miles). Interestingly, the very compact 'ball turret' of the thermal imager is carried on a (presumably detachable) mounting on the port outboard pylon rather than on the forward fuselage, thus restricting the number of pylons available for external stores to three.

Known Air Force of Zimbabwe aircraft of Soviet/Russian origin				
Type	**Serial**	**C/n**	**F/n**	**Notes**
Shenyang FT-5	795	55-****		White overall
Shenyang FT-5	796	55-****		White overall
Chengdu F-7 II	700	?		Crashed 14-11-1994
Chengdu F-7 II	701	?		
Chengdu F-7 II	702	?		Grey/white colour scheme
Chengdu F-7 II	703	?		
Chengdu F-7 IIN	704	?		White overall
Chengdu F-7 IIN	705	?		
Chengdu F-7 IIN	706	?		Grey/white colour scheme
Chengdu F-7 IIN	707	?		White overall
Chengdu F-7 IIN	708	?		White overall
Chengdu F-7 IIN	709	?		White overall
Chengdu F-7 IIN	710	?		Grey/white colour scheme
Chengdu F-7 IIN	711	?		Grey/white colour scheme
Guizhou FT-7BZ	730	?		White overall, later grey/white colour scheme
Guizhou FT-7BZ	731	?		White overall
MiG-23MS	13	?	?	Ex-Libyan Arab Air Force
MiG-23UB	7012	B1037912	?	Ex-Libyan Arab ir Force 7912
Mi-35	01*	?	?	
Mi-35	02*	?	?	
Mi-35P	03	?	?	
Mi-35	04*	?	?	
Mi-35P	05	?	?	

* Existence not proven but likely

Map and Index

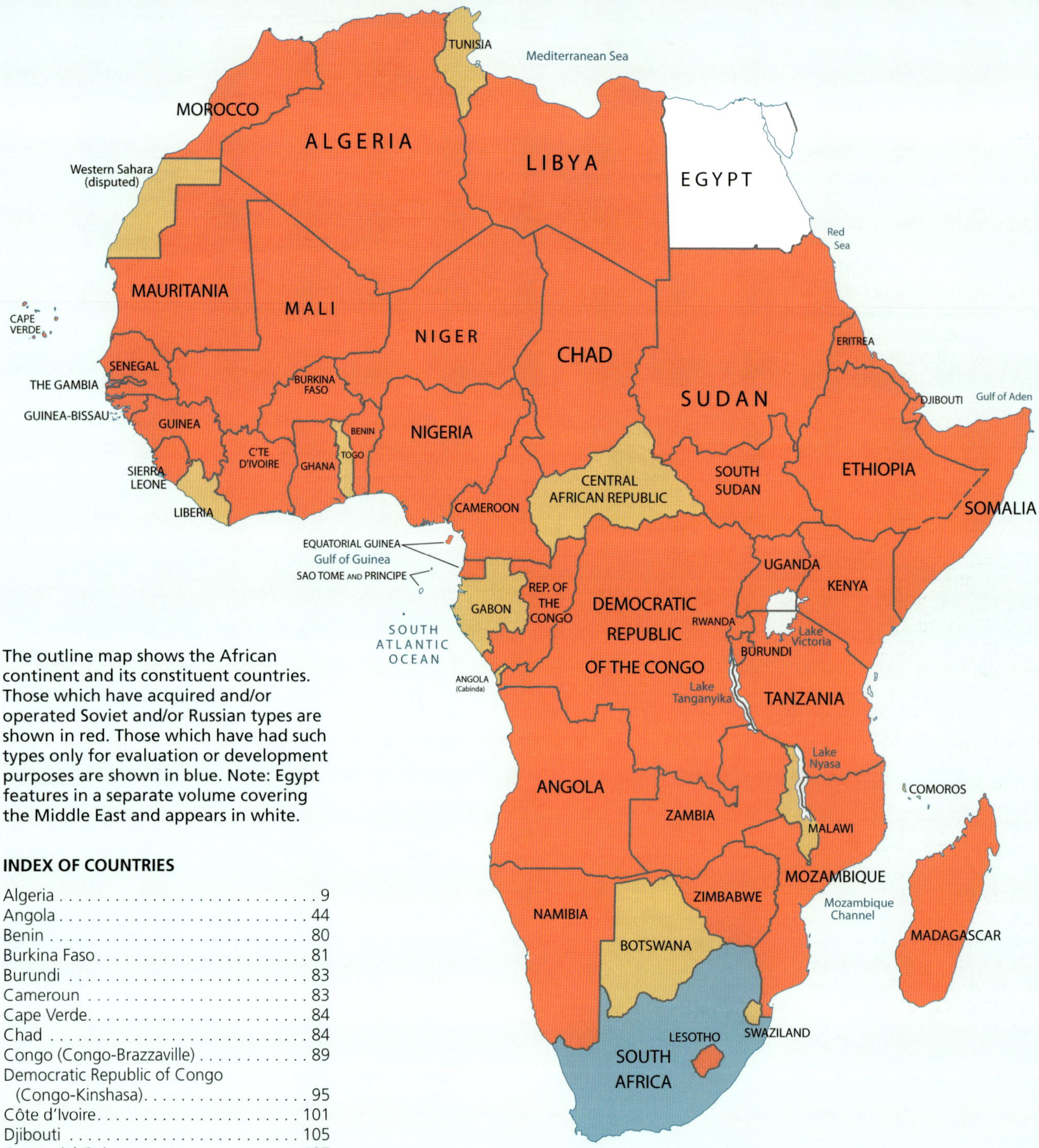

The outline map shows the African continent and its constituent countries. Those which have acquired and/or operated Soviet and/or Russian types are shown in red. Those which have had such types only for evaluation or development purposes are shown in blue. Note: Egypt features in a separate volume covering the Middle East and appears in white.

INDEX OF COUNTRIES

Algeria 9
Angola 44
Benin 80
Burkina Faso 81
Burundi 83
Cameroun 83
Cape Verde 84
Chad 84
Congo (Congo-Brazzaville) 89
Democratic Republic of Congo
(Congo-Kinshasa) 95
Côte d'Ivoire 101
Djibouti 105
Equatorial Guinea 107
Eritrea 110
Ethiopia 117
Gambia 137
Ghana 137
Guinea 138
Guinea-Bissau 141
Kenya 143
Lesotho 143
Libya 144
Madagascar (Malagasy Republic) 189
Mali 191
Mauritania 193
Morocco 194
Mozambique 195
Namibia 200
Niger 202
Nigeria 202
Rwanda 212
Senegal 214
Sierra Leone 217
Somalia 219
South African Republic 227
Sudan 230
South Sudan 249
Tanzania 249
Uganda 253
Zambia 265
Zimbabwe 267

In this book, locations and units are not indexed, as the relevant entries are predominantly contained within each country's individual section and thus quite easy to locate.

AIRCRAFT TYPES

Soviet/Russian
Antonov
An-2 68, 80, 139, 142, 157, 179, 222, 234
An-12 27, 28, 44, 52, 53, 68, 83, 95, 97, 99, 100, 101, 103, 135, 137, 138, 143, 195, 198, 203, 218, 231, 235, 244
An-22 45, 122, 153
An-24 52, 68, 92, 93, 94, 98, 107, 109, 123, 193, 221, 222, 223, 224, 231, 234, 244
An-26 52, 58, 59, 65, 68, 69, 70, 84, 85, 86, 89, 97, 98, 99, 123, 157, 161, 162, 163, 165, 173, 175, 179, 190, 192, 193, 195, 197, 198, 201, 202, 222, 223, 224, 231, 234, 235, 236, 237, 243, 244, 245
An-28 52, 61, 107, 251, 252
An-30 52, 60, 70, 231, 234, 236, 245
An-32 52, 59, 60, 69, 70, 92, 93, 94, 101, 107, 109, 123, 130, 157, 161, 179, 180, 231, 233, 234, 245
An-72 52, 61, 70, 109, 157, 164, 165, 173, 180, 212, 232, 233, 240, 245
An-74 8, 52, 70, 158, 166, 180, 232, 234, 239, 240, 241, 245
An-124 24, 26, 27, 84, 96, 159, 171, 180, 202, 231, 255
An-148 174
Il'yushin
IL-14 27, 92, 94, 139
IL-18 28, 29, 36, 93, 94, 105, 140, 141, 142, 143, 192, 193
IL-28 26, 27, 29, 33, 36, 194, 204, 205, 206, 207, 209, 210, 222, 224
IL-62 137, 159, 167, 180
IL-76 27, 28, 30, 31, 36, 44, 53, 62, 63, 70, 83, 92, 94, 96, 100, 101, 109, 113, 122, 159, 168, 169, 170, 172, 173, 175, 179, 180, 181, 218, 232, 234, 245, 253
IL-78 28, 30, 36, 53, 70, 151, 159, 180, 181
Kamov
Ka-27 33
Ka-29 109, 110
Ka-32 33
Mikoyan-Gurevich
MiG-15 7, 9, 10, 33, 36, 37, 45, 70, 71, 90, 91, 94, 120, 130, 138, 140, 142, 144, 189, 191, 192, 193, 194, 195, 198, 203, 204, 205, 210, 219, 220, 222, 224, 250, 253, 263
MiG-17 9, 11, 25, 33, 37, 44, 45, 47, 55, 63, 71, 81, 82, 89, 90, 91, 94, 120, 124, 131, 138, 140, 141, 142, 189, 190, 191, 192, 193, 194, 195, 196, 197, 198, 202, 203, 204, 205, 206, 207, 210, 219, 221, 222, 224, 225, 231, 236, 250, 253, 254, 259, 261, 262, 263, 265
MiG-21 9, 10, 11, 12, 13, 20, 33, 36, 37, 38, 45, 46, 47, 48, 51, 55, 56, 57, 58, 59, 60, 61, 63, 64, 67, 71, 72, 73, 74, 81, 91, 92, 94, 96, 112, 114, 115, 116, 117, 120, 121, 124, 125, 127, 128, 129, 130, 131, 132, 133, 139, 140, 141, 142, 145, 146, 148, 161, 162, 166, 169, 175, 177, 178, 181, 182, 183, 184, 189, 190, 191, 192, 193, 195, 196, 197, 198, 199, 200, 201, 202, 204, 208, 210, 211, 219, 222, 223, 224, 226, 227, 230, 231, 232, 233, 234, 235, 236, 245, 246, 251, 253, 254, 255, 256, 257, 261, 262, 263, 264, 265, 266
MiG-23 6, 12, 13, 14, 25, 26, 35, 36, 38, 39, 46, 47, 48, 49, 50, 51, 57, 61, 62, 63, 67, 74, 75, 96, 97, 100, 101, 103, 111, 114, 115, 116, 117, 118, 119, 121, 122, 127, 128, 129, 133, 134, 146, 147, 148, 149, 150, 151, 152, 162, 164, 165, 166, 167, 169, 170, 175, 177, 178, 184, 185, 200, 204, 230, 231, 236, 246, 268, 269
MiG-25 12, 13, 15, 16, 17, 25, 39, 149, 150, 151, 166, 167, 169, 185, 236, 246
MiG-29 11, 12, 13, 16, 17, 18, 19, 20, 23, 39, 40, 45, 49, 92, 111, 112, 114, 115, 116, 129, 130, 204, 230, 231, 233, 234, 237, 238, 246
Mil'
Mi-1 29
Mi-2 29, 31, 40, 41, 93, 95, 97, 99, 105, 107, 123, 134, 137, 143, 159, 175, 185, 186, 187, 214, 216, 244
Mi-4 29, 81, 83, 137, 138, 192
Mi-6 31, 41, 54, 123, 140, 141
Mi-8 29, 30, 31, 32, 41, 53, 54, 67, 75, 76, 78, 81, 82, 83, 84, 85, 87, 89, 92, 93, 95, 97, 99, 102, 103, 104, 105, 107, 114, 123, 129, 134, 135, 140, 141, 143, 159, 161, 172, 173, 174, 175, 187, 190, 191, 192, 196, 199, 200, 201, 202, 214, 222, 227, 232, 234, 235, 246, 255, 266
Mi-14 95, 123, 124, 161, 175, 176, 178, 187, 188
Mi-17 12, 30, 31, 32, 33, 41, 53, 54, 64, 65, 76, 77, 78, 81, 82, 83, 85, 87, 89, 93, 95, 97, 99, 102, 103, 107, 110, 114, 116, 123, 127, 128, 135, 136, 137, 138, 140, 142, 143, 159, 175, 202, 212, 213, 214, 215, 216, 217, 218, 219, 232, 233, 242, 243, 246, 247, 249, 255, 259, 260, 264, 265
Mi-19 31, 33, 41
Mi-24 8, 31, 33, 34, 35, 36, 42, 54, 65, 67, 78, 79, 83, 85, 87, 89, 93, 95, 97, 99, 100, 102, 103, 104, 105, 107, 109, 110, 114, 116, 117, 123, 124, 126, 127, 128, 135, 137, 140, 141, 159, 160, 164, 165, 166, 188, 192, 193, 196, 200, 202, 212, 214, 217, 218, 219, 227, 228, 230, 232, 237, 247, 256, 257, 259, 264, 265, 268
Mi-25 31, 54, 66, 67, 82, 139, 160, 164, 165, 166, 188, 196, 197
Mi-26 97, 99, 109, 110
Mi-34 205, 212
Mi-35 31, 54, 66, 67, 82, 92, 93, 95, 106, 107, 114, 116, 123, 126, 127, 128, 129, 130, 135, 160, 174, 175, 177, 178, 188, 205, 208, 209, 211, 212, 214, 215, 216, 232, 242, 243, 247, 257, 260, 261, 262, 268, 269
Mi-171 30, 31, 32, 34, 41, 42, 88, 89, 143, 159, 187, 232, 233, 241, 242, 246, 247
Mi-172 108, 109, 142, 143, 212, 214, 232, 246, 247, 249, 255, 256, 257, 263, 264
Polikarpov
I-15 7
I-16 7
R-5 7
R-Z 7
Sukhoi
Su-20 49, 152, 161
Su-22 49, 50, 51, 53, 63, 64, 79, 80, 152, 153, 154, 155, 156, 164, 167, 168, 169, 175, 188
Su-24 21, 22, 23, 24, 27, 43, 44, 49, 55, 153, 156, 157, 175, 188
Su-25 50, 51, 52, 54, 55, 65, 66, 80, 84, 85, 89, 96, 97, 98, 101, 103, 104, 105, 108, 109, 110, 114, 115, 116, 120, 122, 130, 136, 137, 231, 234, 238, 243, 248
Su-27 23, 48, 56, 57, 65, 80, 112, 113, 114, 115, 116, 120, 121, 122, 129, 130, 135, 161
Su-30 9, 23, 24, 25, 27, 28, 44, 161, 255, 258, 259, 264
Su-35 161
Tupolev
ANT-40 7
SB 7, 34, 35, 42, 230
TB-3 7
Tu-22 153, 154, 155, 156, 157, 158, 160, 162, 164, 165, 166, 251, 263

AIRCRAFT TYPES continued

Tu-95 138
Tu-134 53, 80, 195, 196, 200
Tu-154 123, 135
Yakovlev
Yak-130 26, 27, 44, 161
Yak-40 53, 80, 107, 109, 123, 135, 190, 191, 266

Non-Soviet/Russian
Aeritalia G.222 222
Aermacchi MB.339 129
Aero
CS-102 10, 36, 37, 89, 90, 94, 142, 194, 195, 198, 224
L-29 Delfin 204, 205
L-39 Albatros 57, 80, 151
Aérospatiale SA.342 Gazelle 105
Airbus A300 259
Atlas (Aermacchi) MB.326M Impala 62
Atlas CSH-2 Rooivalk 227, 228
Avia-14 138, 139, 140
BAC Strikemaster 105, 236
Bell
214 268
AH-1 229, 230
Boeing
707 261
747 172
C-97 206
E-3A Sentry 236
RC-135 167
Boeing Vertol CH-47 Chinook 178
CASA CN.235 100
Chengdu F-7 201, 202, 204, 211, 231, 236, 251, 253, 266, 268, 269
Dassault
Mirage IIID 145
Mirage 5D 34, 161, 259
Mirage F.1 57, 58, 59, 61, 71, 148, 151, 166, 167, 169, 178
Rafale 178
Douglas
A-26 Invader 205
DC-3/C-47 205, 206, 224, 254
DC-4 206
DC-6 206
DC-7B 206
Embraer EMB-312 Tucano 59
English Electric Canberra 58, 224
Eurocopter Super Puma 268
Eurofighter Typhoon 178
Fokker
F.27 56, 205
F.28 208
General Dynamics F-111 170, 172
Grumman
A-6 Intruder 169, 170
E-2C Hawkeye 170
EA-6 Prowler 170
F-14A Tomcat 167, 168, 169, 170
Guizhou FT-7 201, 202, 204, 211, 231, 251, 253, 268, 269
LET L-410 61, 107, 151
Lockheed
C-130 Hercules 27, 57, 62, 151, 178, 235, 261, 264
F-104 158
Malmö MFI-9B 205, 210
McDonnell Douglas
F-15 Eagle 93, 236
F/A-18 Hornet 170
McDonnell F-4 Phantom II 160, 236
Nanchang A-5C Fantan 243
North American
B-25 Mitchell 205
F-86 Sabre 7, 125
T-6G Texan 205, 210
Northrop F-5 125, 222, 224, 236
Pilatus PC-7 64
Piper PA-31 Navajo 254
PZL Lim-5 142, 195, 198, 202
Raytheon Beech 1900D 31
Shaanxi Y8 231, 244, 251, 252
Shenyang
F-5 231, 234, 236, 247, 248, 249, 265, 268
F-6 220, 221, 222, 224, 225, 226, 227, 231, 234, 248, 250, 251, 252, 253, 265, 266
FT-2 249, 250, 252
FT-5 231, 247, 248, 249, 250, 252, 265, 267, 268, 269
Transall C.160F 105
Xian Y7H 193

MISSILES
AIM-7M Sparrow 170
AIM-9 Sidewinder 168, 162, 168, 169, 170, 222
Matra R.550 Magic 58, 61
R-3 46, 50, 58, 147, 152
R-13 168, 222
R-23 14, 50, 61, 147
R-24 57, 62
R-27 115, 116, 129
R-40 149, 150
R-60 14, 50, 147, 149, 150
R-73 17, 48, 115, 120, 129, 130
R-77 48
Raytheon MIM-23 Hawk 166

NATO REPORTING NAMES
Beagle 26, 27, 194, 204, 205, 206, 222
Blinder 153, 154, 155, 157, 162, 164, 165, 166, 263
Candid 27, 28, 53, 83, 92, 100, 109, 122, 159, 218, 232
Careless 123
Cash 52, 107, 204, 251
Clank 52, 231
Classic 137, 159, 168
Cline 52, 92, 101, 107, 123, 157, 231
Coaler 52, 109, 157, 158, 212, 232
Cock 23, 45, 122
Codling 53, 107, 123, 190, 266
Coke 52, 92, 107, 123, 222, 231
Condor 24, 84, 96, 159, 202
Coot 28, 29, 93, 140, 141, 143, 192
Crate 27, 92, 139
Crusty 53, 195
Cub 27, 28, 52, 83, 95, 99, 100, 101, 123, 137, 139, 143, 195, 203, 218, 231, 251
Curl 52, 84, 99, 123, 157, 190, 192, 193, 195, 202, 222, 231
Fagot 9, 191, 219
Fencer 21, 27, 49, 50, 153
Fishbed 9, 10, 11, 12, 45, 46, 47, 58, 81, 91, 96, 112, 120, 121, 130, 139, 142, 145, 146, 189, 190, 192, 195, 196, 200, 204, 219, 224, 230, 231, 251, 253, 255, 265, 266
Fitter 25, 33, 34, 49, 51, 152, 153, 168
Flanker 23, 24, 48, 112, 115, 116, 122, 161, 255
Flogger 6, 12, 25, 47, 48, 49, 51, 61, 96, 101, 116, 121, 146, 148, 149, 151, 170, 200, 204, 230, 231, 268
Foxbat 12, 16, 151
Fresco 9, 45, 81, 90, 91, 120, 138, 141, 142, 189, 191, 194, 195, 197, 202, 205, 219, 231, 250, 253, 265, 268
Frogfoot 50, 84, 96, 98, 101, 105, 110, 116, 122, 137, 231
Fulcrum 13, 16, 17, 18, 92, 115, 129, 204, 230, 231, 233
Gainful 166
Gremlin 64
Halo 99, 110
Hare 29
Haze 95, 123, 161
Helix 33, 110
Hermit 205
Hind 31, 54, 66, 67, 83, 85, 89, 93, 99, 100, 102, 107, 110, 116, 123, 128, 130, 141, 160, 164, 166, 192, 196, 202, 205, 212, 216, 218, 219, 227, 232, 233, 256, 257, 259, 261, 268
Hip 30, 31, 53, 81, 83, 84, 93, 99, 100, 102, 107, 109, 116, 123, 138, 141, 143, 159, 190, 192, 196, 202, 214, 216, 217, 218, 222, 232, 233, 255, 260, 266
Hook 31, 54, 123, 141
Hoplite 29, 93, 99, 107, 123, 138, 143, 159, 190, 214
Hound 29, 81, 83, 138, 192
Midas 28, 53, 159
Midget 9, 45, 91, 120, 138, 142, 144, 189, 191, 194, 195, 204, 219, 250, 253
Mitten 27, 161
Mongol 9, 11, 47, 81, 91, 96, 120, 139, 142, 145, 189, 190, 192, 195, 200, 204, 220, 230, 251, 253, 265
Moujik 25